PROPERTY OF : Thomas E. B[illegible]
412 N. [illegible] Ave.
Alhambra CA 91801
213-289-3015

PURCHASED 1979
(McGraw-Hill Book club)

DESIGN AND CORROSION CONTROL

DESIGN AND CORROSION CONTROL

V.R. PLUDEK
M.I.Corr.T., A.C.S., M.N.A.C.E.,
M.C.I.M., M.A.S.M., M.E.C.S.

A HALSTED PRESS BOOK

JOHN WILEY & SONS
New York

First published in the United Kingdom 1977 by
The Macmillan Press Ltd.

Published in the USA by
Halsted Press, a division of John Wiley & Sons, Inc.
New York

Printed in Great Britain

Library of Congress Cataloging in Publication Data

Pludek, V R
Design and corrosion control.

"A Halsted Press book."
1. Corrosion and anti-corrosives. 2. Engineering design. I. Title.
TA418.74.P58 1977 620.1'1223 75-25940
ISBN 0-470-69136-0

Preface

It can scarcely be denied that preventive corrosion control offers the best answer to the costly havoc caused by corrosion. For this reason more and more corrosion-control work must, of necessity, find its way to the desks and drawing boards of various design teams in all branches of industry. A large percentage of expensive service failures attributed to corrosion would not have occurred if proper precautions had been taken at the design stage.

To achieve the requisite global excellence which is required in modern design work, with all its diversified problems, by any single member of a design team is a sheer impossibility. General awareness of corrosion, close co-operation and mutual understanding between individual associates of the team will help to narrow the existing wide gap between corrosion specialists and engineering planners.

Corrosion control does not always fit nicely into the design work pattern and many difficulties, both personal and technical, must be overcome before agreement and reasonable compromise can be reached. Functional designers harbour their own ingrained ideas and consider corrosion control to be an unwanted and unwarranted interference in their domain. Corrosion specialists tend to have little understanding of the functional engineering backbone of design and are prone to over-zealous proselytising of their sometimes very restricted speciality. It is the prime intention of this book to show that there is more to corrosion control than its mere study, more than the functional designer can embody in his creation on his own and also more than the average functional designer can cope with on his own. There is much dynamic power in an orderly and well-organised design team bound together by an integrated corrosion-control plan: power for efficiency of function, power for reliability, continuity and safety and power for financial gain.

It is not the intention of the author to provide yet another corrosion-control manual to tell each member of a design team what to do; there are already many excellent books serving that purpose. Rather, the intention is to show, in a rational and orderly way, how available knowledge can be exploited and how close liaison between various specialists can be streamlined to their mutual benefit, and to point the way to a superiority of product for a corrosion-oriented design force.

The author tries to display in a simple form his ideas, which are based on his long experience in corrosion control and which have been satisfactorily employed by his co-workers. This work does not encompass all facets of corrosion-control procedures in design and is not a panacea for all corrosion problems. It is left to the genius of the particular design team, when solving the actual corrosion problem in the ambient circumstances, to develop these ideas and apply them selectively to the good practical purpose of controlling corrosion effectively and economically.

The sketches and illustrations in the book are typical examples of particular designs and should not be construed as specific and definitive recommendations. The author does not accept responsibility for the use of the methods described—the information is advisory only, and use of the methods is solely at the discretion of the user.

The author wishes to acknowledge, with sincere thanks, the many items of information from other sources that helped to inspire his work and thinking in the past, encouraged him in his experimentation, became absorbed into his work and proved useful during the compilation of this book. He is particularly grateful to the National Association of Corrosion Engineers, Houston, Texas, for their kind permission to reproduce NACE Standard RP 02–72 in Chapter 12 of this book.

1976 V.R.P.

Contents

1 Basic Theory

1.1 Introduction

Rapidly improving standards of living in the western world brought about a changing economic pattern. Not very long ago materials were expensive and labour was cheap, but now we have reached the stage where the cost of labour exceeds the cost of materials. World population is growing and material resources are relatively diminishing. Conservation of raw materials is necessary for sociological reasons and because we simply cannot afford the rising cost of dismantling and replacement of the old and rusty with the new, especially if such replacement can be avoided or where it is not necessary for the maintenance of efficiency.

We live in the technological age. The effect of unchecked corrosion, therefore, is not only confined to the state of the corroding utility itself but it also influences, profoundly, man and his economic and social welfare.

1.2 Scope

It is not the intention of this corrosion primer to turn a designer into a complete corrosion engineer. Initially, that is too vast a subject for him to grasp. He should seek advice and information from laboratories, consultants, corrosion engineers and other specialists.

Nevertheless, a designer should endeavour to gain at least a basic knowledge of corrosion to make himself receptive to the necessity of corrosion control and to understand the advice forthcoming from expert quarters.

1.3 General

(1) All metals are thermodynamically unstable and tend to react with their environment to produce compounds such as oxides or carbonates. This reaction involves the movement of electrons and is called the electrochemical reaction. The readiness with which electrons are lost varies from metal to metal and the greater this readiness the more reactive is the metal.

(2) A guide to the reactivity of a metal can be obtained by its reaction to acids, as indicated in *Table 1.1*, and water may be considered a weak acid. Reference to the table indicates that platinum is the least and sodium the most reactive, whilst zinc gives up its electrons fairly readily and copper not so readily. The positive reactions may be regarded as a form of corrosion since they involve the loss of electrons from the metal. Platinum, gold and silver are known as noble metals because of their low reactivity.

Table 1.1 REACTION OF METALS TO ACIDS

Acid	*Sodium*	*Zinc*	*Copper*	*Platinum*
Cold water	Yes	No	No	No
Diluted hydrochloric acid	Yes	Yes	No	No
Concentrated nitric acid	Yes	Yes	Yes	No

(3) Corrosion, in the normal sense, involves loss of electrons from the metal to the environment (usually water and oxygen) and the formation of corrosion products such as oxides. Since the corrosion products are usually insoluble and may form a protective skin on the metal, the rate of corrosion may not be as rapid as would be expected from the reactivity characteristic of the metal, particularly in the case of aluminium and stainless steel. On the other hand, the corrosion product of mild steel is loose and non-adherent and therefore corrosion continues.

(4) The reaction of metals may be assessed more precisely than by their reaction with acids. If two metals are connected together electrically at one end and their other ends are immersed in a common electrolyte, electrons will flow from the more to the less reactive metal. The force behind this electron flow may be measured by a voltmeter and the higher the voltage the greater is the difference in reactivity between the two metals. With such a combination, known as a galvanic cell (*Figure 1.1*), the two metals are termed electrodes; the one which loses electrons is identified as the anode and the one which gains them, the cathode.

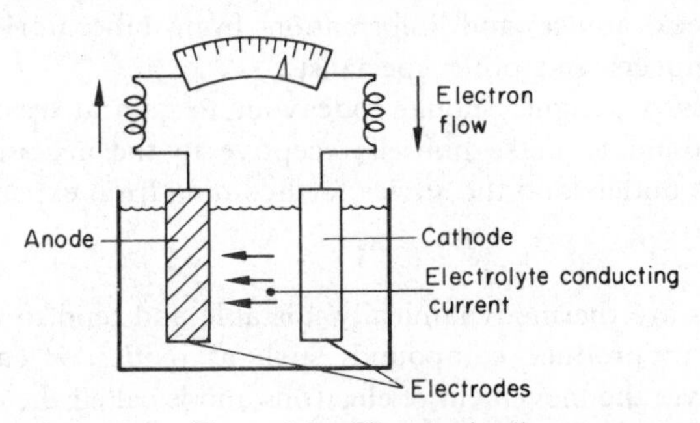

Figure 1.1. The galvanic cell

(5) Since the loss of electrons means conversion of the metal into its compounds, it follows that the anode corrodes and wastes away whilst the cathode remains unchanged (*Figure 1.2*).

(6) The various stages of the corrosion reaction proceed, in detail, as indicated in *Figure 1.3.* Iron in the anodic area loses electrons. These electrons travel through the metal and react with water and oxygen at the cathode to produce hydroxyl ions. The hydroxyl ions from the cathode react with

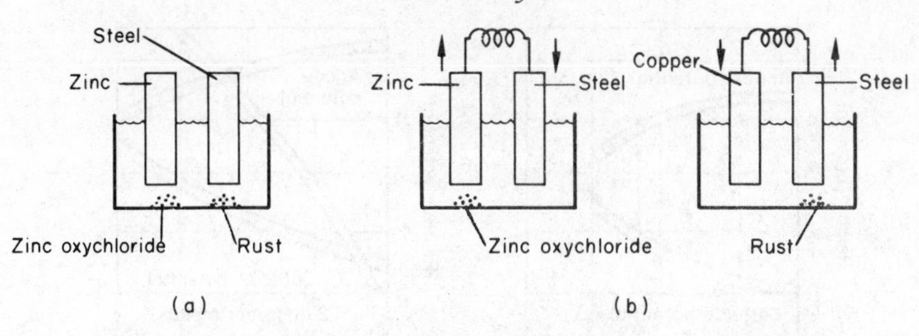

Figure 1.2. Corrosion cells: (a) no electrical connection—corrosion of zinc and steel; (b) with electrical connection—zinc reduces steel corrosion and copper accelerates steel corrosion

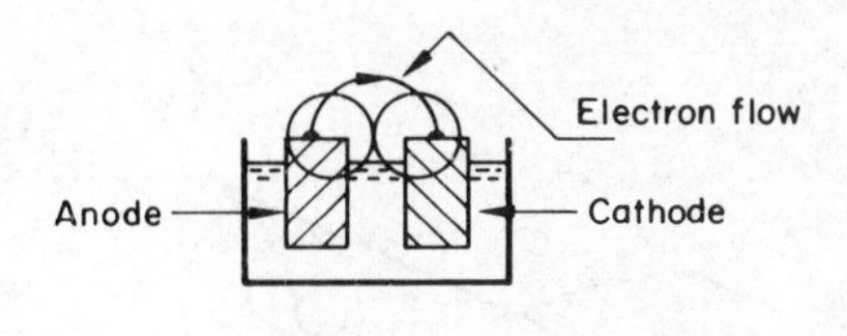

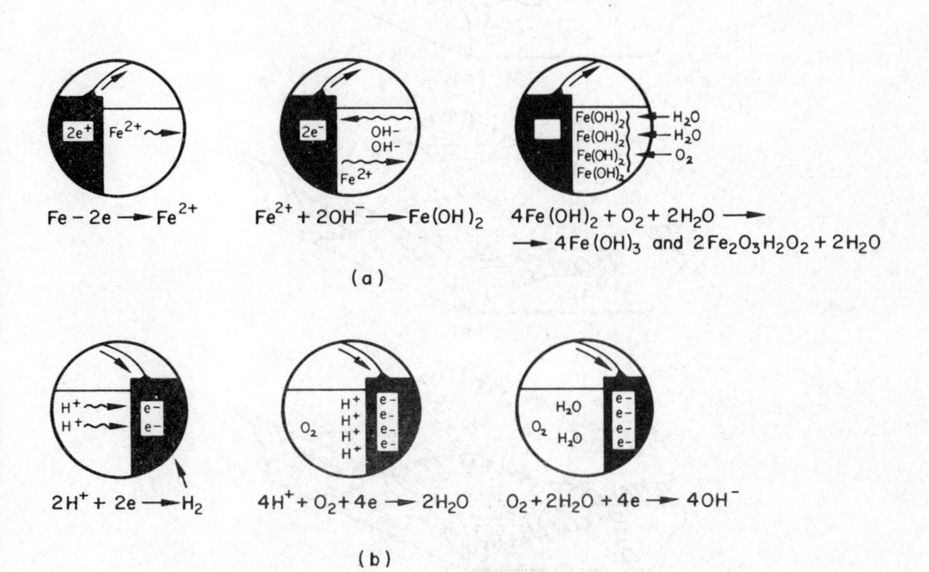

Figure 1.3. Corrosion processes: (a) anodic reaction—oxidation (corrosion) reactions at anode; (b) cathodic reaction—reduction (protective) reactions at cathode

ferrous ions formed at the anode to produce ferrous hydroxide. Ferrous hydroxide oxidises further in the presence of oxygen to produce hydrated oxide of iron, i.e. rust.

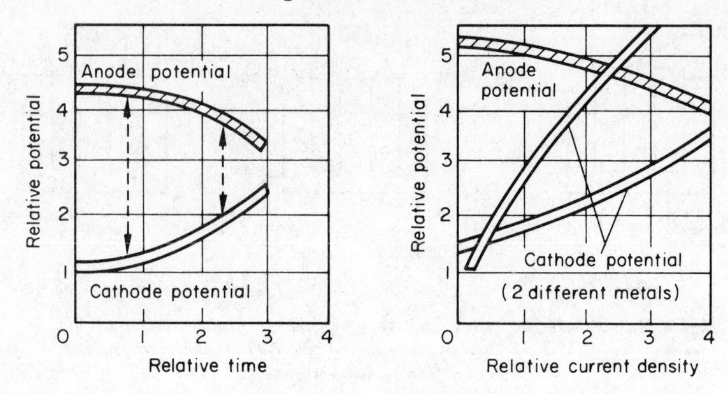

Figure 1.4. Polarisation

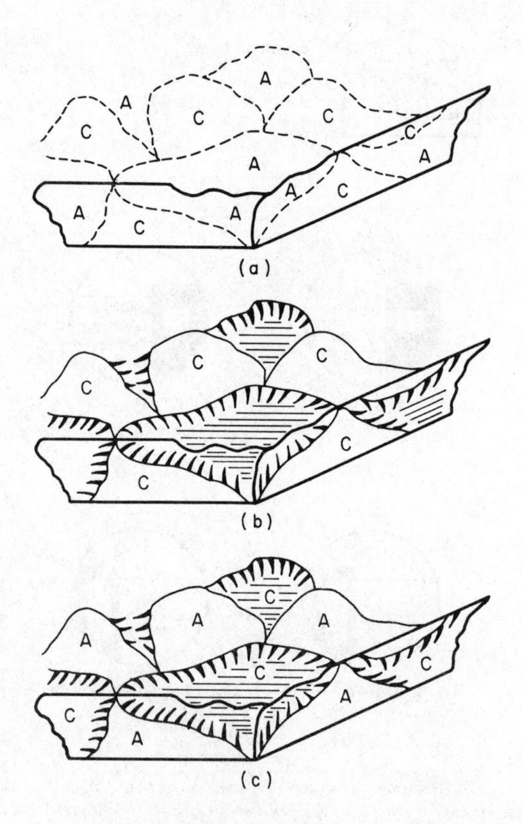

Figure 1.5. Corrosion of a single piece of steel: (a) illustration of crystal boundaries on a corner of a piece of steel (A anodic, C cathodic); (b) corrosion of anodic areas gives rise to pitting and wasting of the steel in these areas; (c) corrosion of steel in anodic areas exposes new crystal boundaries which could be cathodic to uncorroded areas—this leads to corrosion of those areas which were previously protected

(7) As corrosion products build up at anode and cathode, the voltage tends to dip. Anodes and cathode voltages drift towards each other, slowing down the corrosion rate. This voltage change is called polarisation. Polarising at the cathode is usually greater than at the anode. Hydrogen gas can blanket the cathode surface, cutting down corrosion at the anode until such time as it is removed by combining with oxygen to form water (*Figure 1.4*).

1.4 Corrosion of Steel

Steel, when exposed to normal climatic conditions, will corrode even when it is not connected electrically to another metal (*Figure 1.5*). This process conforms with the basic theory since steel is not homogeneous but contains areas of slightly differing compositions. Electropotential differences therefore exist at crystal boundaries and some areas, being anodic to other areas, corrode to protect the cathodic areas—as would be the case with two dissimilar metals.

1.5 Galvanic Series

If one standard electrode is used and all other metals are compared with it, the metals can be arranged in order of their electrode potentials. This is known as the galvanic, or electrochemical, series. *Figure 1.6* shows a number of common metals arranged in order of decreasing electrode potentials.

Steel, being the main metal used in construction, has been placed in the middle of the chart. Under normal circumstances each metal in the scale will be anodic to those listed under it and will, therefore, corrode when connected by an electrolyte. Conversely, each metal is cathodic to those listed above it and will, therefore, be immune from corrosion when connected by an electrolyte.

1.6 Summary

The basic theory of the corrosion process may be summarised as follows:

(1) Oxygen and water (the electrolyte) must normally be present for corrosion of metal to occur (but there are exceptions).

(2) Corrosion involves movement of electrons; thus an electric current flows in the metal.

(3) Corrosion of metal does not take place evenly over the surface; there are local areas of corrosion which give rise to pitting.

(4) The corrosion product, e.g. rust, may form at some distance from the points of origin, namely at a point where anodic and cathodic products combine.

(5) A high concentration of oxygen encourages the formation of a cathode, and the area immediately beyond this becomes the anode and corrodes.

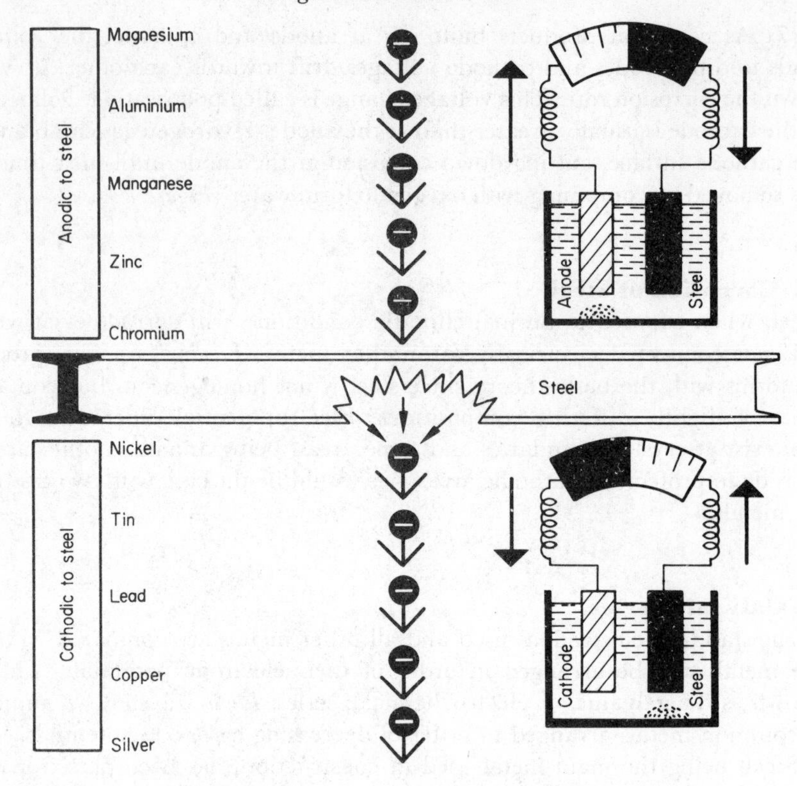

Figure 1.6. Electrode potentials

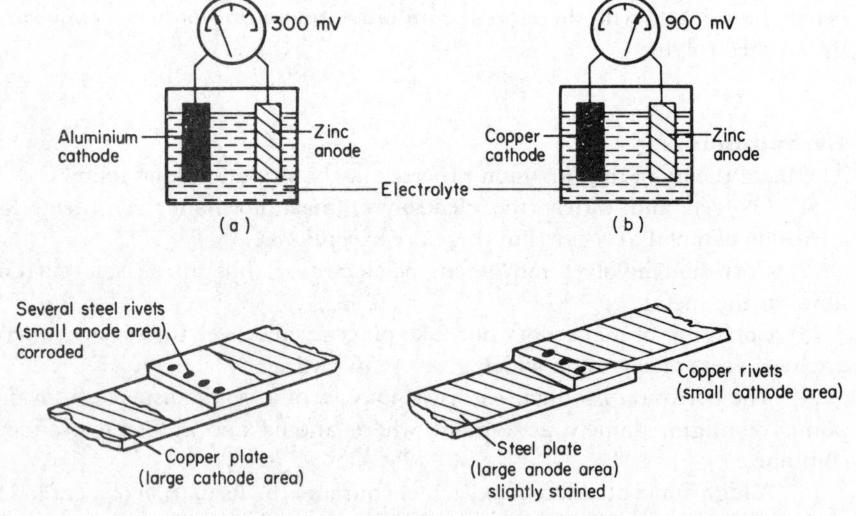

Figure 1.7. Extent of galvanic reaction: (a) less corrosion; (b) more corrosion

Corrosion is accelerated when the oxygen concentration is increased at the cathode or reduced at the anode.

(6) The extent of corrosion reaction between dissimilar metals depends on the emf potential difference between the metals, the electrolyte and the distance apart and relative areas of the two metals (*Figure 1.7*).

(7) Films form on the surface of all metals, giving the metal a degree of corrosion resistance, this resistance depending on the nature of the film. Materials with tightly adherent (insoluble) films have greater corrosion resistance than materials with soluble films. Breakage of the film initiates further corrosion of the bare metal.

(8) Corrosion of metals is influenced by their environment. The degree of this influence is relative to the physical state of the environment, its temperature, concentration and working conditions (e.g. time of exposure, osmosis, effect of cycling).

(9) Corrosion of metals is influenced by their metallurgical composition and microstructure.

2 Glossary

ACTIVE Freely corroding.

ALUMINISING Process for impregnating the surface of a metal with aluminium in order to obtain protection from oxidation and corrosion.

ANCHOR PATTERN/SURFACE PROFILE Shape and amplitude of profile of blast cleaned or grooved steel which influences the bond between metallic or paint films and the substrate.

ANION Negatively charged ion, which migrates to the anode of a galvanic or voltaic cell.

ANODE An electrode at which oxidation of the surface or some component of the solution is occurring.

ANODE POLARISATION Difference between the potential of an anode passing current and equilibrium potential (or steady-state potential) of the electrode having the same electrode reaction.

ANODIC INHIBITOR A chemical substance or combination of substances that prevent or reduce, by physical, physiochemical or chemical action, the rate of the anodic or oxidation reaction.

ANODIC METALLIC COATING A coating, composed wholly or partially of an anodic metal (in sufficient quantity to set off electrochemical reaction) which is electrically positive to the substrate to which it is applied.

ANODIC PROTECTION A technique to reduce corrosion of a metal surface under some conditions, by passing sufficient anodic current to it to cause its electrode potential to enter and remain in the passive region.

ANODISING The formation of hard, corrosion-resistant oxide film on metals by anodic oxidation of the metal in an electrolytic solution.

BASE POTENTIAL Potential towards the negative end of a scale of electrode potentials. See also *Negative Potential*, *Positive Potential* and *Potential*.

BLAST CLEANING Method of cleaning or preparing metal surfaces for painting, galvanising or metal spraying by physical removal of mill scale, rust, old paint or foreign matter, with abrasive propelled through nozzles or ejected by centrifugal force.

BLAST PEENING Treatment for relieving tensile stress by inducing beneficial compressive stress in the surface by kinetic energy of rounded abrasive particles.

BONDERISING A proprietary custom process for phosphatising.

BREAKAWAY CORROSION A sudden increase in corrosion rate, especially in high temperature 'dry' oxidation, etc.

BRUSH PLATING Electrodeposition of coating metal on substrate, in which the electrolyte is held in a pad of cotton wool or other absorbent material, the pad containing an anode.

CATHODE The electrode of an electrolytic cell at which reduction occurs. In corrosion processes, usually the area at which metal ions do not enter the solution. Typical cathodic processes are cations taking up electrons and being discharged, oxygen being reduced and the reduction from a higher to a lower state of valency.

CATHODIC INHIBITOR A chemical substance or combination of substances that prevent or reduce the rate of cathodic reaction by a physical, physiochemical or chemical action.

CATHODIC PROTECTION Reduction of the corrosion rate of a metal surface by passing sufficient cathodic current to it to cause its dissolution rate to become negligible.

CATION Positively charged ion which migrates to the cathode in a galvanic or voltaic cell.

CAUSTIC EMBRITTLEMENT The form of stress corrosion cracking occurring in steel exposed to alkaline solutions.

CAVITATION Refer to Chapter 3, Section 3.3.

CHEMICAL CLEANING Method of surface preparation or cleaning involving the use of chemicals, with or without electrical force, for removal of mill scale, rust, sediments and paint. These chemicals can also be introduced into some systems on-stream while the system is operating.

CHEMICAL CONVERSION COATING A protective or decorative coating which is produced deliberately on a metal surface by reaction of the surface with a chosen chemical environment. The thin layer formed by this reaction may perform several or all of the following functions: protect against corrosion; provide a base for organic coatings; improve retention of lubricants or compounds; improve abrasion resistance; provide an absorbent layer for rust-preventive oils and waxes.

COMPOSITE PLATE An electrodeposit consisting of two or more layers of metals deposited separately.

CONCENTRATION CELLS Refer to Chapter 3, Section 3.4

CONTROLLED GALVANIC SYSTEM Cathodic protection system using sacrificial anodes controlled by means of resistors, fixed or variable.

CORROSION Deterioration of a material, usually a metal, because of a reaction with its environment.

CORROSION–EROSION Refer to Chapter 3, Section 3.5.

CORROSION FATIGUE Refer to Chapter 3, Section 3.6.

CORROSION FATIGUE LIMIT The maximum stress endured by a metal, without failure, in a stated number of stress applications under defined conditions of corrosion and stressing.

CORROSION POTENTIAL The potential of a corroding surface in an electrolyte, relative to a reference electrode.

CORROSION RATE The rate at which corrosion proceeds, expressed by inches of penetration per year (ipy); mils penetration per year (mpy); milligrammes weight loss per square decimetre per day (mdd); microns per year (μm/year) or millimetres per year (mmpy). One micron is equal to 0.0395 mils.

COUPLE Two dissimilar metals in electrical contact.

CREVICE CORROSION Refer to Chapter 3, Section 3.4.

CRITICAL HUMIDITY The relative humidity (RH) at and above which the atmospheric corrosion rate of metal increases markedly.

CURRENT DENSITY Denotes the average current flowing in the electrolyte expressed in amperes per square foot (A/ft^2), amperes per square decimetre (A/dm^2), amperes per square centimetre (A/cm^2) or milliamperes per square centimetre (mA/cm^2) of cathode or, more occasionally, of anode surface.

DEACTIVATION Removal of a constituent of a liquid which is active in causing corrosion. The term is usually applied to the removal of oxygen by physical and/or chemical methods.

DEALUMINIFICATION Refer to Chapter 3, Section 3.14.

DECARBURISATION Refer to Chapter 3, Section 3.14.

DEPOLARISATION The removal of factors resisting the flow of current in a cell.

DEPOSIT ATTACK Localised corrosion under and resulting from a deposit on a metal surface.

DEZINCIFICATION Refer to Chapter 3, Section 3.14.

DIELECTRIC STRENGTH Degree of electrical non-conductance of a material; the maximum electric field a material can withstand without breakdown.

DIFFERENTIAL AERATION The stimulation of corrosion at a localised area by differences in oxygen concentration in the electrolytic solution in contact with metal surface.

DIFFUSION COATING Application of metallic coating, the chemical composition of which was modified by diffusing this at melting temperature into the substrate.

DRIVING VOLTAGE Excess of voltage supplied by anode in relation to the potential of the cathode.

ELECTRODE POTENTIAL The potential of an electrode as measured against a reference electrode. The electrode potential does not include any resistance loss in potential in solution due to the current passing to or from the electrode.

ELECTROGALVANISING Galvanised by electroplating.

ELECTROLESS PLATING Formation of metallic coating by chemical reduction, often catalysed by deposited metal.

ELECTROLYSIS The chemical change in an electrolyte resulting from the passage of electricity.

ELECTROLYTE A chemical substance or mixture, usually liquid, containing ions which migrate in an electric field.

ELECTROLYTIC CLEANING The degreasing/descaling of metal surfaces by electrolysis, the metal being utilised as an electrode.

ELECTRON Indivisible unit of a charge of negative electricity, revolving in definite orbits about the positive nucleus of every atom.

ELECTROPHORETIC PLATING Production of a layer of deposit by discharge of colloidal particles in a solution on an electrode.

ELECTROPLATING Electrodeposition of a thin adherent layer of a metal or alloy of desirable chemical, physical and mechanical properties on metallic or non-metallic substrate.

ELECTROPOLISHING Surface finishing of a metal by making it the anode in an appropriate solution, whereby a specular surface is obtained.

EXFOLIATION/LAMINATION The falling away of metal in layers or leaves.

FILIFORM CORROSION Corrosion which occurs under film in the form of randomly distributed hairlines.

FLADE POTENTIAL The potential at which a metal which is passive becomes active.

FLAME CLEANING/FLAME DESCALING Cleaning of metal with oxyacetylene flame passed rapidly over the surface without overheating, followed by a removal of residual deposits by wire brushing or solvent wash.

FLAME PLATING Deposition of hard metal coating on a substrate by application of molten metal at supersonic velocity (detonation gun).

FLASH CORROSION Intentional or unintentional light surface oxidisation of cleaned metals exposed to the environment for short periods—easily removable.

FOULING Primarily, deposition of flora and fauna on metals exposed to natural waters. This term also applies to any build-up of product soils on metal surfaces caused by the processing or passage of liquids in pipe systems.

FRETTING CORROSION Refer to Chapter 3, Section 3.7.

GALVANIC CORROSION/COUPLE ACTION/BIMETALLIC CORROSION Refer to Chapter 3, Section 3.8.

GALVANISING The accepted term for the coating of iron or steel with zinc by the immersion of the metal in a bath of molten zinc.

GENERAL CORROSION Refer to Chapter 3, Section 3.18.

GREEN ROT Carburisation and oxidation of certain nickel alloys at around 1832°F (1000°C), resulting in a green corrosion product.

HERMETIC SEAL An impervious seal made by the fusion of metals or ceramic materials (as by brazing, soldering, welding, fusing glass or ceramic), which prevents the passage of gas or moisture.

HIGH TEMPERATURE CORROSION Refer to Chapter 3, Section 3.9.

HYDROGEN BLISTERING Refer to Chapter 3, Section 3.10.

HYDROGEN DAMAGE Refer to Chapter 3, Section 3.10

HYDROGEN EMBRITTLEMENT Refer to Chapter 3, Section 3.10.

IMPINGEMENT ATTACK Refer to Chapter 3, Section 3.5.

IMPRESSED CURRENT PROTECTION Cathodic protection of structures, where the cathodic polarisation of metal is secured by electric currents emitted from an independent source.

INHIBITOR See *Anodic Inhibitor* and *Cathodic Inhibitor*.

INORGANIC COATINGS Coatings which contain no organic material after curing. These are coatings based on silicates or phosphates and are usually used pigmented with metallic zinc or ceramic powders.

INTERGRANULAR CORROSION/INTERCRYSTALLINE CORROSION Refer to Chapter 3, Section 3.11.

INTERNAL OXIDATION/SUBSURFACE CORROSION The formation of isolated particles of corrosion products beneath the metal surface, resulting from inward diffusion of oxygen, nitrogen, sulphur, etc.

ION EROSION Deterioration of material caused by ion impact.

IRON ROT Deterioration of wood in contact with iron.

LAMINAR SCALE Rust formation in heavy layers.

LEAD COATING Deposition of lead or lead-rich alloys by hot dipping, electroplating or metal spraying.

LINING Applying protective measures on internal surfaces of objects, e.g. tanks and pipes, with the purpose of improving their corrosion resistance in a particular environment. Miscellaneous materials such as solventless paints, plastics, reinforced plastics, rubber, ceramics, glass and metals can be used.

LOCALISED ATTACK Corrosion in which one area (or areas) of the metal surface is predominantly anodic and another area (or areas) is predominantly cathodic, i.e. anodes and cathodes are separable.

METAL CLADDING/CLAD METAL Combination of two or more metal components bonded metallurgically face to face.

METAL ION CONCENTRATION CELLS Refer to Chapter 3, Section 3.4.

METAL SPRAYING/METALLISING Application of a metal coating to a metallic or non-metallic surface by means of a spray of molten particles. Mechanical deposition of one metal on another.

METALLIC COATINGS Coatings consisting fully or partially of metal applied mechanically, chemically or electrochemically to metals or non-metals, for the purpose of protection, build-up or improvement of properties.

MICROBIAL CORROSION Refer to Chapter 3, Section 3.12.

MILL SCALE An oxide layer on metals or alloys produced by metal rolling, hot forming, welding or heat treatment. Especially applicable to iron and steel.

MIXED POTENTIAL A potential resulting from two or more electrochemical reactions occurring simultaneously on the surface of one metal.

NEGATIVE POTENTIAL Potential towards the positive end of a scale of electrode potentials. See also *Base Potential*, *Positive Potential* and *Potential*.

NOBLE Refers to positive direction of electrode potential.

NOBLE POTENTIAL A potential more cathodic (positive) than the standard hydrogen potential.

ORGANIC ZINC COATING A paint containing zinc powder pigment and an organic (containing carbon) resin.

OXIDATION Loss of electrons by a constituent of a chemical reaction.

OXYGEN CONCENTRATION CELL Refer to Chapter 3, Section 3.4.

PARTING The selective attack of one or more components of a solid solution alloy.

PASSIVATION A reduction of the anodic reaction rate of an electrode involved in electrochemical action such as corrosion.

PASSIVITY A metal or alloy which is thermodynamically unstable in a given electrolytic solution is said to be passive when it remains visibly unchanged for a prolonged period. The following should be noted: (1) during passivation the appearance may change if the passivating film is sufficiently thick (e.g. interference films); (2) the electrode potential of a passive metal is always appreciably more noble than its potential in the active state; (3) passivity is an anodic phenomenon and thus control of corrosion by decreasing cathodic reactivity (e.g. amalgamated zinc in sulphuric acid) or by cathodic protection is not passivity.

PEEN PLATING Deposition of the coating metal, in powder form, on the substrate by a tumbling action in presence of peening shot.

PHOSPHATISING The forming of a thin inert phosphate coating on a surface, usually accomplished by treating with H_3PO_4 (phosphoric acid).

pH VALUE Measure of acidity or alkalinity. A scale ranging from 0 to 14 that is used to express the acidity or alkalinity of aqueous solutions. At 77°F (25°C), a neutral solution has a pH of 7. Values ranging from 7 to 0 denote increasing acidity and from 7 to 14 increasing alkalinity.

PICKLE/PICKLING Form of chemical and electrolytic removal or loosening of mill scale and corrosion products from the surface of a metal in a chemical solution (usually acidic). Electrolytic pickling can be anodic or cathodic depending on polarisation of metal in the solution.

PITTING CORROSION Refer to Chapter 3, Section 3.13.

PLASMA PLATING Deposition on critical areas of metal coatings resistant to wear and abrasion, by means of a high velocity and high temperature ionised inert gas jet.

POLARISATION The deviation from the open circuit potential of an electrode resulting from the passage of current.

POSITIVE POTENTIAL Potential more positive than the potential of a standard hydrogen electrode. See also *Base Potential*, *Negative Potential* and *Potential*.

POTENTIAL Electrical state of a body which promotes the exchange of electrons with a second body. The potential difference is measured in volts. Electrons flow from the negative body to the positive one. See also *Base Potential*, *Negative Potential* and *Positive Potential*.

RASH RUSTING/PEAK SPOTTING Local corrosion due to inadequate coating of the peaks of a rough surface.

REDUCTION An electrochemical term meaning the reverse of oxidation. A chemical change of state in which the substance gains electrons. In the cathodic protection process, reduction takes place at the cathode, where the added electrons result in formation of hydrogen and alkali.

RUST Corrosion product consisting primarily of hydrated iron oxide—the term is properly applied only to iron and ferrous alloys.

RUST CREEP/UNDERFILM CORROSION Corrosive action which develops in damaged or uncoated areas and extends subsequently under the surrounding inert protective coating.

SACRIFICIAL PROTECTION/SACRIFICIAL ANODES/SACRIFICIAL PIECES Pieces of metal which, being anodic to the equipment into which they are introduced, will galvanically corrode and so protect the equipment. Cathodic protection, based on wasting of anodic metal to prevent corrosion of cathodic metal—zinc, aluminium, magnesium, carbon steel, etc.—so protecting steel and other more noble metals.

SCALING The formation at high temperature of thick corrosion product layers on a metal surface. The deposition of water-insoluble constituents on a metal surface.

SEASON CRACKING A term usually applied to stress corrosion cracking of brass.

SELECTIVE ATTACK/LEACHING Refer to Chapter 3, Section 3.14.

SHERARDISING The coating of iron or steel with zinc by heating the product to be coated in zinc powder at a temperature below the melting point of zinc.

STANDARD ELECTRODE POTENTIAL The reversible potential for an electrode process when all products and reactions are at unit activity on a scale in which the potential for the standard hydrogen half-cell is zero.

STRAY CURRENT CORROSION Refer to Chapter 3, Section 3.15.

STRESS-ACCELERATED CORROSION The increased corrosion caused by applied stresses.

STRESS CORROSION CRACKING Refer to Chapter 3, Section 3.16.

SUBSTRATE The basic metal or non-metal whose surface is being protected.

SURFACE PREPARATION Cleaning of surface prior to treatment; hand, mechanical, pickling, blast cleaning, etc., methods are used.

SURFACE TREATMENT Any suitable means of cleaning and treating a surface that will result in the desired surface profile and cleanliness and the required coating characteristics.

SUSTAINED LOAD FAILURE Delayed failure due to the presence of hydrogen in stressed high tensile steels.

TERNE PLATE Deposition of lead–tin alloy on iron or steel sheets by the hot dip process.

THERMOGALVANIC CORROSION Refer to Chapter 3, Section 3.17.

TUBERCULATION The formation of localised corrosion products scattered over the surface in the form of knob-like mounds.

UNIFORM CORROSION/GENERAL CORROSION Refer to Chapter 3, Section 3.18.

VACUUM DEPOSITION/VAPOUR DEPOSITION/GAS PLATING Deposition of metal coatings by the precipitation, sometimes in vacuum, of metal vapour on the treated surface. Vapour may be produced by thermal decomposition, cathode sputtering or by evaporation of the molten metal in air or inert gas.

WEATHER RESISTANCE Ability of a material to resist all ambient weather conditions. These include changes of temperature, precipitation, effect of wind and humidity, sunlight, oxygen and other gases and impurities in the atmosphere, ultraviolet rays, radiation and ozone.

WEATHERING The deliberate exposure of new steelwork in the open for the purpose of loosening mill scale prior to its removal.

WELD DECAY Localised corrosion of weld metal.

WIRE DRAWING Refer to Chapter 3, Section 3.5.

3 Common Forms of Corrosion and their Preventive Measures

3.1 Introduction

Industry still does not fully realise the situation; more and more corrosive effluent is being emitted in a self-poisoning circle and the corrosiveness of atmosphere, soil and natural waters and the complexity of the issuing corrosion are on the increase.

Up to a decade ago various generalisations were the substance of discussions about corrosion. One was supposed to avoid a particular metal or to treat its surface carefully. Data were tabulated to indicate compatibilities between dissimilar metals that should not be placed together or in ionic contact in an electrolytic solution. These generalisations are still valid but one must regard them today as superficial. Any corrosion process is chemically complex; it is not just a simple case of one material being oxidised to an ionic state and the other reduced by gaining electrons. However helpful

Table 3.1 METAL FAILURE FREQUENCY OVER A TWO-YEAR PERIOD (56.9% CORROSION AND 43.1% MECHANICAL)*

Corrosion failures†	%
General corrosion	31.5
Stress corrosion cracking Corrosion fatigue	23.4
Pitting corrosion	15.7
Intergranular corrosion	10.2
Corrosion-erosion Cavitation damage Fretting corrosion	9.0
High temperature corrosion	2.3
Weld corrosion	2.3
Thermogalvanic corrosion	2.3
Crevice corrosion	1.8
Selective attack	1.1
Hydrogen damage	0.5
Galvanic corrosion	0.0

*From the Du Pont Company's Reports.

† The percentages can vary considerably in other industrial locations or environments.

such a model may be, one must consider in the first place the surface effects which initiate a corrosion process and, secondly, the cumulative effect of the individual corrosion attacks.

3.2 Scope

This chapter lays the groundwork to the requisite knowledge of the more common forms of corrosion. It indicates those forms to which materials used in design of structures and equipment are most subjected. This should then enable the designer to connect his records of past failures (*Table 3.1*) with the demands of his new design to secure effective prevention of corrosion. Some of the preventive measures associated with particular types of corrosion are listed and these can be used either individually or in various combinations. Expert consultancy is advisable.

3.3 Cavitation Damage

3.3.1 Definition

Damage of material associated with collapse of cavities in the liquid at a solid–liquid interface (*Figure 3.1*).

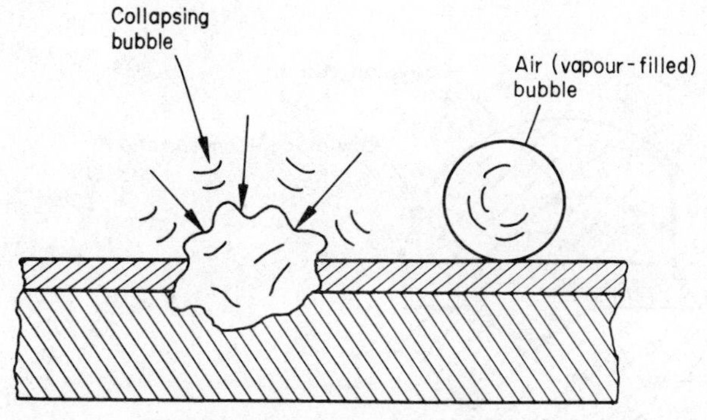

Figure 3.1. Cavitation damage

3.3.2 Cause

Repeated collapse of vapour bubbles on a metal surface can cause mild physical damage to protective films (cavitation corrosion), severe deformation and fracture of surface (cavitation deformation) or a fatigue of metal surface (cavitation fatigue). The low pressure regions are created by flow divergence, water rotation, restrictions met on lines, or by vibration.

3.3.3 Preventive Measures

Avoid conditions which allow absolute pressure to fall below vapour pressure of liquid.

Minimise hydrodynamic pressure differences.
Reduce vibration transfer.
Select suitable geometry or angle of surface to contain formation, amassment or adverse travel of gaseous bubbles in the liquid.
Prevent ingress of dispersed air, if not required for formation of protective scale.
Select resistant material, hardenable by cold-working bubbles.
Specify smooth finish.
Specify resilient surface coating or lining.
Use cathodic protection.
Inject or generate larger air or gas bubbles to buffer cavitation process.

3.4 Concentration Cells

3.4.1 Definitions

Concentration cell: a galvanic cell in which the emf is due to the difference in the concentration of one or more reactive constituents of the electrolyte solution (*Figure 3.2*).

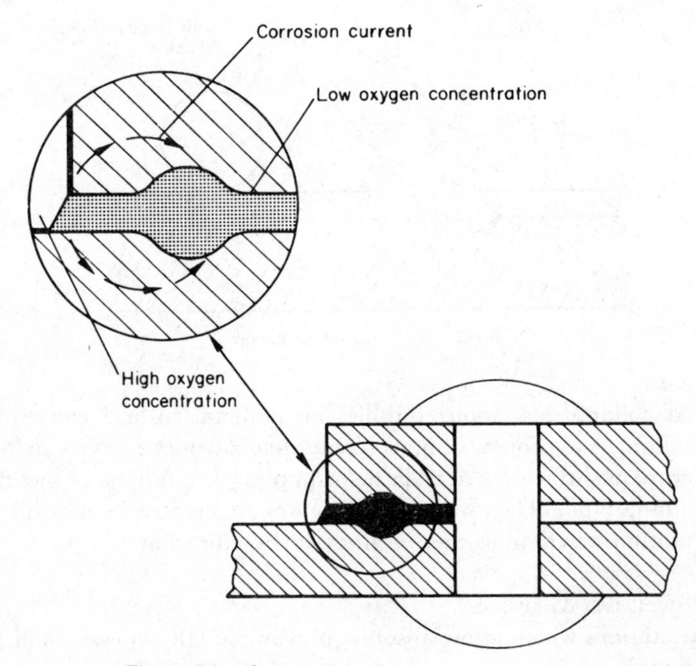

Figure 3.2. Concentration cell

Crevice corrosion: localised corrosion resulting from a crevice formed between two surfaces, one at least of which is a metal.

3.4.2 Cause

Oxygen concentration cells are present in crevices, and also in water lines, adherent deposits and deep recesses, which hinder the diffusion of oxygen and set up differences in solution concentration. The low oxygen areas are anodic and thus corrosion prone.

Metal ion concentration cells, much like their oxygen counterparts, strive to balance out concentration differences. Thus, when the solution over a metal contains more metal ions at one point than another, metal goes into solution where ion concentration is low.

3.4.3 Preventive Measures

Reduce crevices to a necessary minimum, especially in areas of heat transfer or where aqueous environments contain inorganic chemical or dissolved oxygen.

Avoid sharp corners and stagnant areas in design.

Design for complete drainage and provide for uniform environment.

Prevent ingress of corrodant into crevices by improved fit, by use of impervious jointing materials, by encapsulation, by enveloping and by sealing.

Avoid crevice effect between insulation and substrate.

Use design form which facilitates surface cleaning and application of protective coatings.

Specify and assist by design in removal of scale or foreign matter from metal surfaces.

Avoid introduction of foreign matter into media, and settlement of deposits.

Remove solids in suspension—provide filtration.

Use welded butt joints instead of bolted or riveted joints.

Use continuous welds; close crevices in lap joints by continuous welding, caulking or soldering.

Specify sound welding techniques to ensure complete penetration, so avoiding porosity and crevices.

Evaluate individually the crevice effect on each material—select suitable materials.

Avoid fibrous or absorbent packings.

Inhibit environment in crevices or stagnant areas.

3.5 Corrosion–Erosion

3.5.1 Definitions

Corrosion–erosion: a corrosion reaction accelerated by velocity and abrasion; usually accelerated also by presence of solid particles (*Figure 3.3*).

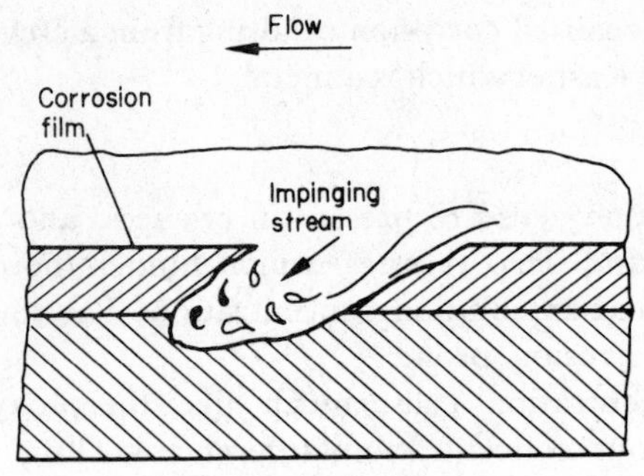

Figure 3.3. Corrosion–erosion

Impingement attack: localised corrosion resulting from the action and/or erosion (separately or conjoint) when liquids impinge on a surface.

Wire drawing: corrosion attack promoted by high velocity wet steam—over 60 m/s (200 ft/s).

3.5.2 Cause

Impingement corrosion is caused by an impinging water stream breaking through corrosion scale and dissolving the metal; the cause of wire drawing is high velocity wet steam. The effect depends mainly on liquid speed and amount of contained air or solids and any factors which affect the rate of formation of protective films.

3.5.3 Preventive Measures

Decrease fluid stream velocity and improve lamellar flow.
Regulate content of air in liquid environment to suit the metal.
Streamline the flow; avoid design which creates turbulence or flow restraint.
Minimise abrupt changes in flow direction.
Align pipe sections.
Streamline inlets and outlets.
Avoid obstructions—real or potential.
Increase thickness of material in vulnerable areas.
Install renewable impingement plates or baffles.
Design for easy renewal by interchangeable parts.
Provide filter for abrasive contaminants in liquids or water trap in steam and compressed air.
Introduce smooth aerodynamic or hydrodynamic surfaces.
Avoid rough texture of surfaces.
Select suitable materials.
Specify suitable coatings or linings.
Protect cathodically.

3.6 Corrosion Fatigue

3.6.1 Definition

Failure by cracking caused by alternating stresses in the presence of a corrosive environment (*Figure 3.4*).

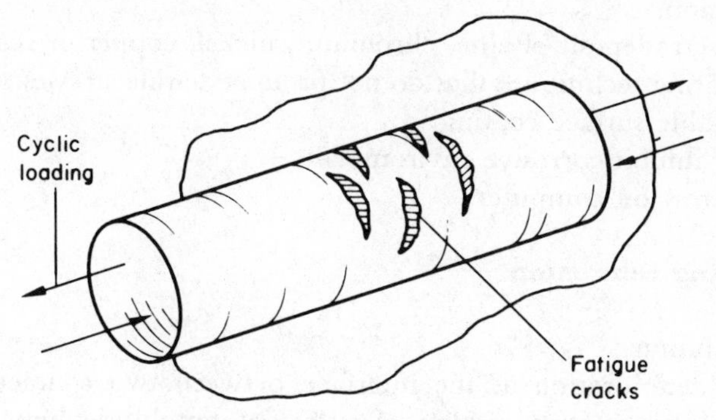

Figure 3.4. Corrosion fatigue

3.6.2 Cause

In much the same way as static stresses link up with corrosion to produce stress corrosion cracking, cyclic loads combine with corrosion to cause corrosion fatigue. This results in metal failure occurring substantially below the fatigue limit for non-corrosive conditions. The combined deteriorating effect of these two—corrosion and fatigue—is greater than the sum of their individual damages.

3.6.3 Preventive Measures

Minimise or eliminate cyclic stressing.
Increase size, bulk or local strength of critical sections.
Reduce stress concentration or redistribute stress.
Streamline fillet design for decrease of stress concentration and improvement of stress flow.
Balance strength and stress throughout the component.
Select the correct shape of critical sections.
Size components by exchange of useless material in non-critical components for stronger critical sections.
Provide for sufficient flexibility to reduce overstressing by thermal expansion, vibration, shock and working of the structure or equipment.
Provide against rapid changes of loading, temperature or pressure.
Avoid internal stress.
Avoid fluttering and vibration-producing or vibration-transmitting design.
Increase natural frequency for reduction of resonance corrosion fatigue.

Improve ductility and impact strength.
Select suitable materials.
Specify stress relieve by heat treatment or by shot peening, swagging, rolling, vapour blasting, tumbling, etc., to induce compressive stresses.
Specify suitable surface finish.
Specify and design for elimination of stress raisers, fretting, scoring and corrosion.
Specify electrodeposit of zinc, chromium, nickel, copper or nitride coatings by plating techniques that do not produce tensile stresses.
Select suitable surface coating.
Change or inhibit corrosive environment.
Analyse stress by computer.

3.7 Fretting Corrosion

3.7.1 Definition

Localised deterioration at the interface between two contacting surfaces, accelerated by relative motion of sufficient amplitude between them to produce slip (*Figure 3.5*).

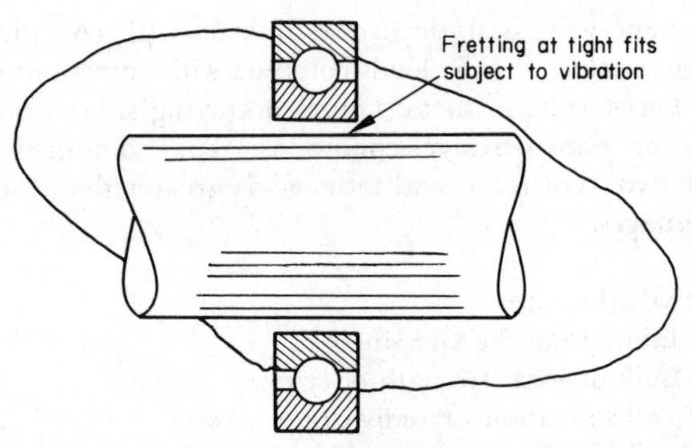

Figure 3.5. Fretting corrosion

3.7.2 Cause

This corrosion occurs between surfaces in close contact, usually under fairly heavy load and subject to very slight relative movement (e.g. minute slippages caused by high frequency vibrations). Differences in elastic strain between surfaces may be sufficient to cause fretting corrosion, which appears to be due to welding of contacting high spots and their subsequent rupture. Local attack may start fatigue cracks, especially where stresses concentrate and mating areas are pitted. Wearing away of surface protective films can initiate galvanic or concentration cell corrosion.

3.7.3 Preventive Measures

Avoid vibration transmitting design.
Introduce barrier between metals which allows slip.
Increase load (but do not overload) to stop motion.
Select suitable materials.
Specify protective coating of a porous (lubricant-absorbing) material.
Isolate moving components from the stationary ones.
Increase abrasion resistance between surfaces, by treating one or both of the surfaces.
Design for exclusion of oxygen on bearing surfaces.
Select compatible metals.
Improve lubrication design—arrange for better accessibility.
Make arrangement for flushing of debris by the motion of lubricant.
Select suitable lubricant.

3.8 Galvanic Corrosion

3.8.1 Definition

Corrosion associated with the current resulting from the coupling of dissimilar electrodes in an electrolyte (*Figure 3.6*).

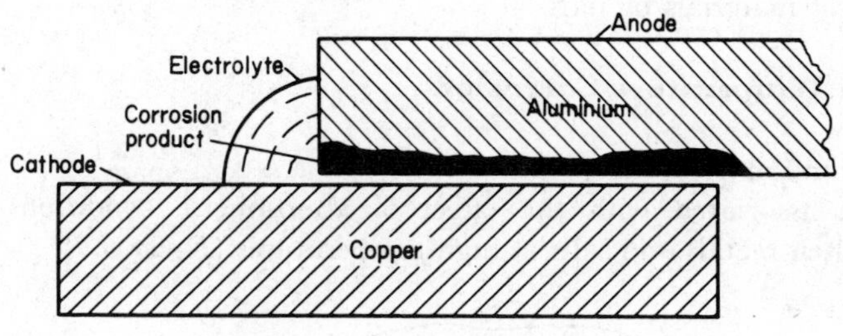

Figure 3.6. Galvanic corrosion

3.8.2 Cause

When two dissimilar metals exposed to an electrically conductive environment are in direct contact, electrically connected by a conductor or by the conductive medium, preferential attack on one, the anodic metal, occurs while corrosion on the other, the cathodic, slows down or stops.

3.8.3 Preventive Measures

Avoid galvanic couples, unless necessary.
If necessary, select from a chart of galvanic series for a particular environment the metals which are compatible or otherwise provide for a *complete dielectric insulation* of one from the other. Select for this insulation suitable and effective materials.

Avoid unfavourable effect of a small anode and a large cathode.
Extend distance between dissimilar metals in conductive medium.
Design for use of readily replaceable anodic parts, or make them thicker.
Change relative position of dissimilar metals to avoid cathodic contamination.
Avoid threaded joints between materials far apart in galvanic series; use brazed, fused or welded joints in preference.
Avoid embedment of dissimilar metal couples in a porous, moisture-absorbing material.
Specify compatible welding and brazing metals.
Specify effective coating of low porosity, *especially on the cathode*.
Do not use conductive coating if it is not compatible with the coupled metals.
Do not rely fully on painting for effective separation of coupled metals.
Use cathodic protection.
Inhibit for reduction of aggressiveness of environment.
Regulate degree of aeration of liquid media to suit coupled metals or induce changes of temperature, movement or chemistry.
Provide for effective ventilation drying of coupled metals.
Prevent access of air and/or water to the bimetallic joint.
Avoid use of non-compatible chemicals for impregnation of materials containing dissimilar metal joints.
Exploit galvanic corrosion in its beneficial form for cathodic protection of critical materials or parts.

3.9 High Temperature Corrosion

3.9.1 Definition

Corrosion associated with the effect of atmospheric conditions, various gases, molten metals and salts at high temperatures (*Figure 3.7*).

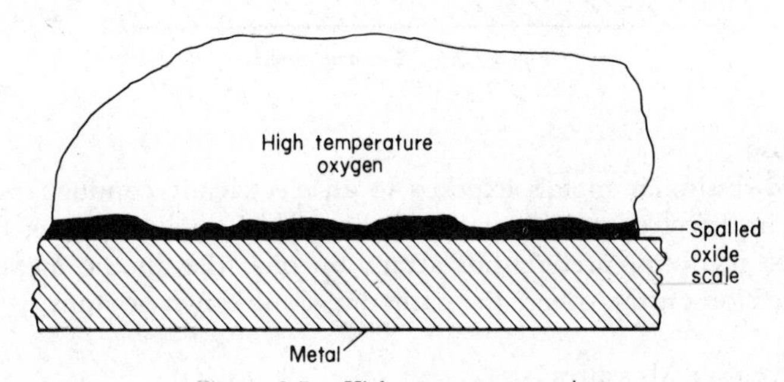

Figure 3.7. High temperature corrosion

3.9.2 Cause

Caused by high temperatures and depends on the composition of the basic metals, composition of the environmental atmosphere, gases, salts or deposited metals, the temperature and the exposure time. It varies considerably. Light metals (those lighter than their oxides) form a non-protective layer that gets thicker as time goes on. This layer forms, spalls and re-forms. Other forms of high temperature corrosion include sulphidation, carburisation and decarburisation.

3.9.3 Preventive Measures
Select stable material.
Adjust temperatures and environment.
Regulate duration of adverse contact.

3.10 Hydrogen Damage

3.10.1 Definition
Reduction of the load-carrying capability by the admission of hydrogen into the metal (*Figure 3.8*).

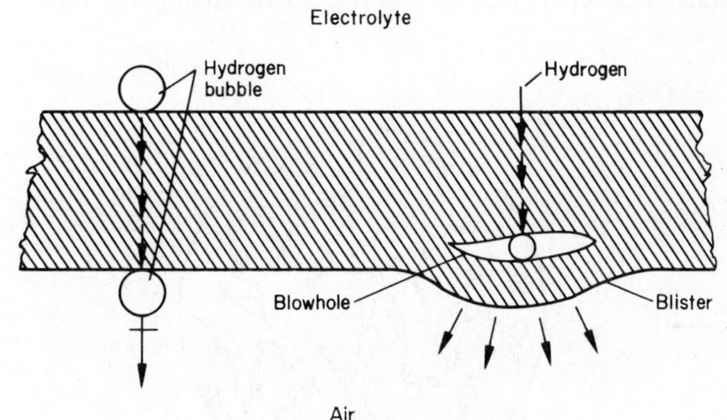

Figure 3.8. Hydrogen damage

3.10.2 Cause
Mechanical damage of a metal caused by the presence of, or interaction with, hydrogen. Hydrogen blistering and hydrogen embrittlement are caused by penetration of atomic hydrogen into metal. Decarburisation is caused by moist hydrogen at high temperatures. Hydrogen attack is a disintegration of oxygen-containing metal in the presence of hydrogen. The origin of hydrogen can be found in the cleaning, pickling, cathodic protection, welding, treatment and operation.

3.10.3 Preventive Measures
Select a *clean* metal.

Select a resistant material—homogeneous or clad.
Select low hydrogen welding electrodes and specify welding in dry conditions.
Select correct surface preparation and treatment.
Avoid incorrect pickling and plating procedures.
Metallise with resistant metal, or use a clad metal.
Induce compressive stresses.
Remove hydrogen from metal by baking (200–300°F; 93–149°C).
Provide for control of media chemistry (e.g. use inhibitors, remove sulphides, arsenic compounds, cyanides and phosphorus-containing ions from environment).
Control cathodic protection potential.
Specify impervious protective coating (e.g. rubber, plastic).
Avoid anodic metallic coatings.

3.11 Intergranular Corrosion

3.11.1 Definition
Preferential corrosion at grain boundaries of a metal or alloy (*Figure 3.9*).

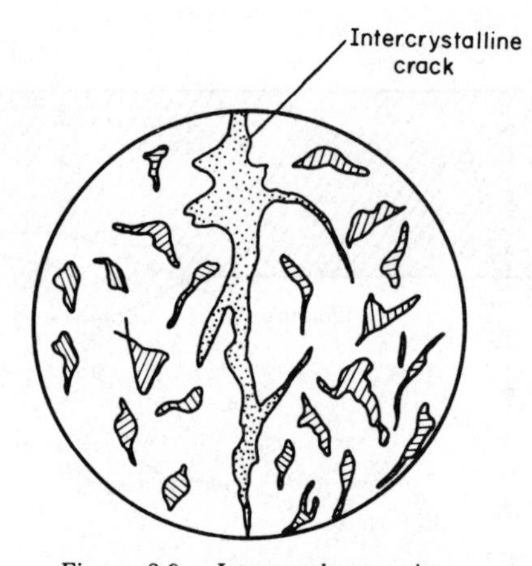

Figure 3.9. Intergranular corrosion

3.11.2 Cause
Due to a number of reasons but generally the result is the same, namely, a selective attack and intercrystalline cracking along the metal's grain boundaries. In some austenitic stainless steels, chromium carbides may precipitate at grain boundaries when cooling from welding temperatures. Corrosion attacks chrome-starved areas.

3.11.3 Preventive Measures
Select materials not susceptible to grain boundary depletion.

Select suitable heat treatment.
Avoid specifying heat treatment or welding in the susceptible range.

3.12 Microbial Corrosion

3.12.1 Definition

Deterioration of materials caused directly or indirectly by bacteria, moulds or fungi singly or in combination.

3.12.2 Cause

Using the term corrosion in the broadest sense, the microbes (bacteria, moulds or fungi) may cause corrosion by:

(*a*) Chemical attack of metals, concrete and other materials by the by-products of microbial life, namely acids (e.g. sulphuric, carbonic or other organic acids), hydrogen sulphide or ammonia.

(*b*) Microbial attack of organic materials (e.g. organic paint coatings, plastic fittings and linings), some natural inorganic materials (e.g. sulphur) or inhibitors.

(*c*) Depassivation of metal surfaces and induction of corrosion cells.

(*d*) Attack of metal by a process in which microbes and the metal co-operate to sustain the corrosion reaction.

(*e*) Attack due to a combination of bacteria.

3.12.3 Preventive Measures

Analyse accurately probabilities of contamination.
Provide for control of media chemistry.
Inhibit or provide for addition of germicide.
Construct non-aggressive surround or secure controlled removal of microbial nutrient.
Select suitable resistant material.
Select suitable protective coating.
Use cathodic protection.
Provide accessibility for frequent cleaning.

3.13 Pitting Corrosion

3.13.1 Definition

Localised corrosion in which appreciable penetration into the metal occurs, resulting in the formation of cavities (*Figure 3.10*).

3.13.2 Cause

When protective film or layers of corrosion product break down, localised corrosion (pitting) occurs. An anode forms where the film has broken, and the unbroken film (or corrosion product) acts as a cathode. The pits form starting points for stress concentration that can cause, or accelerate, stress-corrosion or corrosion fatigue attack.

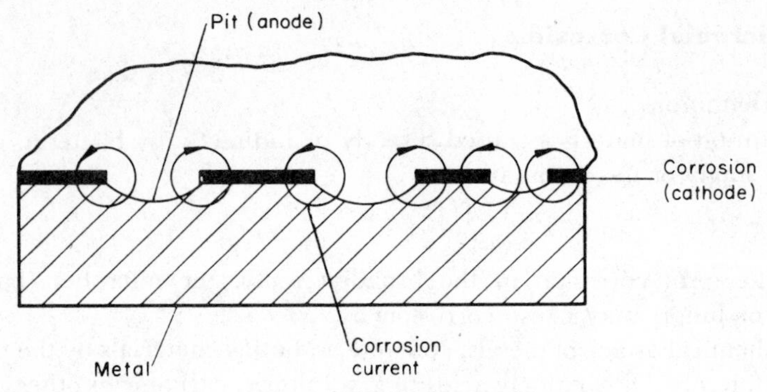

Figure 3.10. Pitting corrosion

3.13.3 Preventive Measures

Select suitable resistant material.

Select suitable geometry to prevent aggressive conditions.

Adjust thickness of material—allow for depth of pits.

Provide for control of media chemistry (do not rely solely on inhibitors unless attack can be thus completely stopped).

Specify protective coating.

Secure formation of continuous and sound protective film.

3.14 Selective Attack (Leaching)

3.14.1 Definition

The process of extraction of a soluble component from an alloy with an insoluble component, by percolation of the alloy with a solvent—usually water (*Figure 3.11*).

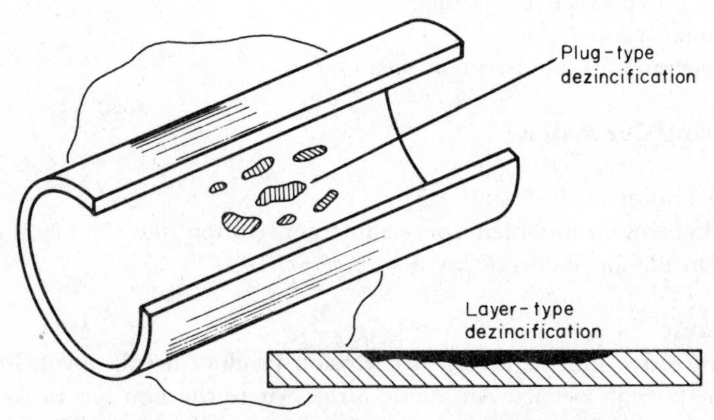

Figure 3.11. Selective attack (leaching)

3.14.2 Cause

Basically, one element of a metal or alloy is singled out for corrosion attack. Common types are dezincification, dealuminification and graphitic corrosion. When, for example, copper–zinc alloys (brasses) containing less than 85% copper are exposed to wet conditions for prolonged periods, zinc may go into solution and the redeposited copper has little mechanical strength. Common cast iron can also act this way inasmuch as, in some corrosives, the iron corrodes out leaving just a porous graphite residue that virtually crumbles.

3.14.3 Preventive Measures

Select resistant material.
Reduce aggressiveness of environment.
Use cathodic protection.

3.15 Stray Current Corrosion (Electrolysis)

3.15.1 Definition

Corrosion resulting usually from direct current flow through paths other than the intended circuit (*Figure 3.12*).

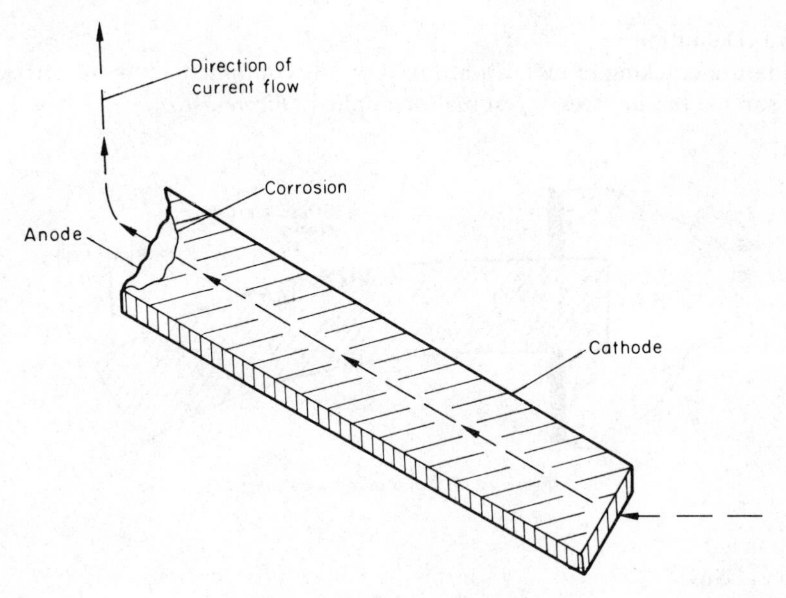

Figure 3.12. Stray current corrosion

3.15.2 Cause

Electrolytic corrosion due to uncontrolled electrical currents (mostly dc or hvdc) from extraneous sources through unintended paths, such as bad earth return on electrical equipment, giving rise to leakage of currents through metal structures and other preferentially conductive paths and causing connected structures and equipment to corrode.

3.15.3 Preventive Measures

Design electrical circuits and equipment so that all exposed parts or panels of conductive materials are at a ground potential at all times.
Insulate electrical cables, components and equipment from structures.
Provide controlled grounding of electrical equipment.
Use non-conducting fluid.
Drain off stray currents with another conductor.
If necessary apply cathodic protection.
Select suitable sites for structures or parts; remove the source of stray current.
Embed structures or parts in inert, non-conducting medium—prevent access of stray currents.
Secure electrical continuity of critical conductors (e.g. pipelines).
Bond metallic structures, normally to an earthing point.
Use expendable *targets* connected to the anodic sides of insulating joints.

3.16 Stress Corrosion Cracking

3.16.1 Definition

Premature cracking of metals produced by the continued action of corrosion and surface tensile stress—residual or applied (*Figure 3.13*).

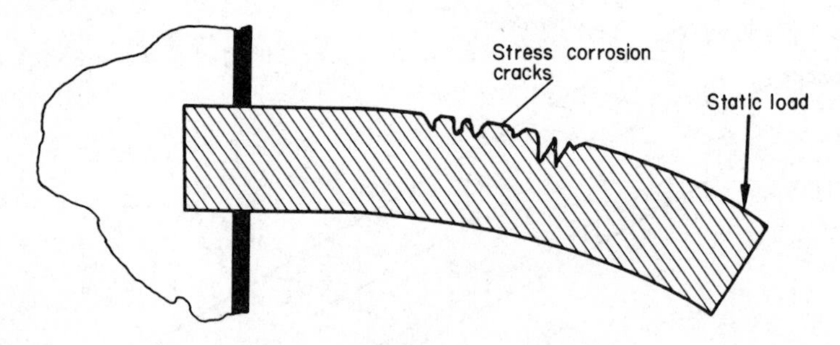

Figure 3.13. Stress corrosion cracking

3.16.2 Cause

Due to a combination of high tensile stress and a corrosive environment (e.g. sea water, perspiration, welding fluxes, cleaning compounds, lubricants, organic solvents and various chemicals, as indicated in *Table 3.2*). Tensile stresses build up under static loading at metal surface. Corrosive action concentrates stresses, causing them to exceed the metal's yield point. Under continual exposure, with the metal corroding and local high stresses concentrations building up, the metal will eventually fail. Non-metals are also subject to similar phenomena.

Table 3.2 ENVIRONMENTS CAUSING STRESS CORROSION

Material	*Environment*
Aluminium	Water and steam; NaCl, including sea atmospheres and waters; air, water vapour
Copper	Tropical atmospheres; mercury; $HgNO_3$; bromides; ammonia; ammoniated organics
Aluminium bronzes	Water and steam; H_2SO_4; caustics
Austenitic stainless steels	Chlorides, including $FeCl_2$, $FeCl_3$, NaCl; sea environments; H_2SO_4; fluorides; condensing steam from chloride waters
Ferritic stainless steels	Chlorides, including NaCl; fluorides; bromides; iodides; caustics; nitrates; water; steam
Carbon and low alloy steels	HCl; caustics; nitrates; HNO_3; HCN; molten zinc and Na–Pb alloys; H_2S; H_2SO_4–HNO_3; H_2SO_4; sea water
High strength alloy steels (yield strength 200 psi plus)	Sea and industrial environments
Magnesium	NaCl, including sea environments; water and steam; caustics; N_2O_4; rural and coastal atmosphere; distilled water
Lead	Lead acetate solutions
Nickel	Bromides; caustics; H_2SO_4
Monel	Fused caustic soda; hydrochloric and hydrofluoric acids
Inconel	Caustic soda solutions; high purity water with few ppm oxygen
Titanium	Sea environments; NaCl in environments 288°C (550°F); mercury; molten cadmium; silver and AgCl; methanols with halides; fuming red HNO_3; N_2O_4; chlorinated or fluorinated hydrocarbons

3.16.3 Preventive Measures

Minimise applied or residual tensile stresses.

Secure sufficient flexibility.

Increase size of critical sections.

Reduce stress concentration or redistribute stress.

Compensate for loss of stiffness produced by penetration.

Avoid misalignment of sections joined by riveting, bolting and welding.

Design simple joints under stress. Avoid lap welding, riveting, bolting. Butt or fillet welding is preferred.

Specify for stressed structures techniques that produce sound welds; also careful preparation and finishing of welds.

Specify and design for elimination of stress raisers.

Select suitable material (metallurgical composition).

Avoid specifying any machining, assembling or welding operations which impart residual tensile stresses.

Use materials in assembly with similar coefficient of expansion.

Secure control of heat treatment and metal working to develop resistant microstructure.

Specify input of compressive surface stresses by suitable treatments.

Specify electroplating or metallising in stressed areas.

Select suitable surface coatings. Specify passive surface films, suitable organic coatings or stoved resin coatings in critical areas.

Use controlled cathodic protection.

Analyse and control environmental conditions. Exclude corrosive environment.

Secure maintenance of low service temperature.

Eliminate possible corrodants from service environments or suitably inhibit.

Prevent by design a repetitive wetting and drying of critical surfaces.

Prevent all types of corrosion in critical spaces by any suitable means.

Analyse stress by computer.

3.17 Thermogalvanic Corrosion

3.17.1 Definition

Corrosion resulting from a galvanic cell caused by a thermal gradient (*Figure 3.14*).

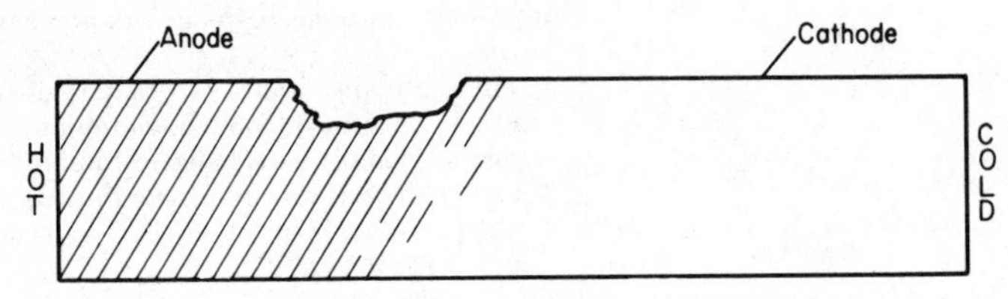

Figure 3.14. Thermogalvanic corrosion

3.17.2 Cause

When a metal is subjected to a thermal gradient by uneven heating or dissipation of heat, this has a similar effect on the metal as galvanic corrosion. The metal is differentially polarised and anodic and cathodic areas are formed, causing preferential attack to develop.

3.17.3 Preventive Measures

Avoid uneven heating, cooling and formation of hot spots.

Design layout to prevent adverse effect of one component on the other.

Provide for continuity of insulation or lining.

Prevent by design access of differentially heated or cooled liquids from exterior sources.

3.18 Uniform Corrosion (General Corrosion)

3.18.1 Definition
Corrosion in which no distinguishable area of the metal surface is solely anodic or cathodic, i.e. anodes and cathodes are inseparable (*Figure 3.15*).

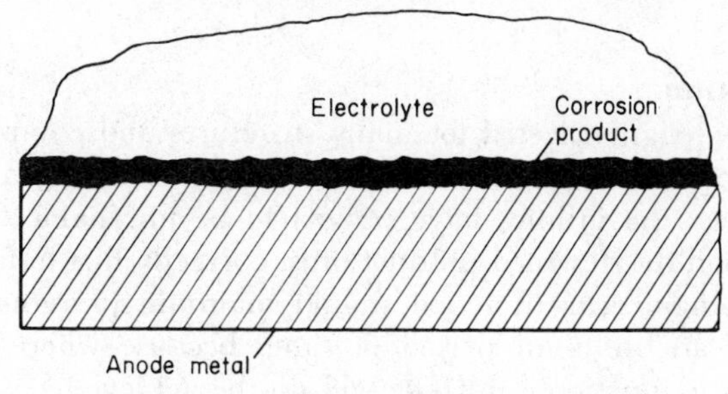

Figure 3.15. Uniform corrosion (general corrosion)

3.18.2 Cause
This corrosion is indicated by a general wasting away of the surface and will normally be found where a metal is in contact with an acid or solution. The presence of such is, however, not essential (e.g. high temperature oxidisation is a form of uniform attack which occurs in a relatively dry atmosphere). The corrosion product may either form a protective layer on the metal and slow down further corrosion or, as in the case of direct chemical attack, the corroded material will dissolve in the corrosive environment.

3.18.3 Preventive Measures
Select suitable resistant material.
Change or inhibit environment.
Specify resistant protective coating.
Use anodic protection.

4 Systematic Analysis

4.1 Introduction

Materials and designs selected for many structures and equipment with the solitary help of functional ratios may not be the most suitable choice. In many such cases the utilities have proved to be unreliable, expensive and unsafe. Although the designed product must correctly first fulfil its functional purpose, this engineering function should maintain its economic and safe continuity for an optimum period of time, because where corrosion can interfere the true functional purpose will not be achieved.

Thus the concerned designer should not concentrate purely on the functional aspects of design, to the total exclusion of other considerations, but must be aware that there are many ways in which corrosion can ruin even the best creation. He should acquaint himself with the basics of corrosion and should be fully aware of his own power and opportunity to ease, retard or stop corrosion in a reasonable and economic way, by selective employment of qualified precautions or by optimal adjustment of his functional design. Sophisticated protective measures should be known to him and he should learn that these require help and intrinsic support from the product he creates. In other words, the corrosion-control measures he takes in his design need tactical, logistic and mainly logical support embodied in the design itself.

The designer should appreciate both his technical formulae and also the corrosion which destroys the function of his product. All this knowledge should be combined in a unified and orderly form of creation.

Basically, the main effort in corrosion control is given to (1) *curative control*—to cure corrosion after it has occurred; and (2) *preventive control*—to prevent or delay corrosion and reduce its harmful effects by taking precautions in advance.

Preventive control deals with pre-production planning, specification of corrosion-control measures and selection of optimal materials, design forms, fabrication and treatment methods to suit the finite environment and conditions of employment. Further, preventive control is concerned with putting these selected measures into effect prior to deployment of the designed structure or equipment and ultimately with the means of securing the appropriate quality of the product's economically extended functionalism.

The cost and degree of efficiency of the embodied corrosion-control measures can be predetermined and their system varied to suit. *The unexpected is more expensive than the planned and predicted.* For this reason preventive control should be the prime consideration of every designer. On the other hand, curative control of the designed utility must not be altogether forgotten and all newly designed products must be made ready for its probable deployment at any appropriate time.

Considering the ultimate responsibility of the designer to the user for the reliability, economy and safety of the designed structure or equipment, it is logical that the designer should make a serious effort to combat corrosion by any available means within his product's design complex. Thus corrosion control should be seriously organised, planned, researched, analysed and executed with the appropriately exacting attention to detail. Generalisation is just not good enough when the combined design team is organised, the information gathered, evaluated and analysed, the merit factors reconciled and the appropriate action planned and detailed.

4.2 Scope

Designers, in their endeavour to secure a unified process of designing a reliable and safe product with an economically sound life-cycle, have to learn from the very beginning how to enrich their information-gathering processes, how to co-operate with their new team-mates and how to adjust their thinking process in parallel to cope with their plan of action, encompassing both the function of the utility and its corrosion control without mutual interference. An attempt is made in this section to indicate summarily any possible precautions that may be taken to cater for the extra work of including corrosion control into the designer's work.

4.3 Organisation of Work

A thorough planning of working sequences and procedures is necessary to secure the requisite efficiency and smooth flow of applying corrosion control in design. It is not sufficient to do this spasmodically at any particular moment in design procedures. Corrosion control should be appreciated concurrently with the whole complex form and function of the utility and at all time—from the moment of conception of the idea until the utility becomes fully operational.

It is to the advantage of the whole design team that proper rules of procedure be arranged and maintained to facilitate co-operation and secure effective results. Existing drawbacks are largely connected with the absence of standardised management methods, and also typical specifications determining the responsibilities of individual specialists within the team for setting up the corrosion-control services. In connection with this, it is desirable to make a comprehensive assessment of such services in each office, in order to find rational forms and approaches which would be technically and economically acceptable to the particular organisation.

Possible extent, sequencing and flow are indicated in a schematic diagram of corrosion control (*Figure 4.1*) and in flow charts (*Figures 4.2–4.4*).

4.4 Teamwork

Corrosion control does not always fit snugly into the overall pattern of design work and many difficulties must be overcome before a reasonable degree of co-operation and mutually acceptable compromise can be reached. Close co-operation between the executive management, designers and corrosion experts is a necessity. In a responsible engineering practice, corrosion control

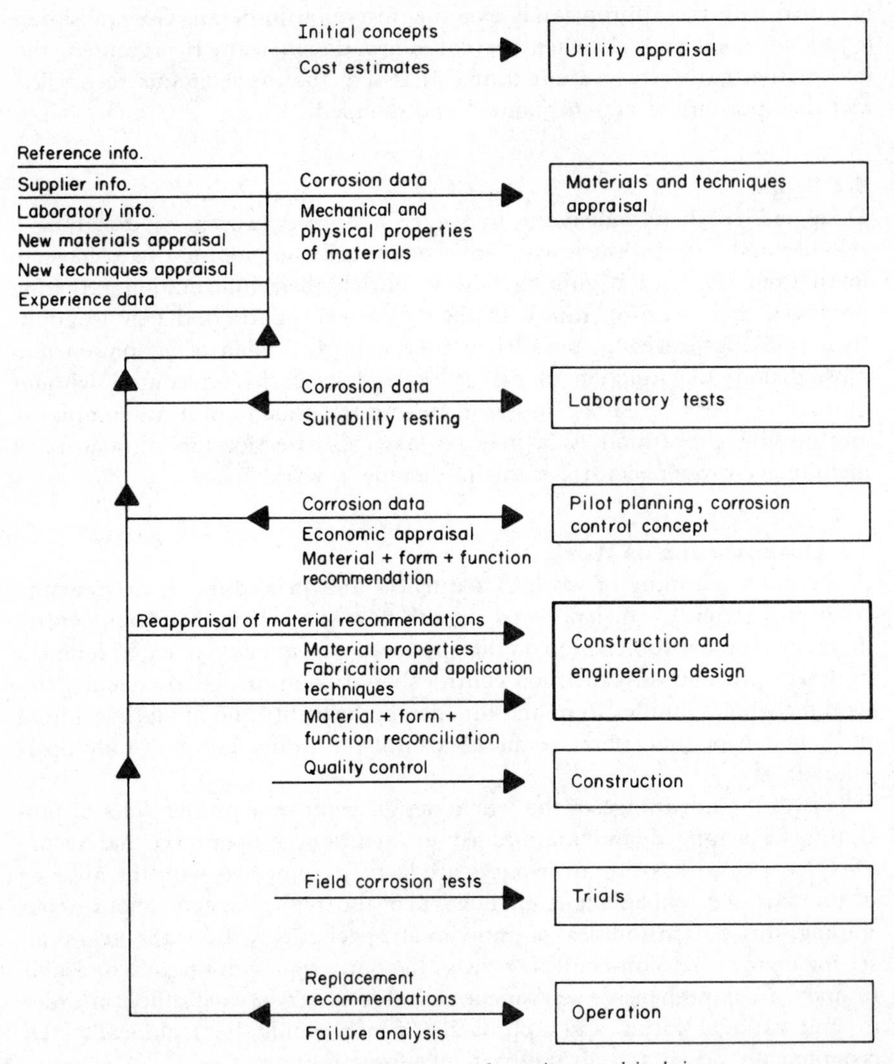

Figure 4.1. Schematic diagram of corrosion control in design

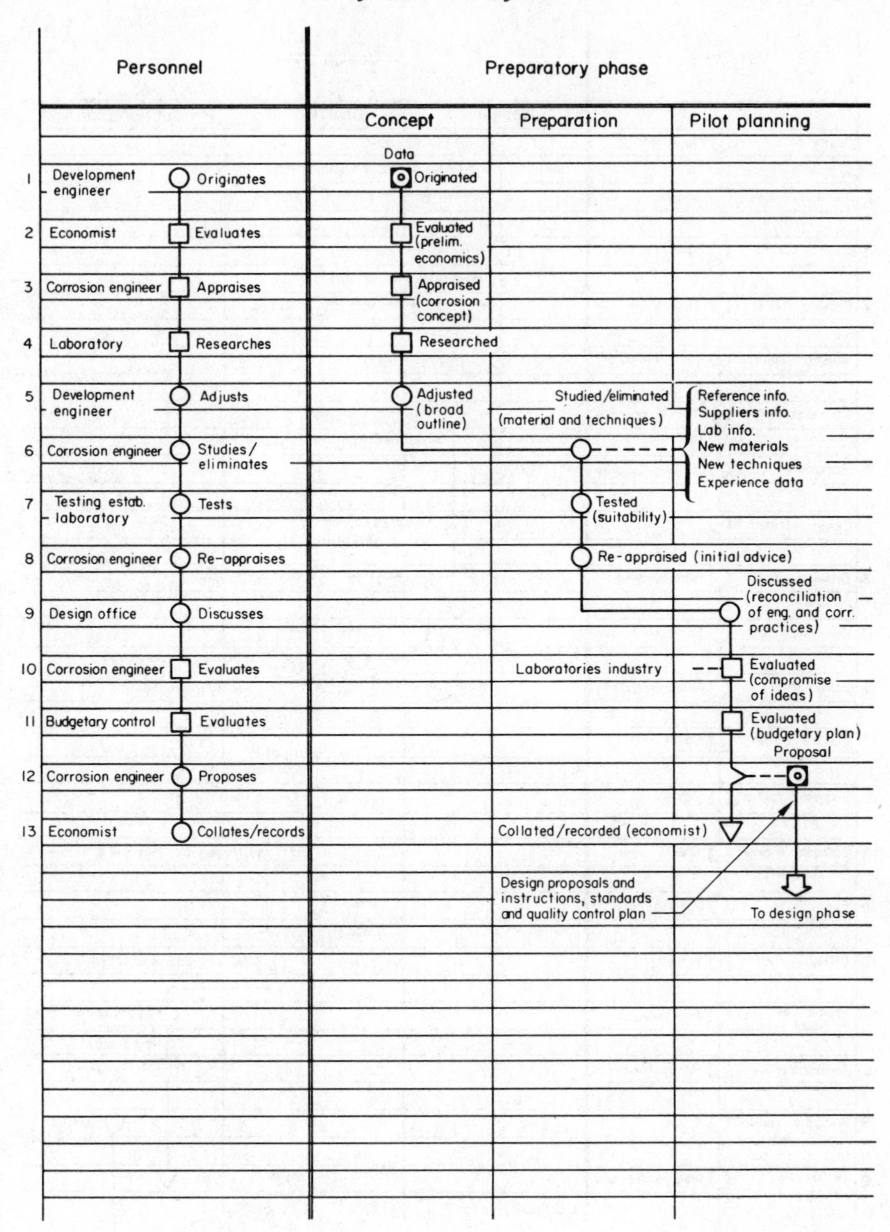

Figure 4.2. Corrosion-control flow diagram—preparatory phase

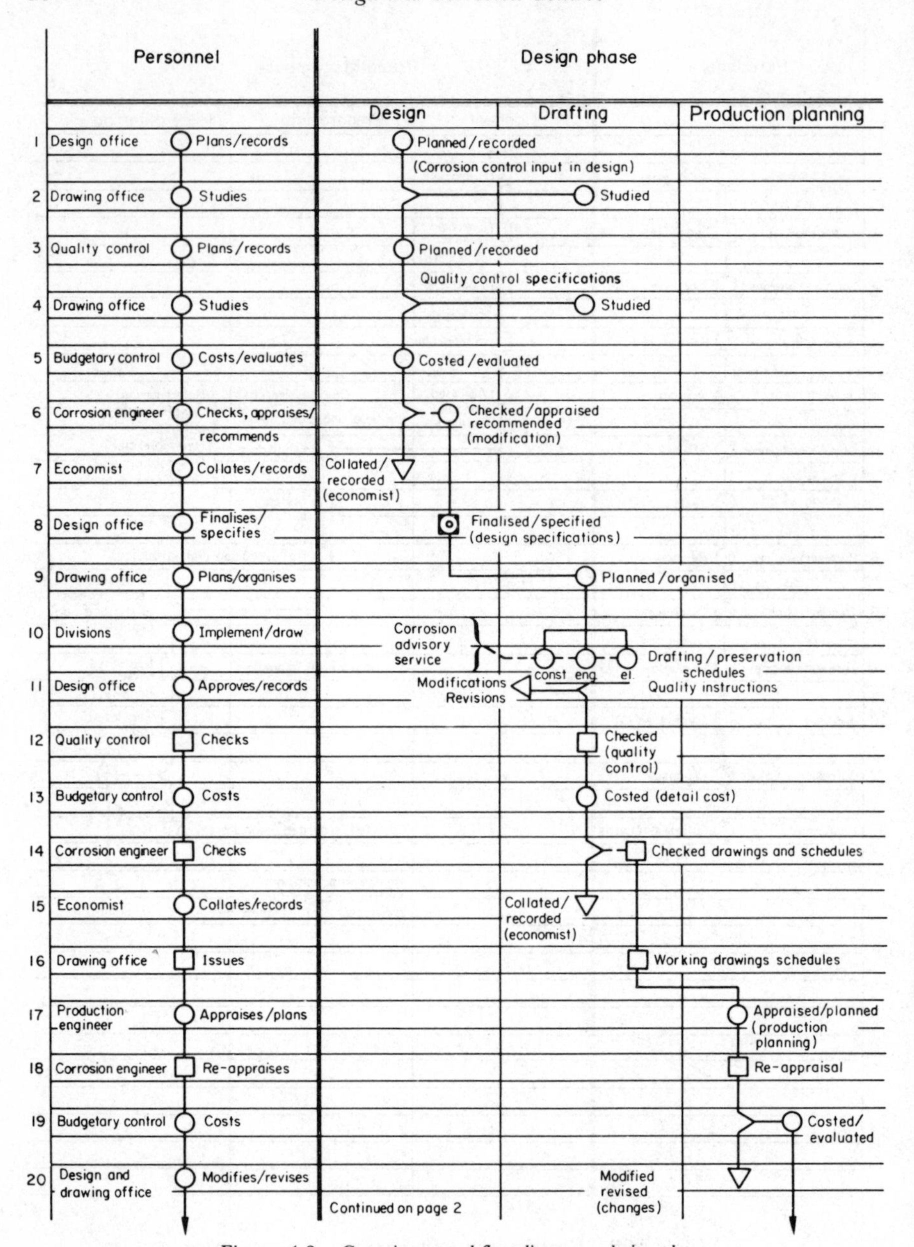

Figure 4.3. Corrosion-control flow diagram—design phase

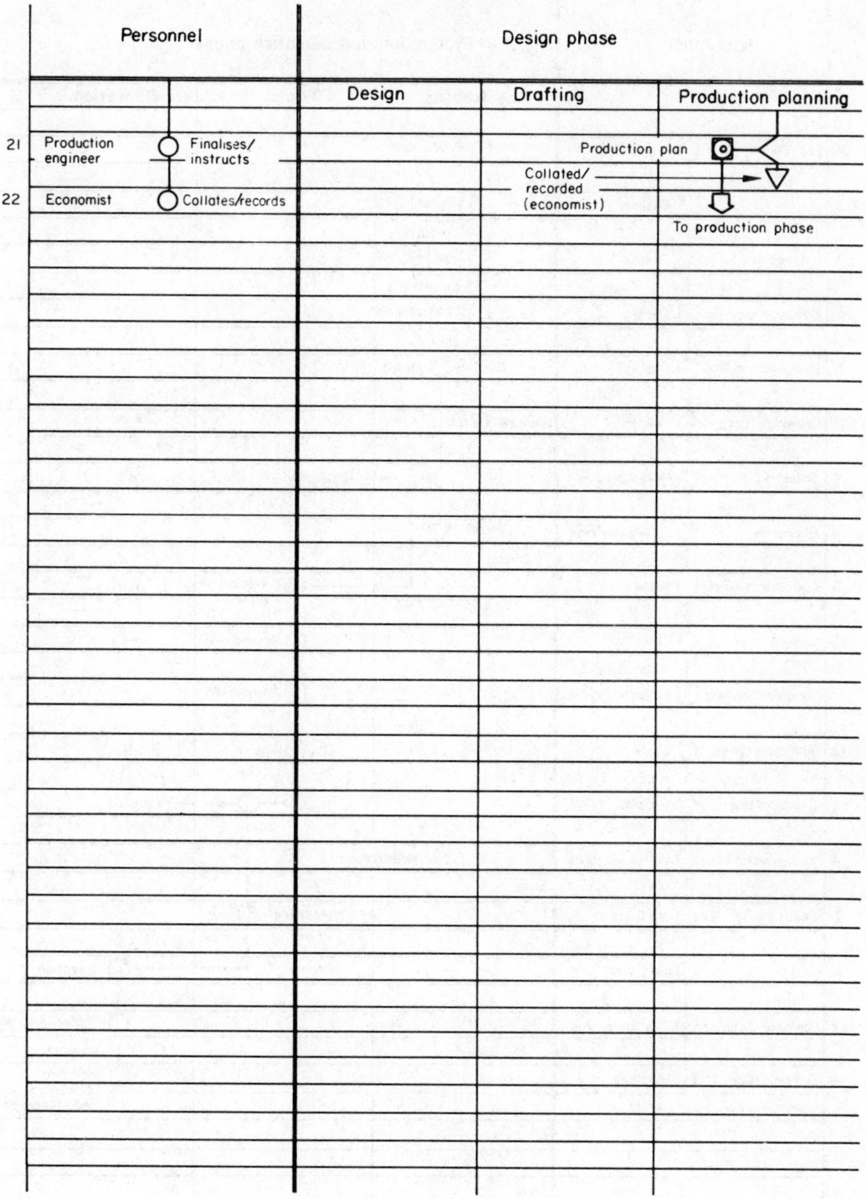

	Personnel		Design phase		
			Design	Drafting	Production planning
21	Production engineer	Finalises/ instructs			Production plan
22	Economist	Collates/records		Collated/ recorded (economist)	To production phase

Figure 4.3 (Contd.)

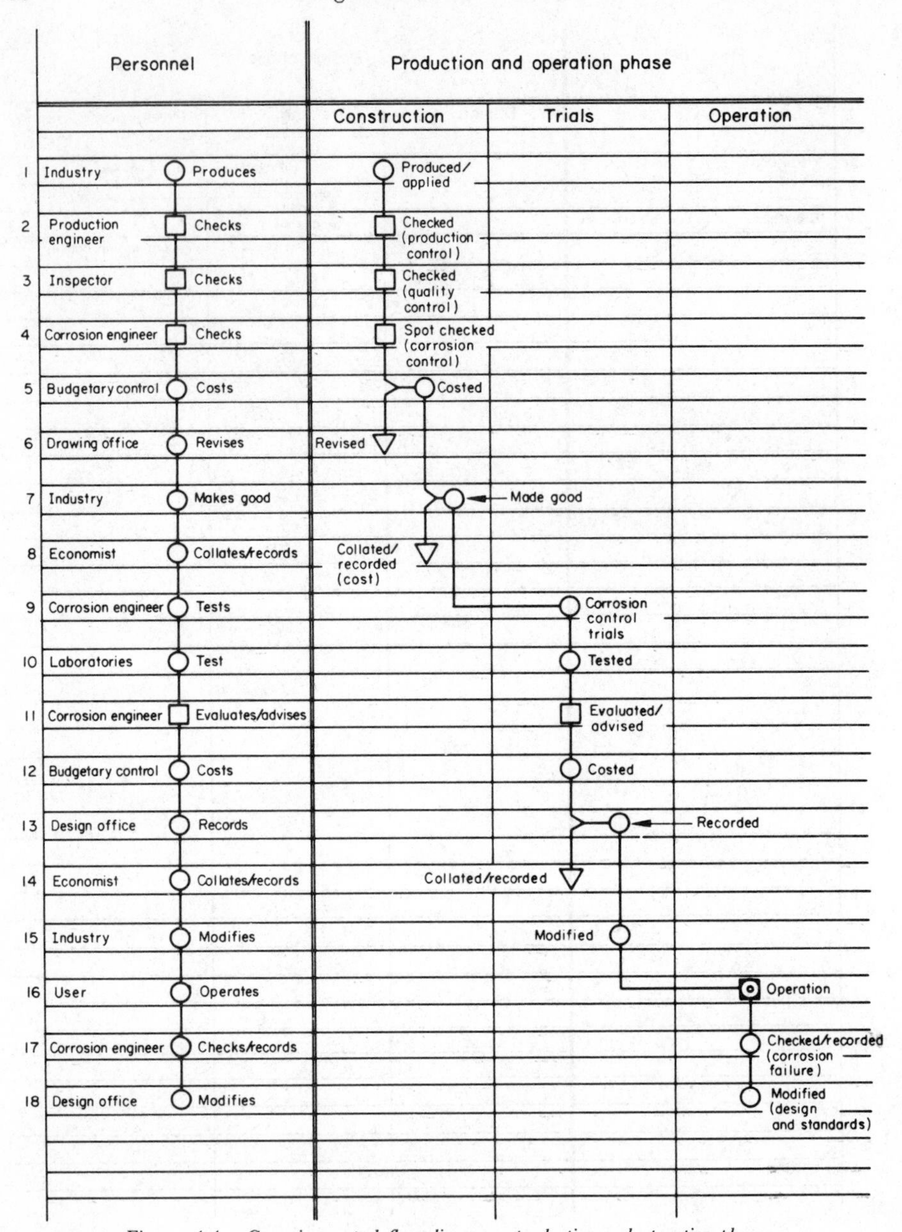

Figure 4.4. Corrosion-control flow diagram—production and operation phase

can no longer be fully subordinate to the others—it should be an equal partner of the team.

Whilst not every project needs to be the same, basically the following organisational units will take part in the process of planning and implementing corrosion control in the design of a product: the development personnel, who conceive the idea; economists and estimators, who evaluate the utility financially; project design personnel, who reduce the idea to an engineering practice; draughtsmen, who illustrate the design in a workable form; production engineers, who convert the idea, design and drawings into production plans; various corrosion experts, who enhance the optimum working life and safety of the product through input of their specialised knowledge; laboratories, who supply the scientific corrosion data and test the soundness of corrosion-control appreciation and practical application; various industries, who supply information, materials and services; and, lastly, the quality control organisations on all levels, who ensure the sound incorporation of anti-corrosion measures into the finished product.

It is in the interests of each of these specialists to know what co-operation one can expect from any other member of his corrosion-control team.

4.4.1 Development Engineer

Informs on overall corrosion involvement within the project utility.

Informs on probable or possible environmental conditions and corrosion and ecological problems created by the product.

4.4.2 Economist

Informs on broad spectrum evaluation of economic feasibility of the product, including its corrosion control.

Instructs on cost limits for implementation of corrosion-control measures.

In latter stages provides budgetary control to prevent corrosion control from running wild and to prevent unnecessary and excessive precautions.

4.4.3 Estimators and Costing Personnel

Compute, financially evaluate, record and report continuously on the cost of anti-corrosion measures at all stages of the design work, to prevent waste.

4.4.4 Designers

Study, consider, reconcile and embody into the design such corrosion-control precautions that materially do not interfere with the engineering function of the utility and serve the purpose of optimal upkeep of its economic function.

Seek relevant information from corrosion specialists and other involved personnel and sources on matters of corrosion-control policy and details.

Supply corrosion specialists with necessary data and allow access to their work for the purpose of study and constructive criticism.

Amend, revise or modify their design to a reasonable degree to suit the demands of corrosion control (changes to be documented).

Implement the design instructions with well-founded quality control rules on matters of corrosion prevention.

Supply required data to budgetary control for costing and evaluation of corrosion prevention.

Initiate procurement or contract documents assisting in optimal corrosion control.

Secure compatibility and prevent corrosive interference within their design between items supplied from various sources.

4.4.5 Draughtsmen

Study and correctly interpret corrosion-control instructions and involvement of guidance drawings, specifications and design and quality control instructions.

Consult with corrosion specialists on inclusion of relative aspects of corrosion control in the working drawings, bills of materials and schedules in the best interest of preventing infusion of corrosion into the product to be fabricated.

Co-operate with production control on adjustment of corrosion-prevention measures to suit both parties and amend working drawings and schedules accordingly.

Supply necessary data to costing personnel.

Assist corrosion specalist in evaluation of drafting work.

4.4.6 Production Control

Secure practical planning of corrosion-control measures to suit the design and the particular production methods and techniques, as well as the application facilities and procedures.

Reconcile design with production.

4.4.7 Corrosion Engineer

Supplies up-to-date information and practical expert advice (discussions, evaluations, advisory worksheets, design instructions, proposals and specifications) on principles and good practices of corrosion control; on the nature and effect of corrosive environments; on structural, metallurgical, physical and mechanical properties of various materials relative to their rate of corrosion, on their availability, fabrication, welding, treatment, their optimum design form, method of application and effective saving in weight.

Advises on substitutions, clad metals, weld overlays, metallising, preservation systems, anodic and cathodic protection, environmental adjustment, etc.

Acts as a clearing house for corrosion information to feed it selectively to the design personnel; to foster their awareness and involvement in corrosion control. Participates in writing specifications, standards and recommended practices instructions on matters of corrosion-control affinity.

In collaboration with laboratories, testing establishments and project officers investigates new corrosion-control materials, processes, equipment and methods consistent with good practices; generates new ideas and investigates changes in design, specifications and standards.

Evaluates economy of individual precautions on demand.

Correlates technical work of design and drawing offices with original corrosion-control specifications, design instructions, manuals, standards and rules of good husbandry in corrosion control; instructs and examines for correct incorporation of corrosion control in all design activities, including guidance and working drawings either in pictorial form or in notes, schedules and bills of materials.

Assists in translation of corrosion aspects of drawings and specifications into practical working instructions to project and production engineers and represents corrosion-control interests in negotiations between the design office and production organisation for sound inclusion of corrosion-control rules, specifications, instructions and quality control stipulations into production planning.

Collaborates with quality control organisation on setting up of appropriate quality control procedures and on maintenance of quality assurance of corrosion-control precautions during design, drafting, production and trial activities.

Represents management and design office on corrosion and failure committees for interpretation of applied corrosion-control measures and proposes revisions.

Co-operates with ordering channels on corrosion-control suitability of bought-out items and contents of appropriate instructions contained in contracting documents.

4.4.8 Laboratories and Testing Establishments

Report on pure research of corrosion phenomena and applied research of corrosion-control materials and methods.

Test at various stages of the design programme or on request the performance and suitability of materials and methods to assure optimum use, application and design form in the given conditions.

Participate on pilot and trial runs for evaluation of efficiency or merit of tested corrosion-control precautions.

Install and operate scientific testing and recording apparatus for evaluation of failures and non-destructive testing.

Participate in quality assurance.

Assist design organisation in avoidance of guesswork in preparation of design and in establishing a more stable scientific basis for engineering decisions.

4.4.9 Industry

Supplies accurate and complete corrosion data on their own products, methods, techniques and facilities. Collaborates on applied corrosion research and testing relevant to their products.

Supplies correct materials and services in accordance with specifications, design, schedules and working drawings and maintans their uniform quality.

Trains and supplies efficient advisory staff, approved applicators and site inspectors for securement of effective corrosion-control measures (materials and work) when arranged.

4.4.10 Quality Control Organisation

Assures that quality control measures are maintained on all levels of planning, design, drawing and fabrication.

Plans and organises quality control for individual corrosion-control systems and procedures.

Composes written or drawn instructions and quality assurance specifications overall or in detail for individual tasks.

Performs practical inspection in co-operation with laboratories and corrosion specialists.

Indicates modes of enforcement of quality assurance.

4.5 Sources of Information

Before project analysis can commence the basic common concepts of the project utility should be known to all personnel engaged both in functional design and corrosion-control work, as well as the basic philosophy of the utility complex and the principles of working and flow sequences of all mechanical, chemical and electronic components which constitute the utility.

Further, the design personnel should make a total effort to collect from all available sources accurate data and information relating to corrosion-control requirements of the project, in their most comprehensive form, to allow the designers to analyse and select proper measures and appreciate accurately their probabilities.

4.5.1 Environmental Conditions

Effective corrosion-control analysis needs a preliminary and accurate assessment of the environmental conditions, to which the utility or any of its parts may be subjected on exterior and/or interior surfaces. The appreciation should not be based on vague guesses; the most accurate description or prediction that can be obtained should be used and its annotations should be presented for analysis in the standard units of measurement, wherever possible.

Prior assistance of experts (e.g. environmental, corrosion and material engineers) should be asked for, to interpret the more complex conditions, as well as their effect on the analysed design materials, forms and other items.

The extent of the appreciation may cover the following environmental parameters:

4.5.1.1 Atmosphere

Climatic influences; basic nature of atmospheric conditions (marine, industrial and countryside); dry or wet conditions (rain and dewing); temperature range; relative or critical humidity; frequency, duration and pH of exposure to rain, snow and sleet; exposure to condensation; exposure to sun (including ultraviolet and infrared parts of the spectrum) and impact of ionised particles; exposure to ozone; direction and velocity of prevalent winds; ecological consistency of air (airborne solid and fluid pollution); etc.

4.5.1.2 Natural Waters

Composition of waters; organic matter content; inorganic mineral and metallic constituents and impurities; dissolved mineral salts, calcium carbonate, etc.; entrained and dissolved oxygen and other gases; contents of mercury and ammonia; constancy of composition or mix; chlorinity; salinity; temperature gradients; pH value; rate of flow and velocity; stratification; density; fouling organisms and agents; type and continuity of exposure (immersion, wash, spray, spillage, residual, condensation); electrical conductivity and resistivity of water; type of bottom; etc.

4.5.1.3 Soils

Chemical composition; texture and structure; clay function; aeration and oxygen diffusion; gravitational, capillary and free ground water; pH reaction; soluble salts content; microbiological and organic content; differences in soil potential; current induction; atmospheric lighting effect; electrical conductivity and resistivity; pick-up of stray currents; etc.

4.5.1.4 Chemicals

Type and composition of chemical; physical state (solid, liquid, gaseous); toxicity; purity; concentration/dilution; pH value; continuity and type of exposure (cycling, immersion, spillage, fumes); maximum and minimum temperatures; fluid velocity; aeration and oxygen content; effect of corrosion products on the chemical; catalytic effect; probability of osmosis; etc.

4.5.1.5 Liquid Metals

Temperature and chemical gradient; metal transfer; non-metal transfer; etc.

4.5.1.6 Dry Heat or Cold Exposure

Maximum and minimum temperature; temperature gradient; temperature spread; frequency of variations; hot spots; etc.

4.5.1.7 Abrasion Exposure
Degree; duration; concentration; etc.

4.5.1.8 Microbiological Influence
Type of microbial life; direct or indirect effect; medium; temperature; periodicity of exposure; etc.

4.5.1.9 Shock and Vibration
Source; strength; frequency; concentration; transfer path; etc.

4.5.1.10 Atomic Radiation
Type; exposure; continuity; temperature; etc.

4.5.1.11 Absorbent Materials
Type (mortar, concrete, brick, floor compositions, wood, plastics, insulating and gasket materials, etc); thickness; pH value; consistency; porosity; evaporation rate; absorbence rate; conductivity and resistivity; etc.

4.5.1.12 Atmospheric Environment Classification—Aircraft Design

Type	*Name*	*Example*
1	Inert	Outer space, continuous vacuum system Airtight dehumidified nitrogen, argon or helium pressurised system, oil sumps
2	Humidity controlled (below saturation) Crit. hum. = 50% RH—clean test environment—and under 50% with airborne hygroscopic particulates	Heated and/or air conditioned buildings
3	Interior (uncontrolled humidity)	Interior of unsheltered vehicles
4	Exterior	Exterior of unsheltered vehicles

4.5.2 Case Histories and Technical Data Records
Historically documented cases of corrosion behaviour of the same or similar product and of the effectiveness of corrosion-control methods applied in similar environmental and operational circumstances are very useful for comparative evaluation of corrosion control in design. Such information, however, should be studied and considered with caution, taking into account the possibility of many variations and combinations of conditions, from which errors and misconceptions could arise. Ultimately, each design case could be

appreciated as unique and then no individual case history may be accepted as an unquestionable dictate.

4.5.2.1 *Failure Reports*

The negative information contained in these documents should be recorded in a comprehensive form (object, materials, fabrication, treatment, operational data, locality, description and cause of failure), evaluated either by corrosion experts or a failure board and filed for easy reference by all design personnel.

The reports can either be filed individually or together. Where a number of failure reports on a related subject can accumulate, a corrosion failure index dealing with various sections of the problem or various parts of utility is preferred. Where a considerable number of failures of a comparatively restricted and repetitive nature is expected, it is desirable to record such information by electronic data processing. It is important that such information, in accurate form and preferably converted into a useful summary, be distributed as soon as possible to all interested personnel to be used either for design revision or maintenance programming.

4.5.2.2 *Materials and Treatment Records*

The positive information recorded in an index form, and accurately updated, can illustrate the whole development and progress of corrosion-control application in design of a particular project or its part and may become a source of valuable information for corrosion-control design analysis, specifications, working drawings, schedules, standards and procedures.

4.5.2.3 *Reference File*

No person engaged in corrosion control should be without access to a filing system which covers accurately all corrosion control in his particular enterprise. The volume of information required is too extensive to memorise.

This can be achieved either by a well-organised personal file or through access to a large-scale or computerised filing system.

The volume of such a file, in so far as it depends on the extent of activities, should not be static but altogether dynamic; it should grow in size and utility with the demand and progress of corrosion science and art and be immediately usable to cover the need of the moment and so allow an easy literature search.

4.5.2.4 *Comparative Index*

When the extent of the reference file becomes too unwieldy for a quick search, or where several materials of the same generic group are often evaluated for preferential use, a comparative index can prove of value.

4.5.2.5 *Standards*

Setting of corrosion engineering standards helps the speeding up of design analysis, so far as repetition of environmental and operational conditions will allow their use. In this way a designer may be able to cut down his work, but he will not be absolutely absolved from appreciation of the finer points of corrosion control which may vary considerably.

4.6 Analysis

Once the preparatory stage of corrosion-control work in design (i.e. setting up a suitable organisation and assembling ample information) is complete, the design team may commence with a step-by-step appreciation and evaluation of corrosion-control data, requirements, rules and other relevant information in a suitable and systematic manner. The results of this analysis will be followed by a reconciliation of the arising corrosion-prevention ideas with the product of the designers' functional engineering appreciation, in accordance with their merits. Finally an overall decision will be made and also a plan of action compatible with the requirements of a rational securement of the planned function, the utility's economic life and the safety of the utility or its parts appointed.

One may however consider, after a detailed examination of the schematic design control analysis (*Figure 4.5*) and the following individual sections, that by a total separation of the mentioned two efforts some work will be duplicated and valuable time wasted. Thus it may be left to the discretion of individual organisations in general or to necessities of individual projects in particular, whether by a judicious combination of items, at least in some of the opposing sections of analysis, a method of parallel thinking can be developed and unnecessary repetition avoided.

Each individual item of the two main parts of design analysis is important, by the degree of its merit for securement of the final results, and should not be forgotten or neglected. For this reason the combined analytical effort should be systematically suited to the project and systematically followed without fail.

One can mention here that the reasoning obtained by the corrosion-control analysis does not absolve the designer from implementing the basic engineering requirements vested in the utility itself, and a correct corrosion-control decision must not obstruct the product's engineering function. Both of the mentioned efforts are, however, so closely knitted together that they should be considered of equal importance, although on a selective basis—it is not good policy to consider only one branch of design analysis and neglect the other.

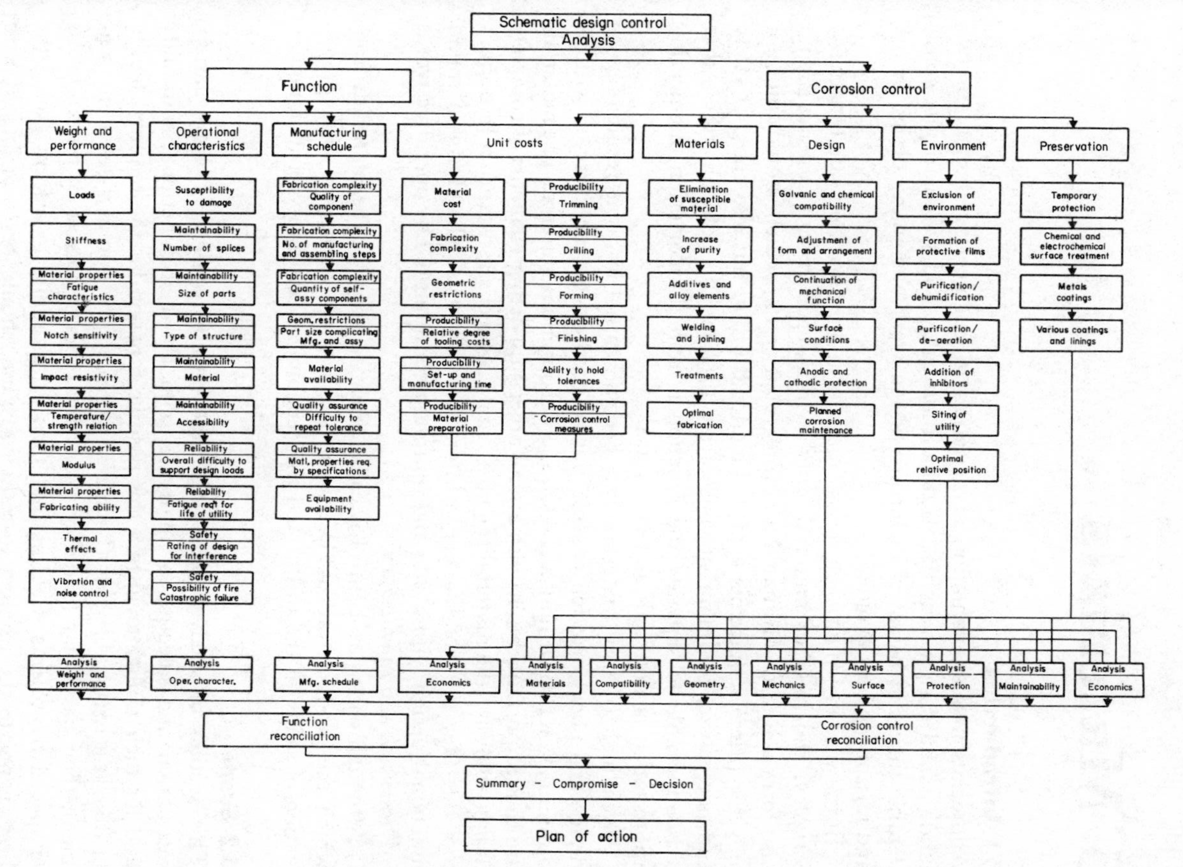

Figure 4.5. Schematic design control analysis

5 Materials

5.1 Introduction

Appreciation and evaluation of materials in engineering design are two fundamental design activities. If done thoroughly and systematically considerable time can be saved in further design work and many design errors and misconceptions avoided.

In the way in which each design must form a unified and safe functional complex, the bulk of materials used within each design system, should also form a well co-ordinated and integrated entity, which should fulfil not only the prime requirements of its functional utility but also those of safety and necessity of optimum survival of the product. It is obvious, then, that it is impossible to separate the functional from the corrosion engineering appreciation of materials and vice versa.

Even high quality materials can become casualties as a result of poor design and poor corrosion control, but no good design can carry on with its optimum function without a good support of optimum material. The materials are thus important, but so also are all other supporting elements of complete design analysis; the material scientist's and engineer's assistance is important too, but they are not the sole arbiters deciding over what forms a sound functional design with a safe and optimum continuity. The operative words are *teamwork* and *co-ordination*.

Although corrosion control is called upon to preserve the materials from which the products are made, its purpose is more concerned with preservation of safety of the utility and the economical continuity of its function.

5.2 Scope

This section of analysis deals with the parallel system of appreciation, evaluation and selection of materials both for their functional suitability and for their ability to sustain this positive function for the required length of time at a reasonable cost. Their resistance to corrosion in a given environment, their particular tendency to specific types of corrosion, requirements of special treatments, welding and their adjustability to a form giving the best chance to the product to resist corrosion, should be co-ordinated with functional parameters. Strength, fabricability and ease of production, appearance, availability and cost should also be appreciated.

Considering the profusion of materials and material-oriented literature, this description is, of necessity, schematic. It attempts only to indicate the system of parallel evaluation of various materials and the few guidelines that are presented are of a general nature.

The knowledge of materials is so vast and far reaching that close co-operation of the designer with metallurgists, material engineers, corrosion engineers and other materials specialists is stressed.

5.3 General

(1) Materials should be selected with due consideration to their functional suitability and ability to maintain their function safely for an economical period of time at a reasonable cost. The particular material selected should be accurately specified.

(2) The whole material complex should be considered as an integrated entity, rather than each material separately. The more highly resistant materials should be chosen for the critical components and where relatively high fabrication costs are involved. It may be necessary to compromise and sacrifice some mechanically advantageous properties to satisfy corrosion requirements and vice versa.

(3) Where corrosion rate is either very low or very high, the choice of materials is simple; where it is moderately low, a thorough analysis of all aspects is required.

(4) In dry environments and carefully controlled fluids, many materials can be used—and these often may be left unprotected. Under atmospheric conditions, even polluted atmospheres, such metals as stainless steels and aluminium alloys may be left unprotected. Also copper and lead have a long life. In a more severe wet environment, for example in marine conditions, it is generally more economic to use relatively cheap structural materials (mild steel) and apply additional protection, rather than use the more expensive ones. For severest corrosive conditions it is preferable in most cases to use materials resistant to the corrosive, than to use cheaper material with an expensive protection.

(5) Materials more expensive than absolutely necessary should not be chosen unless it is economical in the long run and necessary for safety of personnel or product or for other important reasons. Using fully corrosion-resistant materials is not always the correct choice—a balance between first cost and cost of subsequent maintenance should be found over the full estimated life of the designed utility.

(6) Certain combinations of metal and corrosive are a *natural choice*:

(*a*) aluminium—non-staining atmospheric exposure;
(*b*) chromium-containing alloys—oxidising solutions;
(*c*) copper and alloys—reducing and non-oxidising environments;
(*d*) hastelloys (chlorimets)—hot hydrochloric acid;
(*e*) lead—dilute sulphuric acid;
(*f*) monel—hydrofluoric acid;

(*g*) nickel and alloys—caustic, reducing and non-oxidising environments;
(*h*) stainless steels—nitric acid;
(*i*) steel—concentrated sulphuric acid;
(*j*) tin—distilled water;
(*k*) titanium—hot strong oxidising solutions;
(*l*) tantalum—ultimate resistance.

(7) Composition of alloy alone does not ensure quality of the product. Evaluation of resistance to corrosion in a given environment, adverse effect of corrosion products on utility or contents, susceptibility to a specific type of corrosion and fouling, and tendency to corrosion failure due to fabrication and assembly processes such as welding, forming, machining, heat treatment, etc., are of prime importance for selection of material and design.

(8) Due consideration should be given to special treatments required to improve resistance to corrosion, e.g. special welding techniques, stress relieving, blast peening, metallising, sealing of welds, etc. Also to any fabrication or assembly methods which would aggravate any tendency of the material to corrosion failure.

(9) Alloys in as highly alloyed a condition as necessary should be used when the cost of fabrication is higher than the cost of basic material. Proportional cost of material in some multishaped or complicated components is much less than in the simple ones.

(10) An alloy or temper should be selected which is free of susceptibility to the critical corrosion under the respective general and local environmental conditions in the utility and that meets the strength and fabrication properties required for the job.

It is sometimes better to use a somewhat weaker but less sensitive alloy, than to use the one which does not lend itself to reliable heat treatment and, due to this, whose resistance to a particular corrosion is poor.

(11) If heat treatment after fabrication is not feasible, materials and method of fabrication chosen should give optimum corrosion resistance in the as-fabricated condition. Materials prone to stress corrosion cracking should be avoided in environments conducive to failure, observing that stress relieving alone is not always a reliable cure.

(12) When corrosion or erosion is expected an increase in wall thickness of the structure or piping should be provided over that required by other functional design requirements. This allowance in the judgement of the designer should be consistent with the expected life of the structure or piping.

The allowance should secure that various types of corrosion or erosion (including pitting) do not reduce the thickness of structure or piping below the thickness that is required for mechanical stability of the product. Where no thickening can be allowed or where lightening of product is contemplated a proportionally more resistant alloy or better protection should be used.

In general, and subject to mechanical requirements, wall thickness is made double the thickness that would give desired life (*Figure 5.1*).

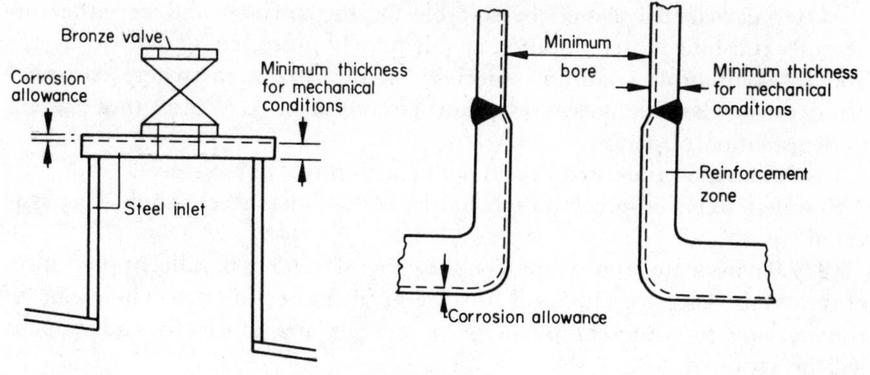

Figure 5.1

(13) Short-life materials should not be mixed with long-life materials in non-reparable subassemblies. Materials forming thick scale should not be used where heat transfer is important.

(14) Where materials could be exposed to atomic radiation it is necessary to consider whether the effect will be derogatory or beneficial, observing that some controlled radiation may enhance the property of a metal.

(15) Not only the structural materials themselves but also their basic treatment should be evaluated for suitability (e.g. chromate passivation, cadmium plating, etc.) at the same time.

(16) Non-metallic materials complying with the following requirements are preferred: low moisture absorption, resistance to fungi and microbes, stability through temperature range, compatibility with other materials, resistance to flame and arc, freedom from out-gassing and ability to withstand weathering.

(17) Flammable materials should not be used in critical places. Their heat could affect corrosion stability of structural materials. Toxic materials producing dangerous volumes of toxic or corrosive gases when under fire or high temperature conditions should not be used.

(18) Fragile or brittle materials which are not, by design, protected against fracture should not be used in corrosion-prone spaces.

(19) Materials which produce corrosion products that can have an adverse effect on the quality of contents should not be used, especially when the cost of wasted contents exceeds the cost of container.

(20) All effort should be made to obtain from the suppliers of equipment an accurate detailed description of materials used within their products.

(21) The following should be noted with regard to electrical equipment. The use of hygroscopic materials and of desiccants should be avoided. The latter, when their use is necessary, should not be in contact with an unprotected metallic part.

Fasteners should be of a well-selected corrosion-resistant material, or materials better protected than the parts which they join together.

Materials selected should be suitable for the purpose and be either inherently resistant to deterioration or adequately protected against deterioration by compatible coatings, especially in problem areas where corrosion can cause low conductivity, noise, short circuits or broken leads, thus leading to degradation of performance.

Insulation materials used should not be susceptible to moisture.

Stainless steels or precipitation hardening stainless steel should be passivated.

(22) To indicate, approximately, the general trend of parallel appreciation of materials, selective check-off lists are given in Section 5.4. These can, of course, vary for different materials or designs and a selective adjustment will be required.

It is obvious from the contents of these selection lists that a thorough expert knowledge is required, both in engineering and in corrosion control, to complete and evaluate the required data. Only very seldom, and then mostly in simple or repetitive projects, can this task be left to an individual—normally, close co-operation of designer and corrosion or material engineer is needed as mentioned previously, and each will have to bring into play his overall and specialised expertise.

The data obtained in such a selection list, after appropriate evaluation and comparative appreciation, should serve as a base for decision as to whether the appreciated conglomerate of material and its fabrication methods are suitable for the considered purpose. Although in some cases a clear-cut confirmation of suitability may be secured, in many more cases several materials and methods may be evaluated before the optimal one is found. Even then such materials will not always satisfy all required properties and under such circumstances the most satisfactory compromise should be accepted.

5.4 Material Appreciation

5.4.1 Metals: Selective Check-off List

5.4.1.1 Physical Character of Material

(a) General

Ballistic properties
Chemical composition (%)
Contamination of contents by corrosion products
Corrosion characteristics in:
  atmosphere
  waters
  soil
  chemicals

gases
molten metals
other environments
Creep characteristics × temperature
Crystal structure
Damping coefficient
Density (g/cm^3)
Effect of cold working
Effect of high temperature on corrosion resistance
Effect on strength after exposure to hydrogen
Effect on strength after exposure to higher temperature
Electrical conductivity (mho/cm or S/m)
Electrical resistivity (ohm/cm)
Fire resistance
Hardenability
Hydraulic permeability
Magnetic properties (Curie point)
Maximum temperature not affecting strength (°C)
Melting point (°C)
Position in electromotive series
Rapidity of corrosion – corrosion factor
Susceptibility to various types of corrosion:
general
hydrogen damage
pitting
galvanic
stress corrosion cracking
corrosion fatigue
fretting
concentration cell/crevice
corrosion/erosion
cavitation damage
intergranular
selective attack
high temperature
others
Thermal coefficient of expansion ($°C^{-1}$)
Thermal conductivity (W/(m.°C))
Toxicity
Wearing quality:
inherent
given by heat treatment
given by plating

(*b*) *Strength or mechanical*

Above strength properties:
at elevated temperatures and for long holding times at temperature
at room temperature after exposure to elevated temperatures
Bearing ultimate* (N/mm^2 or kN/m^2 or kg/mm^2)
Complete stress–strain curve for tension and compression
Compression modulus of elasticity (kg/mm^2)
Compression yield* (N/mm^2 or kg/mm^2)
Fatigue properties:
S–N curve
endurance limit
Hardness (Vickers)
Impact properties (Charpy kg/cm^2 at 20°C):

notch sensitivity
effect of low temperature
maximum transition temperature (°C)
Poisson's ratio
Response to stress–relieving methods
Shear modulus of elasticity (kg/mm²)
Shear ultimate* (Pa)
Tangent modulus curves in compression—with and across grain
Tension modulus of elasticity (Pa)
Tension—notch sensitivity
Tension yield*

*Typical and design values, variability—with and across grain.

5.4.1.2 Design Limitations

Restrictions

Size and thickness
Velocity
Temperature
Contents
Bimetallic attachment
Geometric form
Static and cyclic loading
Surface configuration and texture
Protection methods and techniques
Maintainability

5.4.1.3 Fabrication Character of Materials

(*a*) *Brazing and soldering*

Compatibility
Corrosion effect
Flux and rod

(*b*) *Formability at elevated and room temperatures*

Ageing characteristics
Annealing procedure
Corrosion effect of forming
Heat treating characteristics
Quenching procedure
Sensitivity to variation
Tempering procedure
Effect of heat on prefabrication treatment

(*c*) *Formability in annealed and tempered states*

Apparent stress × local strain curve
Characteristics in:
bending
dimpling
drawing
joggling

shrinking
stretching
Corrosion effect of forming
Elongation × gauge length
Standard hydropress specimen test
True stress–strain curve
Uniformity of characteristics

(*d*) *Machinability*

Best cutting speed
Corrosion effect of:
drilling
milling
routing
sawing
shearing
turning
Fire hazard
Lubricant or coolant
Material and shape of cutting tool
Qualitative suitability for:
drilling
milling
routing
sawing
shearing
turning

(*e*) *Protective coating*

Anodising
Cladding
Ecology
Galvanising
Hard surfacing
Metallising
Necessity of application for:
storage
processing
service
Paint adhesion and compatibility
Plating
Prefabrication treatment
Sensitivity to contamination
Suitability
Type of surface preparation optimal

(*f*) *Quality of finish*

Appearance
Cleanliness
Grade
Honing
Polishing
Surface effect

(*g*) *Weldability*

Arc welding
Atomic hydrogen welding
Corrosion effect of welding
Cracking tendency
Effect of prefabrication treatment
Electric flash welding
Flux
Friction welding
Heat zone effect
Heli-arc welding
Pressure welding
Spot welding
Torch welding
Welding rod

(*h*) *Torch cutting*

Cutting speed

5.4.1.4 Economic Factors

(*a*) *Availability*

In required quantities:
- single
- multiple
- limited
- unlimited

In different forms:
- bar
- casting (sand, centrifugal, die, permanent mould)
- extrusion
- forging
- impact extrusion
- pressing
- sintered
- powder pressing

In metallised and pretreated forms:
- galvanised
- plastic coated
- plated
- prefabrication treated

In cladded forms
Uniformity of material
Freedom from defects
Time of delivery:
- present
- future

(*b*) *Cost in different forms*

Bar, shape or plate
Casting (sand, centrifugal, die, permanent mould)
Extrusion
Forging

Impact extrusion
Pressing
Sintered
Powder Pressing

(*c*) *Size limitation in different forms*

Gauge
Length
Weight
Width

(*d*) *Size tolerances for different forms*

5.4.2 Plastics: Selective Check-off List
To be adjusted for generic groups

5.4.2.1 Physical Character of Material

(*a*) *General*

Anisotropy characteristics—main and cross-direction
Area factor ($in^2\ lb^{-1}\ mil^{-1}$)
Burning rate (in/min—in sec or cm/min—in sec)
Bursting strength (1 mil thick, Mullen points)
Change in linear dimensions at 100°C for 30 min (%)
Clarity
Colour
Contamination of contents (decomposition and extract)
Creep characteristics × temperature—creep-apparent-modulus (10° lbf/in^2 or kg/mm^2)
Crystal structure
Crystalline melting point
Damping coefficient
Decay characteristics in:
  atmosphere
  alcohol
  chemicals
  gases
  high relative humidity
  hydraulic oils
  hydrocarbons
  lubricants and fats
  solvents
  sunlight
  water
  other environments
Deflection temperature (°F or °C)
  264 (lbf/in^2) fibre stress
  66 (lbf/in^2) fibre stress
Density (g/cm^3)
Dielectric constant (60 c/s, 10^3 c/s, 10^6 c/s, 10^9 c/s)
Dielectric strength—1 to laminations (v/mil or v/mm):
  short time
  step by step
Dissipation factor (1 MΩ)

Effect of high temperature on decay
Effect of low temperature on decay
Effect on strength after exposure to temperature
Electrical conductivity (mho/cm or S/m)
Electrical properties at room temperature per thickness
Electrical loss factor (1 M Ω)
Electrical power factor (60 c/s, 10^3 c/s, 10^6 c/s)
Electrical resistivity:
  arc/sec
  insulation (96 h at 90% RH and 35°C) MΩ
  volume (Ω/cm—50% RH and 25°C)
Evolvement of combustion products
Evolvement of corrosives
Fillers
Fire resistance
Flammability (ignited by flame)
Folding endurance
Fuming on application
Gas permeability ($cm^3/100\ in^2$/mil thick/24 h/atm at 25°C):
  CO_2, H_2, N_2, O_2
Generic group and composition
Gloss
Haze (%)
Heat distortion temperature at 264 lbf/in^2 (°F)
Heat expansion (in^{-1}°F) – ($10^{-5}\ in^{-1}\ in^{-1}$°C)
  (coefficient of linear thermal expansion)
Heat degradation
Heat insulation (thermal conductivity) (btu/$ft^{-2}h^{-1}$°F in^{-1})
Heat sealing temperature range
Hydraulic permeability
Intensity of colour
Light transmission, total white (%)
Maximum service temperature (°F or °C)
Maximum temperature not affecting strength (°C)
Melt index (dg/min)
Migration of plasticisers
Minimum temperature not affecting strength (°C)
Opacity
Rate of water vapour transmission ($g/100\ in^2$/24 h—mg/m^2/24h/mil) at 37.8°C
Reflectance
Refractive index (n_d)
Reinforcement
Softening temperature (°C)
Specific gravity (25°C)
Specific heat (cal/°C g^{-1} or btu lb^{-1}°F)
Stability of colour
Stiffness—Young's modulus
Susceptibility to various types of deterioration:
  general
  cavitation/erosion
  erosion
  fatigue
  fouling
  galvanic (metal-filled plastics)
  impingement
  stress cracking and crazing
  others
Taste

Thermal conductivity (W/(m.°C))
Toxicity
Transmittance (%)
Unit weight (m^3/lb)
Water absorption (24 h/1⅛ in thick/%)
Wearing quality:
  inherent
  given by treatment

(*b*) *Strength or mechanical*

Abrasion resistance
Average yield (in^2 lb^{-1} mil^{-1})
Bearing ultimate* (lbf/in^2 or N/mm^2)
Bonding strength (lb/thickness or kg/thickness)
Brittleness
Bursting pressure (lbf/in^2 or kg/mm^2)
Complete stress—strain curve for tension and compression
Compressive strength:
  flatwise (lbf/in^2 or N/mm^2)
  axial (lbf/in^2 or N/mm^2)
  at 10% deflection (lbf/in^2 or N/mm^2)
Deformation under load
Elongation (%)
Elongation at break (%)—75°F (24°C)
Fatigue properties
  S–N curve
  endurance limit
Flexibility and flex life
Flexural strength* (lbf/in^2 or N/mm^2)
Hardness (Rockwell)
Impact strength, Izod (ft lb^{-1} in^{-1} of notch)*
Inherent rigidity
Modulus of elasticity (lbf/in^2 or kg/mm^2):
  in compression (10^5 lbf/in^2 or kg/mm^2)
  in flexure (10^5 lbf/in^2 or kg/mm^2)
  in tension (10^5 lbf/in^2 or kg/mm^2)
  in shear (10^5 lbf/in^2 kg/mm^2)
Resistance to fatigue
Safe operating temperature (°C)
Shear, ultimate* (Pa)
Tear strength:
  (g/mil)—propagating
  (lb/in)—initial
Tensile strength (lbf/in^2 or kg/mm^2)
Vacuum collapse temperature

*Typical and design values, variability—with and across grain

5.4.2.2 Design Limitations

Size and thickness
Velocity
Temperature
Compatibility with adjacent materials:
  at ambient temperature

at elevated temperature
Geometric form
Static and cyclic loading
Surface configuration and texture
Protection methods and techniques
Maintenance

5.4.2.3 Fabrication Character of Materials

(*a*) *Moulding and injection*

Compression ratio
Compression moulding pressure (lbf/in^2)
Compression moulding temp. (°F or °C)
Injection moulding pressure (lbf/in^2)
Injection moulding temp. (°F or °C)
Moulding qualities
Mould (linear) shrinkage (in/in or cm/cm)
Specific volume, etc. (lb^3)

(*b*) *Lamination*

Laminating pressure (lbf/in^2)
Laminating temperature (°F or °C), etc.

(*c*) *Formation at elevated temperatures*

(*d*) *Machinability*

Adverse effect of:
  drilling
  milling
  sawing
  shearing
  turning
Best cutting speed
Fire hazard
Machining qualities
Material and shape of cutting tool, etc.

(*e*) *Protective coating*

Cladding
Painting
Plating
Sensitivity to contamination
Suitability
Type of surface preparation optimal

(*f*) *Quality of finish*

Appearance
Cleanliness
Grade
Polishing
Surface and effect

(g) Joining

Adhesive joining
Bonding
Cracking tendency
Heat zone effect
Welding, etc.

5.4.2.4 Economic Factors

(a) Availability

Cladded forms
Forms available
Maximum width (in)
Thickness range
Treated forms
Uniformity of material
Freedom from defects

(b) Cost of different forms

(c) Size limitations

(d) Size tolerances

5.4.3 Natural and Synthetic Elastomers: Selective Check-off List

5.4.3.1 Physical Character of Material

(a) Uncured properties

Colour
Consistency
Shelf life
Viscosity

(b) Cured properties

Chemical composition
Compression set
Contamination of contents by contact
Creep
Drift (room temperature)
Elasticity
Elastic memory
Flexing
Hardness range
Linear shrinkage (%)
Permeability to gases
Permanent set
Resilience

Self-damping
Specific gravity
Swelling resistance (in various environments)

(*c*) *Thermal properties*

Coefficient of thermal expansion × 10^{-3} per °F or °C
Drift at elevated temperature
Elongation % at elevated temperature
Flame resistance
Heat ageing
Low temperature—brittle point (°C)
Low temperature—stiffening (°C)
Maximum service temperature range (°C)
Tensile strength—elevated temperature (Pa)
Thermal conductivity (btu h^{-1} ft^{-3} °F ft^{-1})

(*d*) *Electrical properties*

Dielectric constant
Dielectric strength
Dissipation factor
Electrical resistivity (Ω/cm)
Volume resistivity

5.4.3.2 *Mechanical Properties*

Crack resistance
Cut growth
Elongation % plain
Elongation % reinforced
Resistance to abrasion
Resistance to impact and shock
Resistance to wear
Tear resistance
Tensile strength—reinforced (lbf/in² or mPa)
Tensile strength—unreinforced (lbf/in² or mPa)

5.4.3.3 *Environmental Properties*

Chemical resistance in:
- water
- acid
- alkali
- aliphatic hydrocarbons
- aromatic hydrocarbons
- halogenated hydrocarbons
- alcohol
- oil and greases
- synthetic lubricants (e.g. diester)
- hydraulic fluids—silicates, phosphates

Light
Oxidation

Radiation
Weather

5.4.3.4 *Subjective Properties*

Odour
Staining
Taste

5.4.3.5 *Fabrication—Character of Materials*

Bonding to rigid materials
Limits of geometric form
Moulding limits
Processing characteristics
Sagging

5.4.3.6 *Corrosion Effect on Substrate*

Fumes in combustion
Physical contact
Vapour on setting and set

5.4.3.7 *Fabrication Character of Materials*

5.4.3.8 *Economic Factors*

5.4.4 Adhesives: Selective Check-off List

5.4.4.1 *Physical Character of Material*

(*a*) *General*

Chemical composition
Colour
Contamination of contents by adhesive
Characteristics in:
  atmosphere
  waters and humidity
  soil
  chemicals
  lubricating oil
  hydraulic fluid
  degreasing solvent
  gases
  other environments
Creep characteristics and temperature
Density

Effect of forming on substrate
Effect of organic vapours on substrate
Effect of high temperatures
Effect of low temperatures
Effect on strength after exposure to higher temperature
Electrical conductivity
Electrical resistivity
Fire resistance
Flammability

(*b*) *Strength or mechanical*

Load on glue line:
  cleavage (lb/in of width)
  peel (lb/in of width)
  shear (lb/in^2)
  tension (lb/in^2)

5.4.4.2 Design Limitations

Utility:
  development
  field repair
  production
  prototype
Application:
  bonding
  reinforced joint
  restriction on joined area
  restriction on pressure
  scaling
  type of drying
Surface types:
  base metal
  finish
Type of joint:
  edge
  filled
  lap
Hydraulic permeability
Maximum temperature not affecting strength
Melting point
Odour
Smoke development on fire
Surface effect (corrosion):
  metals
  non-metals
  coatings
Susceptibility to environments:
  outdoor sheltered
  outdoor direct exposure
  outdoor sunlight only
  indoor room condition
  indoor controlled
  stored sealed
  stored sealed (opened for test)
  stored outdoor sheltered

stored outdoor direct exposure
stored in warehouse
stored in room
stored in controlled atmosphere
others
Thermal coefficient of expansion
Thermal conductivity
Toxicity

5.4.4.3 Fabrication Character of Materials

5.4.4.4 Economic Factors

5.5 Guidelines for Selection of Dielectric Separation Materials

(1) The main purpose of separation sealing is to secure discontinuity of conductance between dissimilar metals in physical contact and to prevent ingress of water and air to their faying surfaces.

(2) Dielectric materials are good electric insulators. Choice of materials subject to their dielectric strength should be graduated in relation to the environment and function, i.e. the more aggressive the conditions, wider emf potential and more critical functional conditions the higher is the required ohmic resistance.

(3) Various shapes and sizes of faying surfaces demand diverse materials; whilst an extended linear surface may require sealant in a form of sealant tape, multiform or small connection may benefit from an elastomeric sealant or caulking compound, applied by a gun or spatula, dielectric gasket or washer.

(4) Separation materials should not be porous to such a degree that the absorbed water or other electrolytes will cause uninterrupted conductivity between the metals of the bimetallic couple. For example, in heavily wetted areas dielectric asbestos or other suitable gaskets with dielectric effect and low water absorption should be used, instead of porous asbestos gaskets.

(5) Composition, working consistency, pot life, adherence to surfaces and application procedures should suit the practical field application, this without excess effort and without impeding a work undertaken in the vicinity.

(6) Separators should be suitable for local environmental conditions, i.e. their resistance to environment, heat, pollutants and spillage (fuels, oils, fumes) should be considered.

(7) Separators should be suited to the functional requirements of the joint. For example, only such separators as allow free movement should be used where the connection is not fixed and mobility between the faying surface is required.

(8) Separators should be of sufficient thickness or allow build-up of thickness to a ratio serving the requisite separation under prevalent environmental conditions. The thickness can vary to suit and should be stipulated in design.

(9) Separators should not adversely affect the materials of the couple by their chemical contents and composition, e.g. the effect of free sulphur content in vulcanised rubber. Graphite content of gaskets can seriously aggravate corrosion of joined metals under wetted marine conditions.

(10) Dielectric materials should be in the range acceptable to fire safety, i.e. they do not necessarily need to be fire resistant unless in direct proximity of fire exciters. The degree of fire retardance required depends on the quantity of material, critical position and exposure to heat.

(11) Paint coating of faying surfaces between dissimilar metals subject to wetting is not a substitute for dielectric separation.

(12) Where design or functional requirements do not allow the separation indicated above, steel or aluminium components should be metallised on faying surfaces with aluminium or zinc.

5.6 Guidelines for Selection of Fasteners

(1) Fasteners should maintain their function of connecting safely two pieces of metal in a way easy to disassemble.

(2) Fasteners should not adversely affect the materials of the basic components and should not be affected reciprocally by them.

(3) In aggressive conditions fasteners should not be made of metal anodic to both metals of the joint.

(4) Fasteners should preferably be made in a metal compatible to both metals in the connection, i.e. slightly cathodic.

(5) If, for design and economic reasons, fasteners made of a metal anodic to the cathodic member of the joint are used, and dielectric separation cannot be incorporated, the fastener should be metallised with anodic metal (i.e. with aluminium, zinc or cadmium) or otherwise equivalently protected.

(6) Fasteners should preferably be made of material economically resistant to the environment (or so protected) and no material subject to catastrophic failures through hydrogen embrittlement, stress corrosion cracking, etc., should fasten critical connection.

5.7 Request for Material Information

Date: ....

(1) CORROSION DATA WORK SHEET

Please give complete information.

CORROSION FACTORS

1. PROCESS OR OPERATION: ....
2. PROCESS UNIT : ....
3. CORROSIVES (principal substances, impurities, composition, concentration, pH, etc.). Whenever possible also give their common names, i.e. 'black liquor', 'spent pickle', 'spinning bath', etc:
   ....

4. TEMPERATURE: Max: .... Min: .... Aver: ....
5. OPERATING PRESSURE: .... or Vacuum: ....
6. AERATION: Air-Free .... Moderate: .... Complete: ....
7. AGITATION, or velocity of solution: ....
8. CONSISTENCY (viscosity, stiffness, etc.) ....
9. ABRASIVES or SUSPENDED SOLIDS (nature and quantity) ....
10. METALLIC CONTAMINATION (what, if any, must be considered and limits) ....
11. GALVANIC FACTORS: If more than one metal or alloy is involved:
    (a) Which, if any, are or will be in immediate galvanic contact? ....
    (b) What are their approximate area relationships? ....

EQUIPMENT EXPERIENCE

12. METAL, alloy or other material used: ....
13. THICKNESS of sheet, plate, or tubing used: ....
14. SERVICE LIFE of material used: ....
15. FAILURE (general corrosion, pitting, liquid level attack, vapour attack, cracking, abrasion etc.): ....

NAME .... TITLE ....
COMPANY ....
ADDRESS ....

Use reverse side for sketches or for additional details including any special factors not covered by this sheet.

(2) COATING ENGINEERING DATA

Please give complete information. Use reverse side of sheet for sketches or other details of problem.

(a) INTENDED PURPOSE OF COATING

( ) Preliminary Design Study	( ) Production	( ) Cost Estimating
( ) Research & Development	( ) Plant Maintenance	( ) Government Usage
( ) Pilot Production	( ) Field Maintenance	( ) Industrial Usage
( ) General Information	( ) Product Improvement	( ) Commercial Usage

(b) GENERAL DESCRIPTION OF ITEM TO BE COATED

....
Estimated area to be coated: .... Length of part: .... Height of part: ......
Width of part: .... Estimated number of units to be coated: ....
Test & Evaluation: .... Production: ....

(c) BASIC FUNCTIONS OF COATING

( ) Corrosion Protection		( ) Waterproofing	( ) Sealant
( ) Wear Resisting		( ) Fire Retarding	( ) Identification
( ) Weather Proofing		( ) Fire Proofing	( ) Decorative
( ) Insulating	( ) Thermal	( ) High Temperature	( ) Release Agent
	( ) Dielectric	( ) Low Temperature	( ) Adhesive
( ) Conducting	( ) Thermal	( ) Optical (High Reflectivity)	
	( ) Electrical	( ) Optical (Low Reflectivity)	

(d) SURFACE TO BE COATED

( ) Metal (Alloy) ....
( ) Part suitable for forced drying
( ) Low Temp. ( ) Med. Temp. ( ) H.T.
( ) Other Materials ....
( ) Coating now in use ....

(e) SURFACE CONDITION

( ) As Machined ( ) As Cast .... RMS
( ) Metallised ....
( ) As Wrought ( ) Polished .... RMS
( ) Blast-cleaned ( ) Pickled..............
( ) Primed or Coated.....................

(f) METHOD OF APPLICATION DESIRED

( ) Spray ( ) Dip ( ) Brush
( ) Electrostat. Spray ( ) Airless Spray
( ) In-plant Facility ( ) Approved Applicator
( ) Outside Production (Vendor)

(g) METHOD OF FORCED DRYING AVAILABLE

( ) Air-dry (only)
( ) Oven Bake ...... °C/Part max .... °C
( ) Gas ( ) Electric ( ) Infrared
( ) Ambient In-Service Operation Cure

(h) ENVIRONMENTAL REQUIREMENTS

( ) Heat—Range Min ..... °C Max ..... °C—Rise ..... °C per ( ) Sec ( ) Min ( ) Hr
( ) Ambient Operating Temp.—Interior ...... °C—Exterior ...... °C
( ) Thermal Shock ...... °C to ..... °C for .... cycles/min
( ) Maximum Continuous Operating Temp ..... °C for ..... ( ) Sec ( ) Min ( ) Hrs ( ) Days
( ) Immersion—Type of Liquid .... ( ) Continuous ( ) Intermittent
Temperature Range Min ..... °C Max ..... °C—Concentration .... %—pH .....
( ) Gases—( ) Steam—Type of Gas ...... Temperature—Range Min ...°C Max .... °C
Velocity Min...cm/s Max...cm/s—Pressure Min...kg/mm^2 Max..kg/mm^2—Conc...%
( ) Dielectric Strength .... Volts/mil
( ) Humidity ( ) Salt Atmosphere ( ) Salt Spray ( ) Vacuum ( ) Ultra-High Vacuum
( ) Cryogenic ( ) Solar Radiation ( ) Nuclear Radiation ( ) Traffic Surface/Abr.
( ) Non-Traffic Surface/Non-Abrading ( ) Others ( ) Accessible

(i) CHEMICAL RESISTANCE

( ) Solvent............( ) Cont. Immers. ( ) Inter. I. ( ) Spill. ( ) Fumes ( ) Splash
( ) Fuel/Oil ....( ) Cont. Immers. ( ) Inter. I. ( ) Spill. ( ) Fumes ( ) Splash
( ) Acid...............( ) Cont. Immers. ( ) Inter. I. ( ) Spill. ( ) Fumes ( ) Splash
( ) Alkali..............( ) Cont. Immers. ( ) Inter. I. ( ) Spill. ( ) Fumes ( ) Splash

(j) DESIRED COATING CHARACTERISTICS

( ) Colour .... Match No .... STD. ( ) Flat/Matte ( ) Semi-Gloss ( ) Gloss
Thickness ...... mils Tolerance .... in.

6 Compatibility

6.1 Introduction

In an endeavour to satisfy themselves on the functional requirements of individual components, some shortsighted designers fail to consider the structure and the equipment as a whole. Often, for them, each individual item stands on its own in splendid isolation.

Unfortunately, in the same way that bad relations between individuals can play havoc with human society, so badly conceived relations between individual materials of a complex can ruin even the best design. Thus it is imperative that all inter-material influences are properly appreciated and evaluated before any final decision in design is taken, whether these are caused by direct contact between dissimilar metals or induced by changes of polarity, transfer of electrolysis through a medium, carry of metallic particles in the stream, adverse influence of stray currents or by any other derogatory effect arising from the near proximity of materials (e.g. chemical, thermal or radiation) selected to form the required entity.

In complex structures and equipment, process streams and piping arrangements, different metals, alloys or other materials are frequently used in corrosive or conductive environments within an easy reach of each other; in practical applications the contact between dissimilar materials cannot be totally avoided. It is up to each individual designer to create benign conditions of contact between the various materials and units implanted into the designed project, and to take proper precautions for avoidance of the consequences of the probable and less optimal selections enforced by predominant functional requirements. These precautions will mainly consist of selecting compatible materials, designing effective electric separation and of adjusting environmental media.

Compatible materials are those which, although used together in a particular medium in appropriate relative sizes and compositions, will not cause an uneconomic breakdown within the utility. Materials do not only influence each other by virtue of their inherent or induced difference of electric potentiality (electrochemically), but also by their composite chemistry. These adverse chemical influences may be caused by materials in ambient state or induced by changes in materials caused by variations of environmental conditions. All the above-mentioned possibilities will bear influence on the designer's appreciation of the problem.

Not all considerations may, however, be given to the adverse effects of proximity of materials. There is a number of outlets where, by judicious choice of dissimilarity between materials, beneficial results can be obtained (e.g. sacrificial cathodic protection, cleaning of metals).

6.2 Scope

This part of analysis is concerned with various types of inter-material relations met in engineering design and some of the means of controlling their adverse consequences or exploiting their benefits. Cathodic protection, which in fact could appear here, is dealt with in Chapter 10.

Considering all possible involvements of compatibility in design, one can only indicate some of the many available means within the designer's reach; for the rest it is left to the ingenuity of each designer to apply selectively the suggested precautions to suit the function and, at the same time, the particular environmental and other conditions appertaining to the product.

This reconciliation may depend largely on the following relevant parameters :

(*a*) component metals or other materials;
(*b*) difference of emf;
(*c*) distance between dissimilar materials;
(*d*) degree of exposure to corrosive environment;
(*e*) relative sizes of anode to cathode or a contaminator to the affected material;
(*f*) conductivity of environment versus conductivity of materials;
(*g*) resistivity of environment versus resistivity of materials;
(*h*) temperature gradients and spread;
(*i*) fluid current strata, directions and velocity;
(*j*) criticality of resulting failures;
(*k*) contents of cathodic metals or aggressive materials in solvent waters, other liquids or atmospheres;
(*l*) sources of dc stray currents and their conductive paths;
(*m*) development of corrosive fumes in specific conditions;
(*n*) nature of the effect—beneficial or derogatory, etc.

6.3 General

(1) Dissimilar metals in intimate contact or connected by conductive path (e.g. water, condensation, electrolyte or conductor) should be applied only when the functional design or other important considerations render this unavoidable.

(2) If the use of dissimilar metals in a conductive environment is necessary, an attempt to select metals which form *compatible couples or groups* should first be made (for the marine environment a useful *Galvanic Corrosion Indicator* is published by International Nickel Company Ltd; for other environments

Table 6.1 GROUPING OF COMPATIBLE MATERIALS—AIR-SPACE ENVIRONMENT

Type 1	Inert environment	All materials are compatible
Type 2	Humidity controlled heated and/or air-conditioned building	Platinum, gold, graphite and silver are not compatible with low alloy steel, aluminium, magnesium, copper and cadmium—other combinations are compatible
Type 3	Interior of unsheltered vehicles uncontrolled humidity	I, magnesium II, beryllium, zinc, clad and non-clad aluminium alloys, cadmium III, steel (except corrosion resist.), lead, tin IV, 12% Cr 400 series steels, pH corrosion-resistant steels, 18% Cr 400 Series steels, chromium, brass, bronze, copper, beryllium copper, aluminium bronze alloys, 300 series stainless steels, monel, Inconel, nickel alloys, titanium alloys V, silver, graphite, gold, platinum *Notes* : Each material is compatible with other members of the same group but not with materials of a different group with the following exceptions: titanium fasteners installed in aluminium alloys are considered similar titanium is similar to group V metals tin is similar to group II alloys graphite composites are considered similar to group V metals and the last five members of group IV
Type 4	Exterior of unsheltered vehicles	Titanium alloys, nickel-base and cobalt-base alloys (Inconel), 300 series stainless steels, gold, platinum and graphite are compatible with each other but not with other materials

different indicators may be required (*Table 6.1*), and contact with an expert laboratory to obtain a suitable indicator is recommended).

(3) The *Scales of Galvanic Potentials* used at present are meaningless unless the amount of current flowing between dissimilar metals is known.

(4) The designer should make proper arrangements to obtain accurate information on material composition of all bought-out items within the design complex; they belong to and form an integral part of overall corrosion-safe design.

(5) For accurate evaluation of compatibility an exact engineering description of all materials or their metallurgical composition is required—general descriptions (e.g. mild steel) do not provide an adequate base to guarantee compatibility in conductive media.

(6) Connections between same or compatible metals should not normally cause any galvanic problems; one cannot, however, rely on this altogether (see Chapter 3).

(7) Galvanic corrosion of dissimilar metals can be avoided or at least reduced by preventing the extended presence of humidity (e.g. condensation)

at such bimetallic joints; continuously dry bimetallic connections do not corrode (see Chapter 7).

(8) Bimetallic connections in the proximity of fumes from combustion generators should be avoided.

(9) When joining of non-compatible metals (e.g. copper alloy to aluminium) is unavoidable, their effective dielectric separation is imperative, even in environments of lesser conductivity.

(10) Connections between stainless steel and steel, or stainless steel and aluminium components in conductive environments are normally considered to be bimetallic couples and adequate selective precautions against galvanic action should be taken.

(11) Faying surfaces of dissimilar metals should be separated completely effectively (*Figure 6.1*).

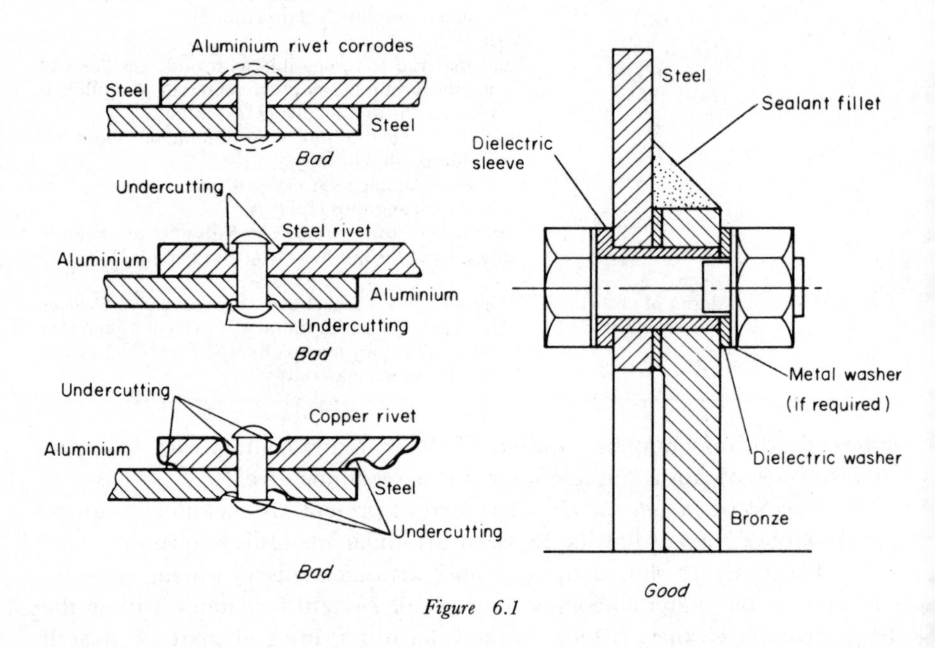

Figure 6.1

(12) Where, however, complete dielectric separation cannot be implemented any possible increase of electrolyte path would be of advantage (*Figure 6.2*).

(13) Dielectric separation can be provided in miscellaneous ways (*Figure 6.3*): (*a*) insulating gaskets (e.g. synthetic rubber, ptfe and other non-porous materials) for shaped contacts; (*b*) butyl tape (minimum 0.020 in thick) for linear extended contacts; (*c*) spreadable sealant (two coats of approved sealing compound to each surface, the first being allowed to dry before applying the second one) for multiform or small size contacts, etc.

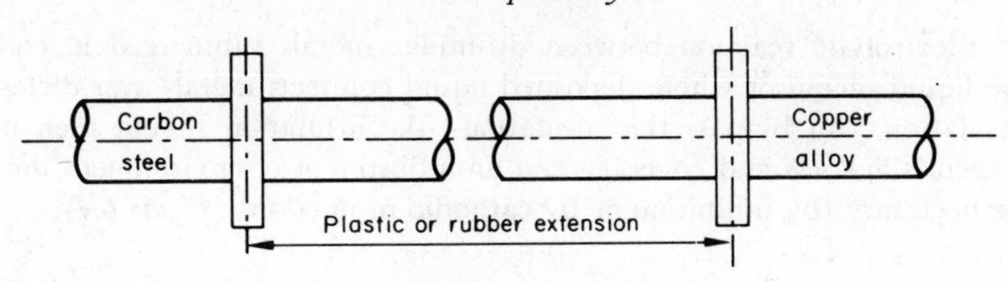

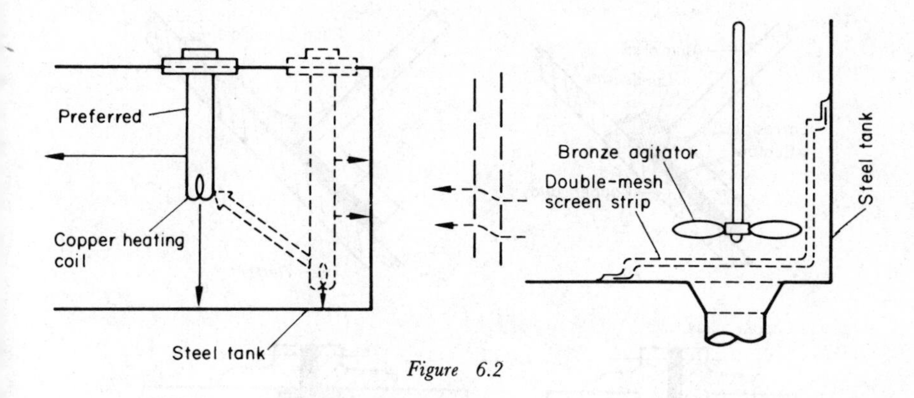

Figure 6.2

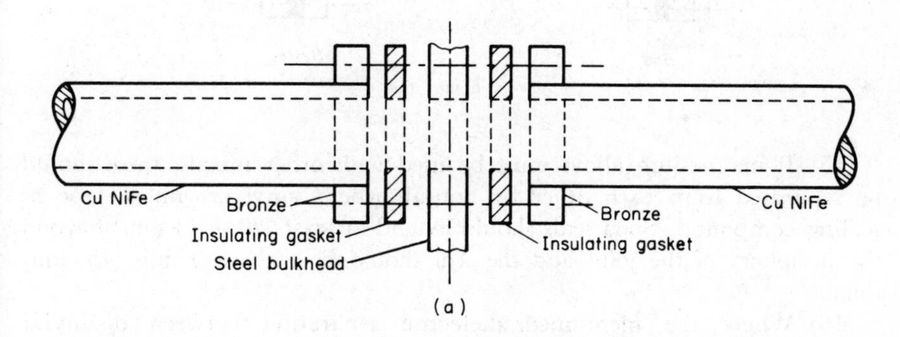

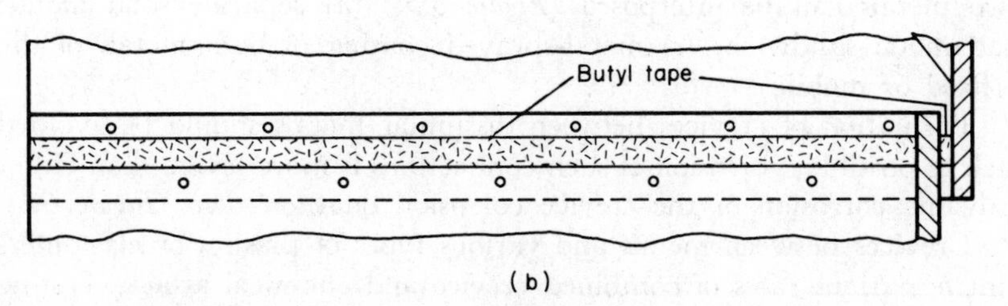

Figure 6.3

(14) Electrolytic reaction between dissimilar metals submerged in conductive liquid media or where deposited liquid connects metals over dielectric insulation, can by-pass this insulation—the insulation should then be of sufficient thickness and coverage and an adjustment of environment may also be necessary (by inhibition or by cathodic protection) (*Figure 6.4*).

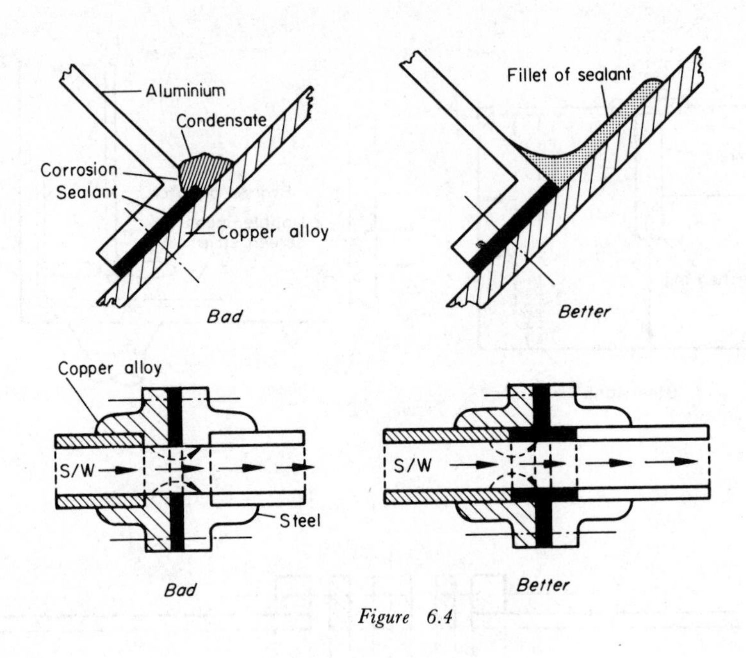

Figure 6.4

(15) If magnesium alloys must be used with other metals, these should be separated from each other by non-absorbent vinyl or rubber tape or sealing compound. Such seals should extend at least 1/8 in (3 mm) beyond the periphery of the joint and the seal should be at least 3 mils (75 μm) thick.

(16) Where the mentioned dielectric separation between dissimilar metals cannot be used, a metal which reduces the potential difference between the two metals can be interposed (*Figure 6.5*): (*a*) separate solid metal; (*b*) clad metal sandwich; (*c*) metal-sprayed coating of both metals of the joint (fixed or mobile).

(17) Formation of crevices between dissimilar metals should be avoided as much as possible; corrosion of such connections is more severe than either the galvanic corrosion or the crevice corrosion on their own (*Figure 6.6*)

(18) Crevices between metals and various types of plastics or elastomers may incite various rates of combined crevice and chemical attack. Testing is recommended prior to inclusion into design.

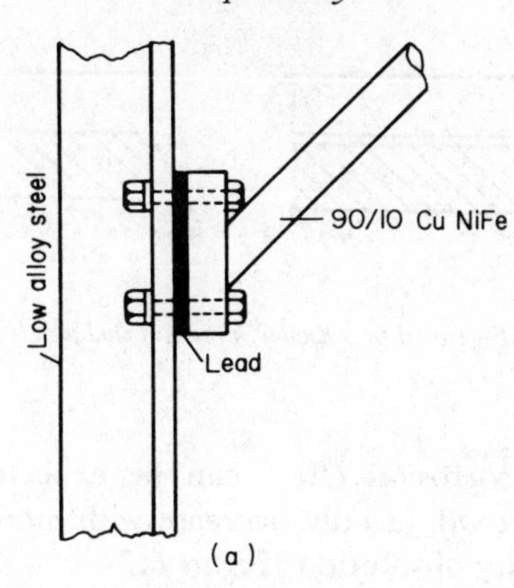

Low alloy steel
90/10 Cu NiFe
Lead
(a)

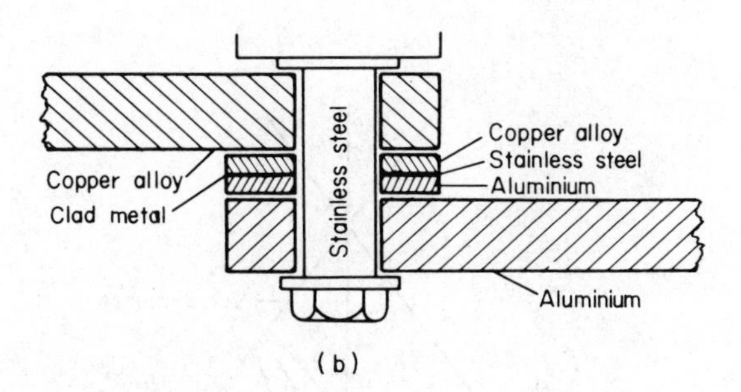

Stainless steel
Copper alloy
Stainless steel
Aluminium
Copper alloy
Clad metal
Aluminium
(b)

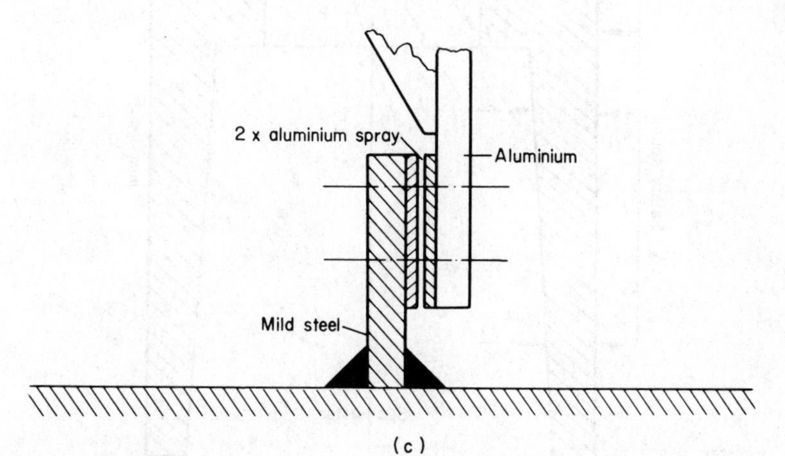

2 x aluminium spray
Aluminium
Mild steel
(c)

Figure 6.5

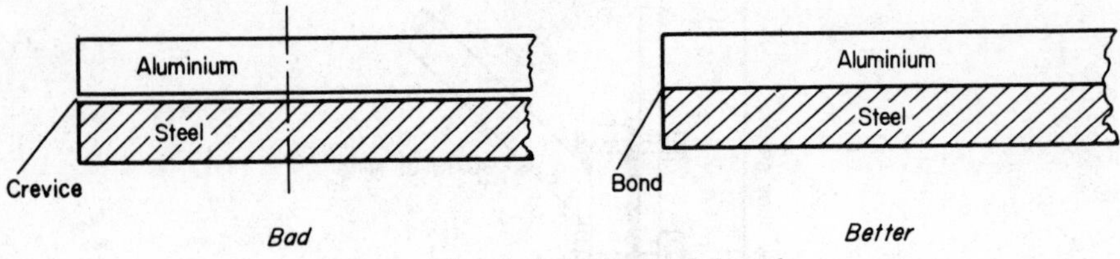

Figure 6.6. Explosion-bonded clad metals

(19) The greatest corrosion effect can be expected at the junction of dissimilar metals and it will usually decrease with increasing distance, which depends on conductivity of solution (*Figure 6.7*).

(20) In marine and other conductive atmospheres the adverse effect of galvanic coupling is apparent within approximately 2 in (5 cm) around the contact. Dielectric separation within this range should be effective or

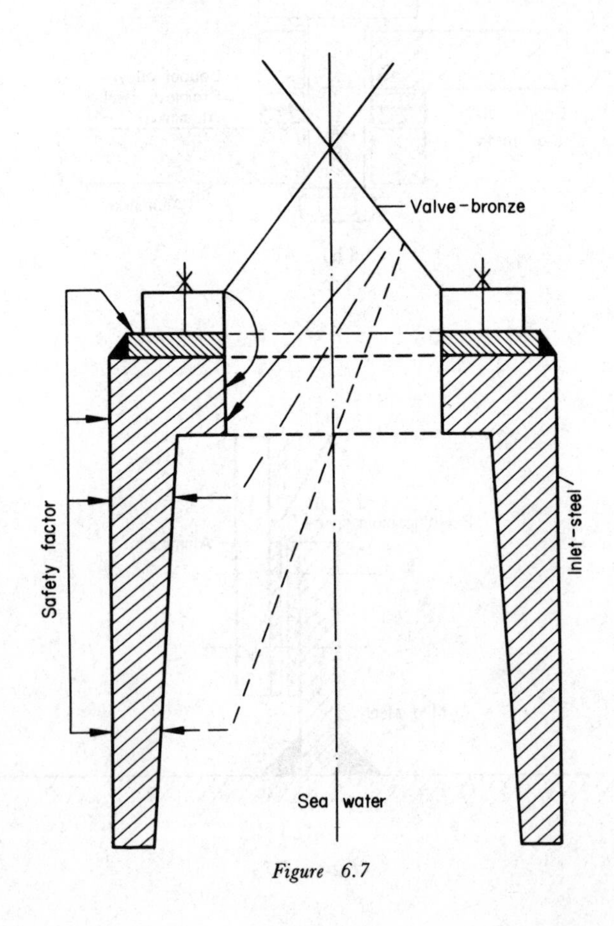

Figure 6.7

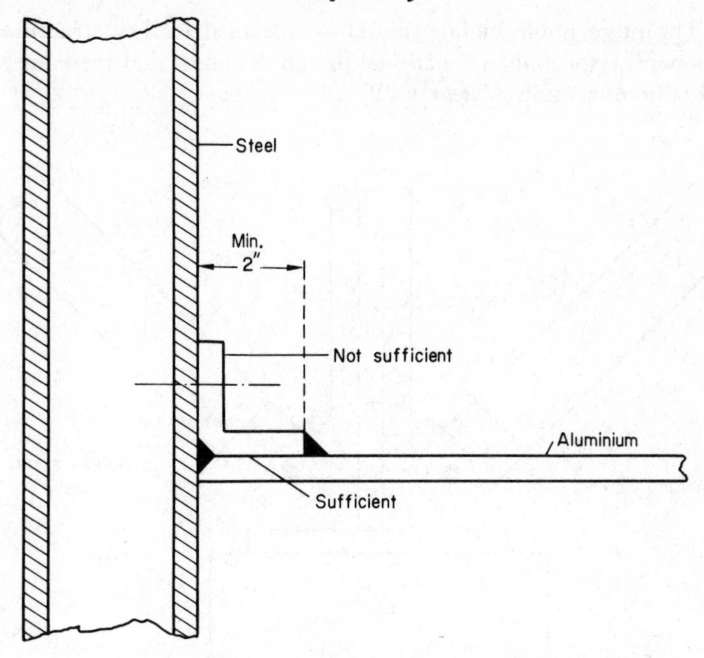

Figure 6.8

appropriate compensation for weight/strength loss should be made (*Figure 6.8*).

(21) All precautions should be taken against formation of rust in direct vicinity of galvanic couples. Rust, due to its hygroscopic properties (it absorbs 40–50% humidity), can aggravate galvanic attack.

(22) Every effort should be made to avoid the unfavourable area effect of small anode and large cathode. Corrosion of relatively small anodic area may be 100–1000 times greater, in comparison with the corrosion of bimetallic components which have the same area submerged in conductive medium (*Figure 6.9*).

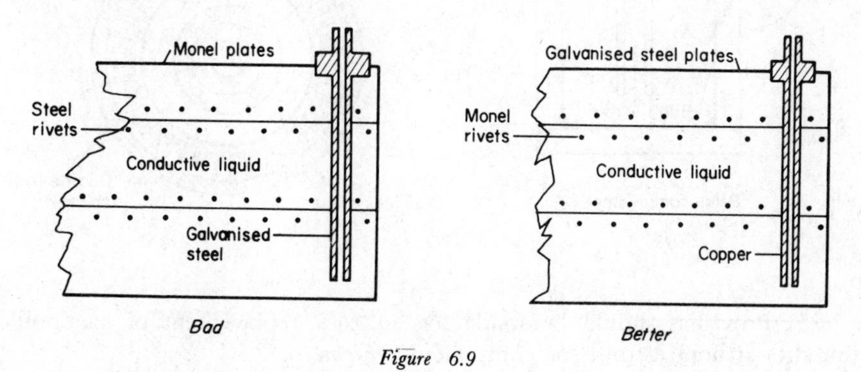

Figure 6.9

(23) The more noble metals should be specified for key structural units or components, especially if functional design demands that these are smaller than the adjoining units (*Figure 6.10*).

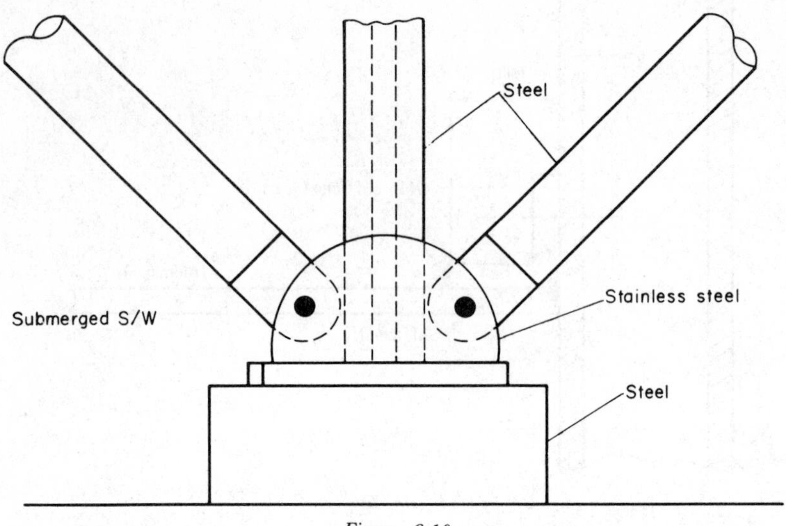

Figure 6.10

(24) Less noble (anodic) components should be made larger or thicker to allow for their corrosion (*Figure 6.11*).

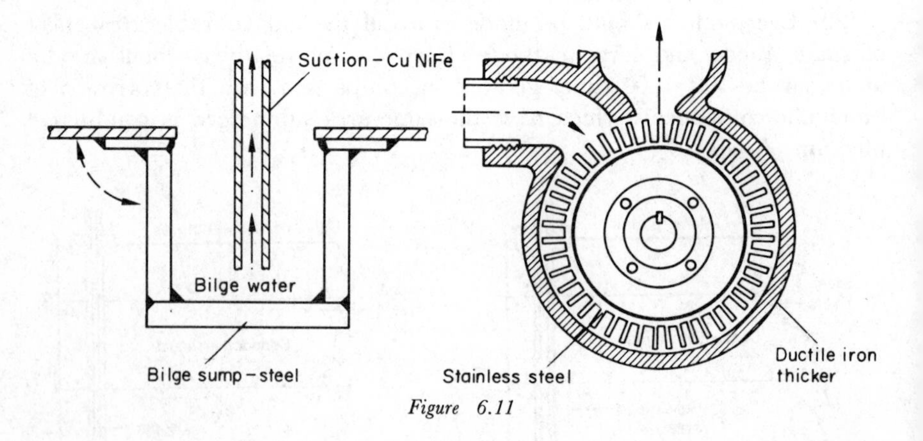

Figure 6.11

(25) Provision should be made for an easy replacement of less noble (anodic) structural units or components (*Figure 6.12*).

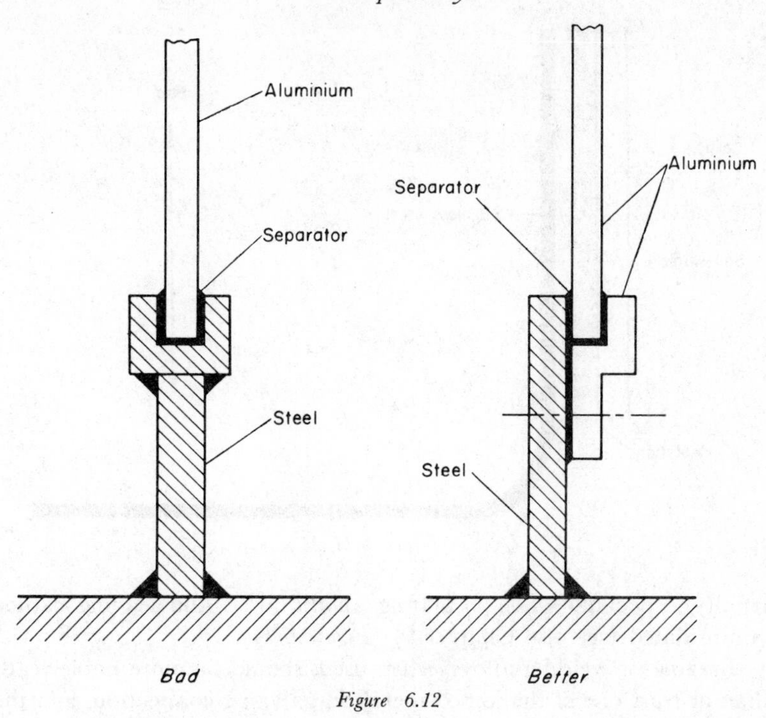

Figure 6.12

(26) No less noble part should be inserted in a conductive environment haphazardly into an otherwise unified system (*Figure 6.13*).

(27) Insulation of bimetallic connection should not be used haphazardly

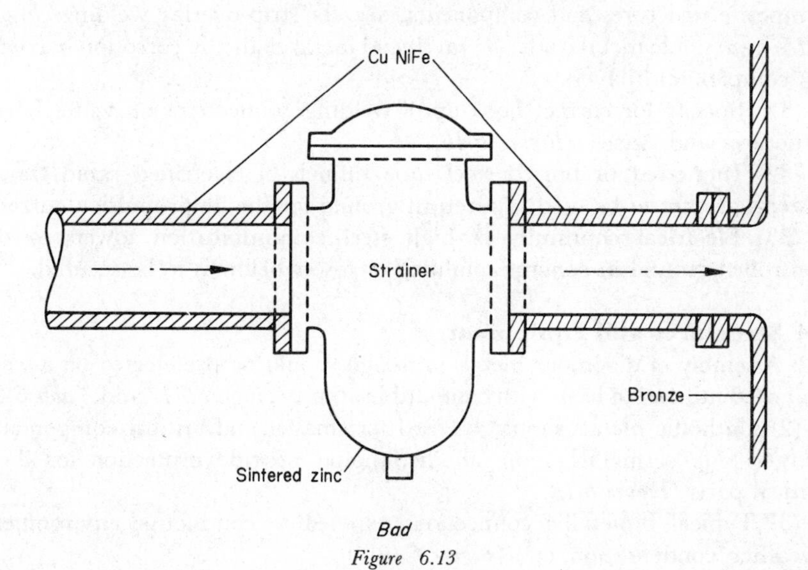

Figure 6.13

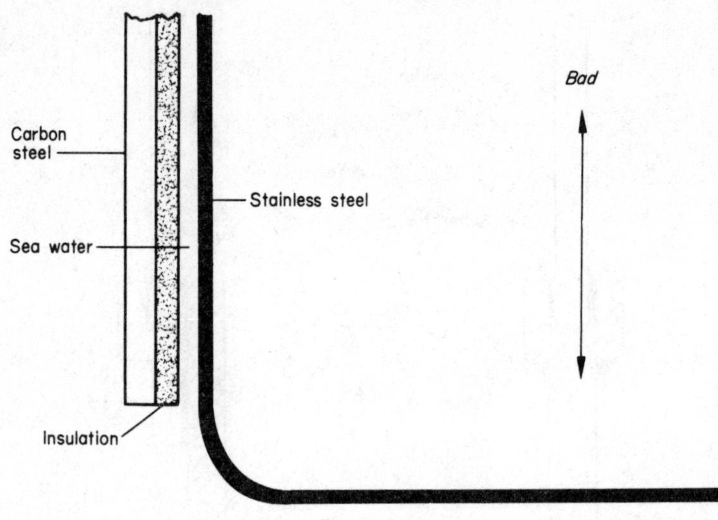

Figure 6.14

or partially, especially when insulating anodic areas and leaving cathodic areas uninsulated (see also Chapter 9) (*Figure 6.14*).

(28) Brazing or welding alloys, when used, should be more noble (cathodic) than at least one of the joined metals in galvanic connection, and these alloys should be compatible to both of them.

(29) Adhesives, when used for joining dissimilar metals, should be of low conductivity (insulating).

(30) Laminar composites may be used in marine environment for bimetal composite structures and components; also for strip overlay welding (*Figure 6.15*) : (*a*) noble metal clads; (*b*) sacrificial metal clads; (*c*) corrosion barriers; (*d*) complex multilayers.

(31) Specify for correct hook-up of welding connections on water borne structures and vessels (*Figure 6.16*).

(32) Immersed or buried steel tube tunnels of electrified rapid transit systems are not to be used as a return ground for the dc propulsion current.

(33) Electrical continuity of high steel communication towers to the controlled ground in concrete building or tower below is to be secured.

6.4 Structures and Equipment

(1) Assembly of dissimilar metals in design should be preselected on a well-balanced utilitarian basis with compatible affinity (*Figure 6.17* and *Table 6.2*).

(2) Cathodic metals should be used for smaller and critical components. Larger bulk of metals, more anodic, should provide protection for these critical parts (*Figure 6.18*).

(3) Typical bimetallic connections exposed to conductive environment (weather, condensation, etc) (*Figure 6.19*).

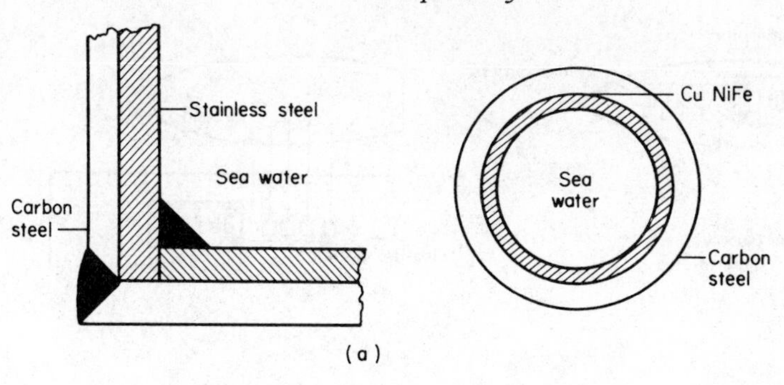
Stainless steel
Sea water
Carbon steel
Cu NiFe
Sea water
Carbon steel
(a)

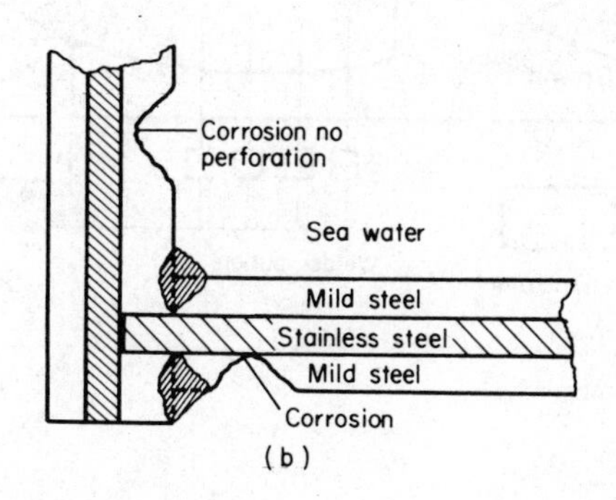
Corrosion no perforation
Sea water
Mild steel
Stainless steel
Mild steel
Corrosion
(b)

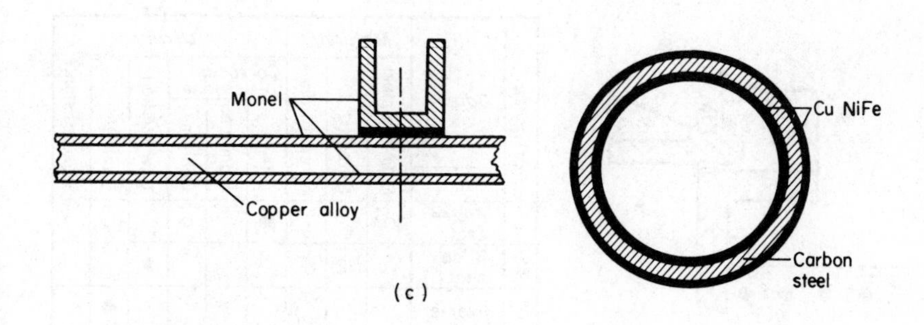
Monel
Copper alloy
Cu NiFe
Carbon steel
(c)

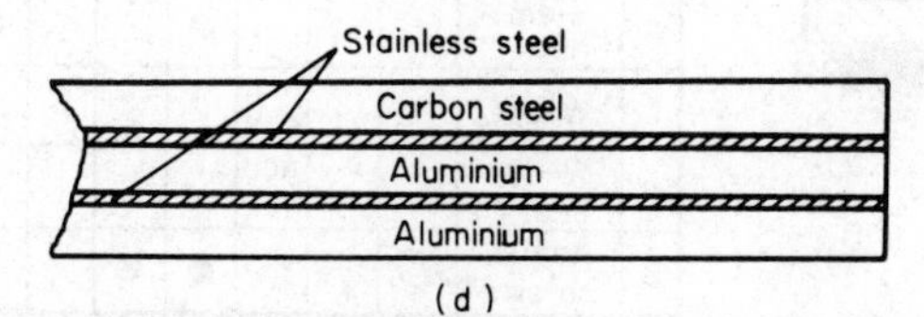
Stainless steel
Carbon steel
Aluminium
Aluminium
(d)

Figure 6.15

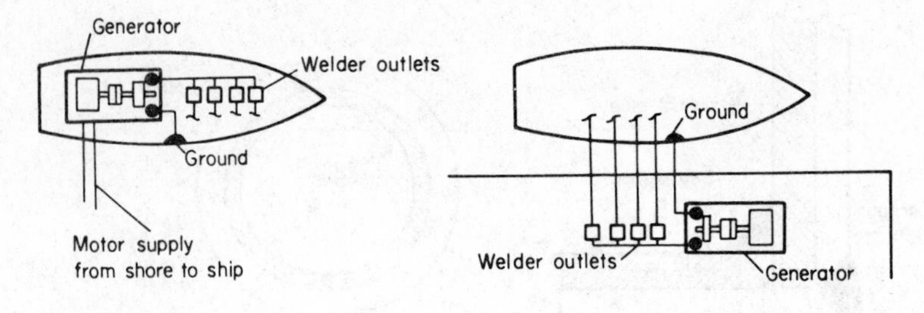

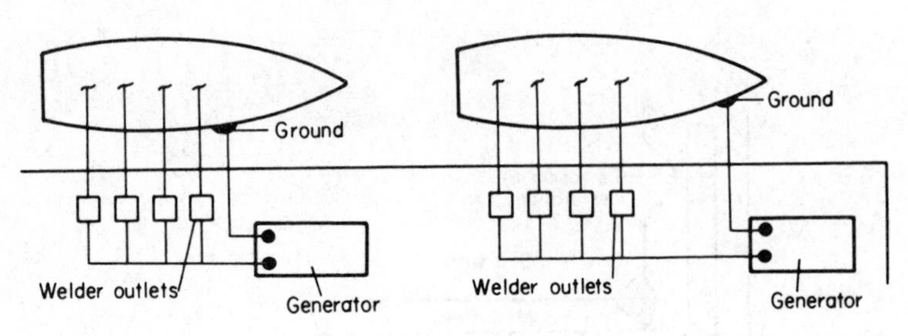

Figure 6.16

Table 6.2 (See also *Figure 6.17*)

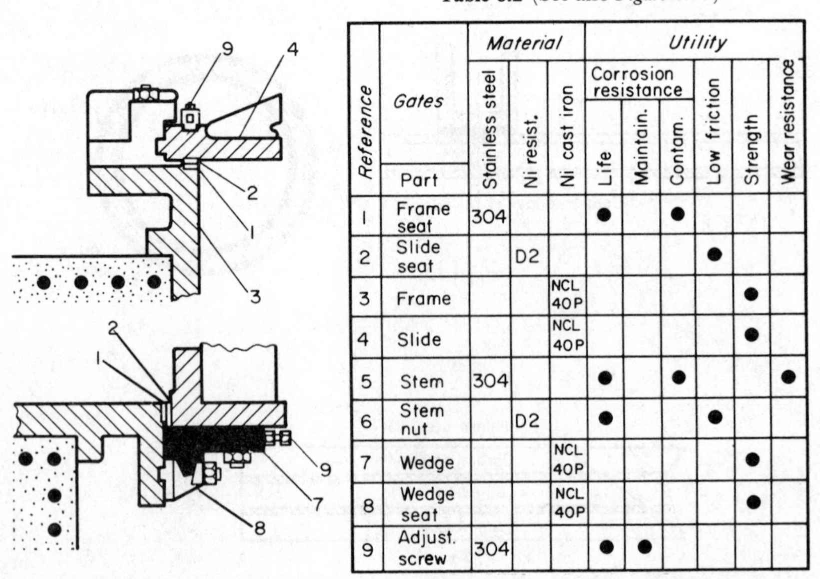

Reference	Gates	Material			Utility					
					Corrosion resistance					
	Part	Stainless steel	Ni resist.	Ni cast iron	Life	Maintain.	Contam.	Low friction	Strength	Wear resistance
1	Frame seat	304			●		●			
2	Slide seat		D2					●		
3	Frame			NCL 40P					●	
4	Slide			NCL 40P					●	
5	Stem	304			●		●			●
6	Stem nut		D2		●			●		
7	Wedge			NCL 40P					●	
8	Wedge seat			NCL 40P					●	
9	Adjust. screw	304			●	●				

Figure 6.17. Plating: structural steel CSA G40.8 (see also Table 6.2)

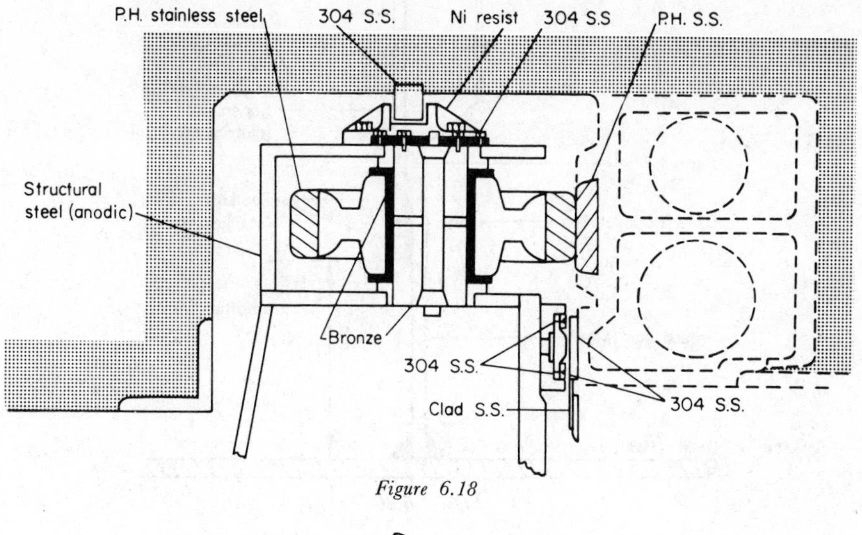

Figure 6.18

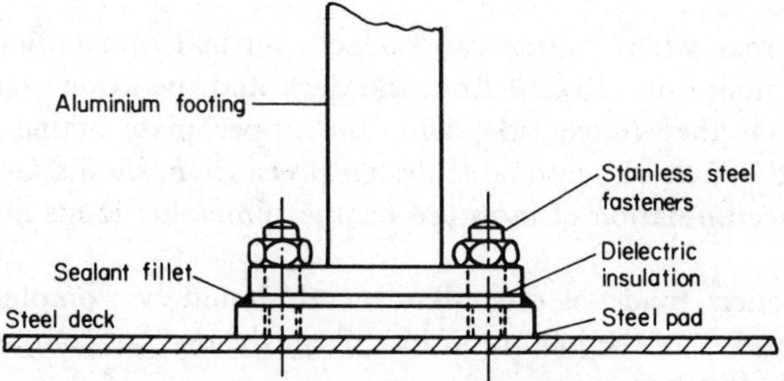

Bimetallic equipment joint

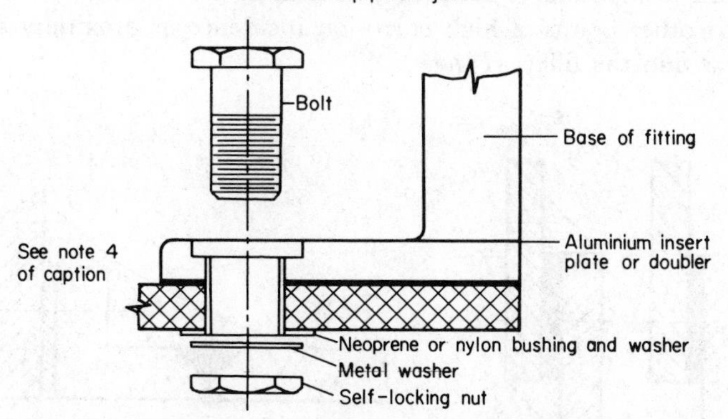

Figure 6.19. The following should be noted: (1) bolt and washer material must be compatible with fitting; (2) aluminium and steel faying surfaces to be primed with lead-free zinc chromate primer; (3) two coats butyl rubber sealant to be applied—first coat to set before second coat is applied; (4) excess sealant to be formed into fillet around periphery of fitting

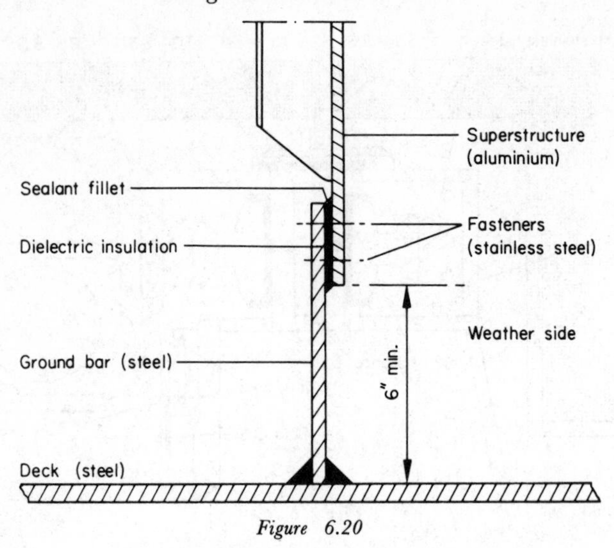

Figure 6.20

(4) In areas where water can collect, vertical bimetallic joints should not be less than 6 in (15 cm) from the deck and the lower edge of the joint should be on the wetted side, with the upper plate acting as a flashing (*Figure 6.20*). *Note*: all reasonable design precautions should be taken against extended accumulation of moisture on the bimetallic joints and their interface.

(5) Fasteners made of dissimilar metal should be completely insulated from both metals of the joint or at least from the one not compatible with the metal of the fastener.

(6) The excess insulation compound squeezed out of the joint, augmented with sealing compound if necessary, should be formed into sealing fillets. Welds and other points of high corrosion incidence in proximity should be included within the fillets (*Figure 6.21*).

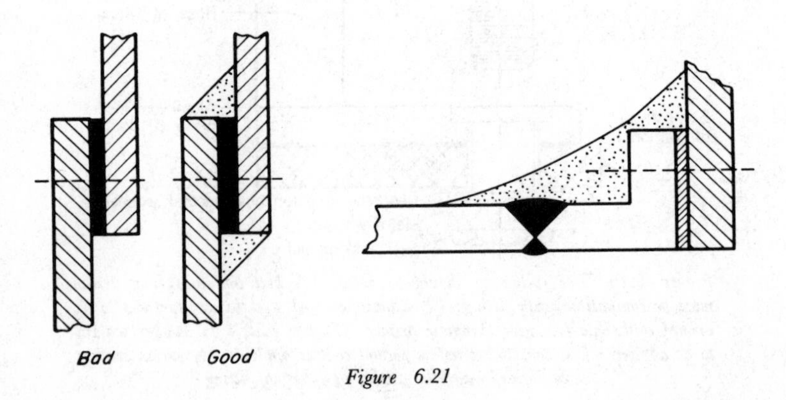

Figure 6.21

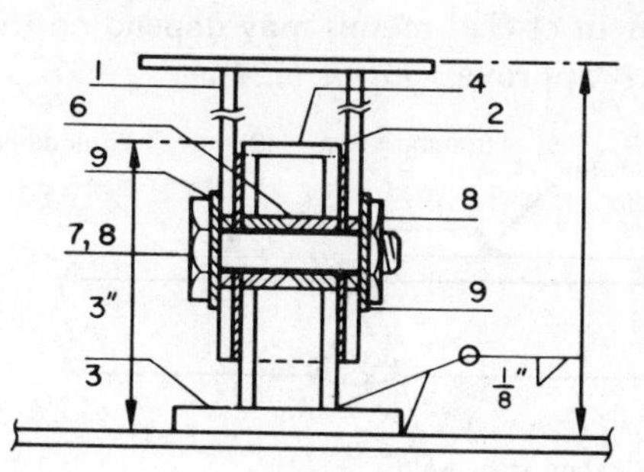

Table 6.3 (See also *Figure 6.22(a)*)

Item no	Description	Material
1	Tube $1\frac{1}{2}$ in. sq. x 0·120 in. wall	Steel or alum
2	Tube 1 in. sq. x 0·120 in. wall	" "
3	Flat $2\frac{1}{2}$ in. x $\frac{1}{4}$ in. x $2\frac{1}{2}$ in. lg	" "
4	Flat 1 in. x $\frac{1}{8}$ in. x 1 in. lg	" "
5	Tube $1\frac{1}{4}$ in. sq. x 0·125 in. wall	Syn rubber
6	Tube $\frac{5}{8}$ in. O/D x $\frac{7}{16}$ in. I/D	Nylon
7	Screw M/C hex. HD $\frac{3}{8}$ in. 16 U.N.C.-7A	St steel
8	Nut hex. $\frac{3}{8}$ in. 16 U.N.C.-2B	" "
9	Bush	Nylon

Aluminium tube

metallic pad

$\frac{1}{8}$"

$\frac{1}{8}$"

(b)

Table 6.4 (See also *Figure 6.22(b)*)

Item no.	Description	Material	
1	Bar, flat (dimensions as required)	Aluminium	Select suitable shape
1	Angle (dimensions as required)	Aluminium	
1	Tee (dimensions as required)	Aluminium	
1	Tube, round, hollow (dimensions as required)	Aluminium	
1	Pad, bi-metallic $\frac{1}{2}$ in. thk. ($\frac{1}{4}$ in. aluminium $\frac{1}{4}$ in. steel) length and width as required	Aluminium / Steel	

Figure 6.22. (a) Normal bimetallic joint (see also Table 6.3); (b) transition joint for the same purpose (see also Table 6.4)

(7) Structural connections between dissimilar metals can be facilitated by a design involving transition joints (*Figure 6.22* and *Tables 6.3* and *6.4*). These joints are normally provided by clad metals with a sound high strength bond, not adversely affected by the used method of joining (welding, brazing). Note also the saving of labour costs.

(8) Combination of metals in the bimetallic pad best suited or compatible with the attached members (structures or equipment) should be used (*Figure 6.23*).

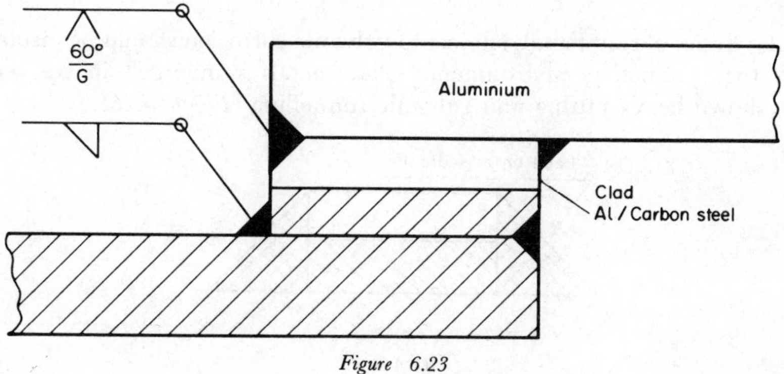

Figure 6.23

(9) Utility, fabrication and attachment of clad metals may depend on the technique of bonding; ask for suppliers' approval (*Figure 6.24*).

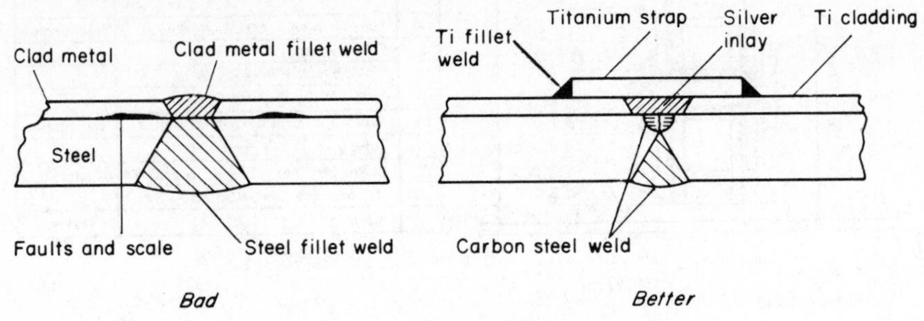

Figure 6.24

(10) Clad metals may be subject to galvanic corrosion along exposed edges, if the metals are far apart according to a galvanic corrosion indicator (e.g. copper/aluminium clad to aluminium) (*Figure 6.25*).

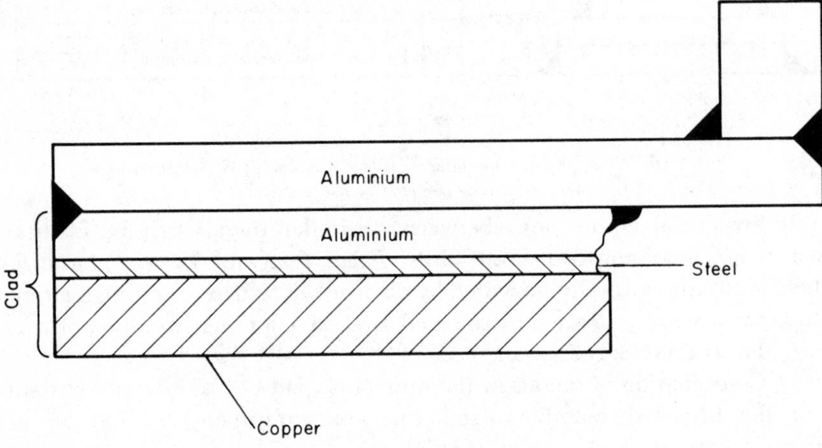

Figure 6.25

(11) Edges of clad metals subject to galvanic corrosion should be insulated from the conductive environment. Clad metals submerged in sea water have shown heavy pitting and galvanic tunnelling (*Figure 6.26*).

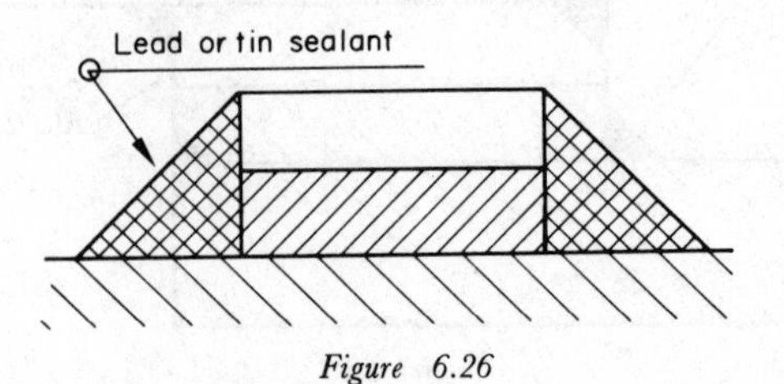

Figure 6.26

(12) The size, thickness (each metal in sandwich) and shape (plane and peripheral) of clad metal connections (bimetallic pads) should be adjusted to the requirements of their utility, shapes of the joined surfaces, strength of the connection and method of attachment. Round or oval pads are able to avoid the adverse effects of sharp corners—stress risers (*Figure 6.27*).

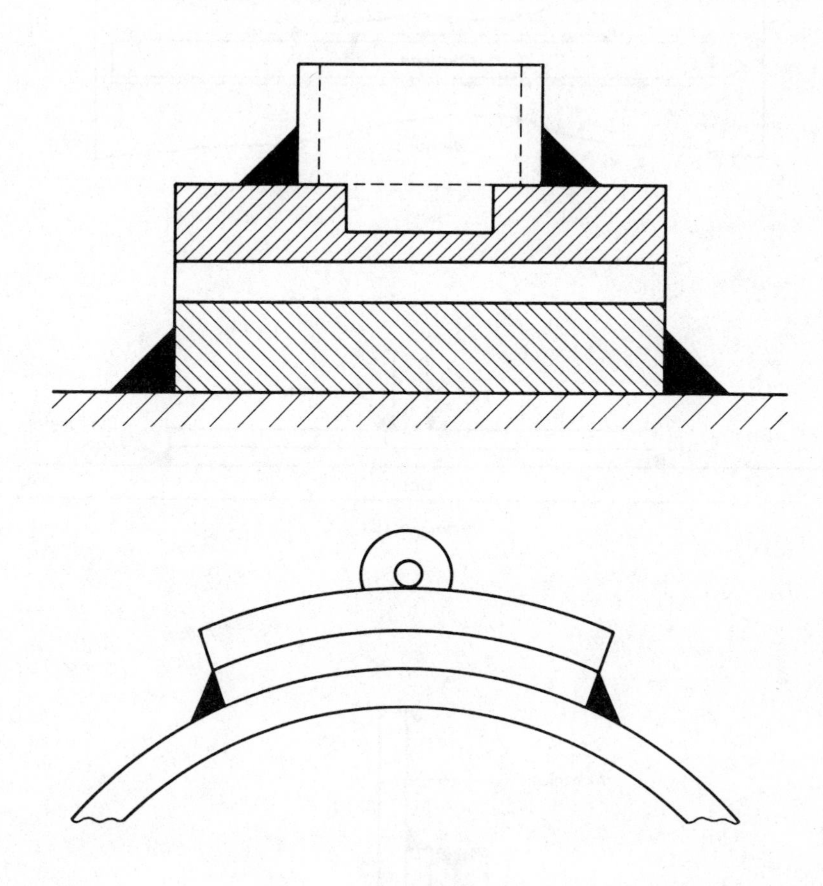

Figure 6.27

(13) Correct system and sequences of welding attachment of bimetallic pads should be specified to avoid distortion and input stresses (*Figure 6.28*).

(14) Fixed attachment of bimetallic pads or strips in dry or moderately damp environmental conditions (*Figure 6.29*).

(15) Fixed attachment of bimetallic strips or pads in corrosion-prone conditions—transition joint (*Figure 6.30*).

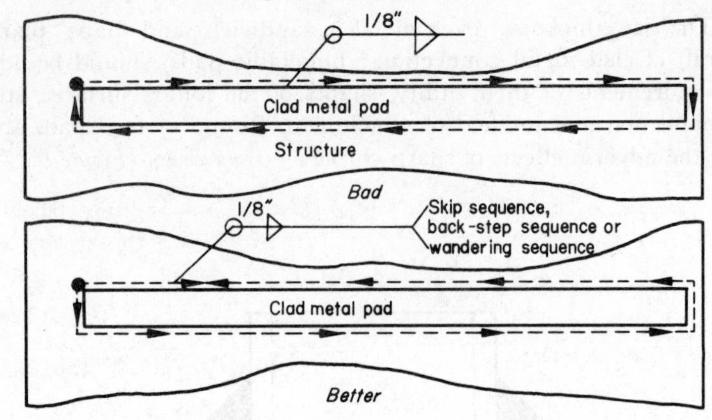

Figure 6.28

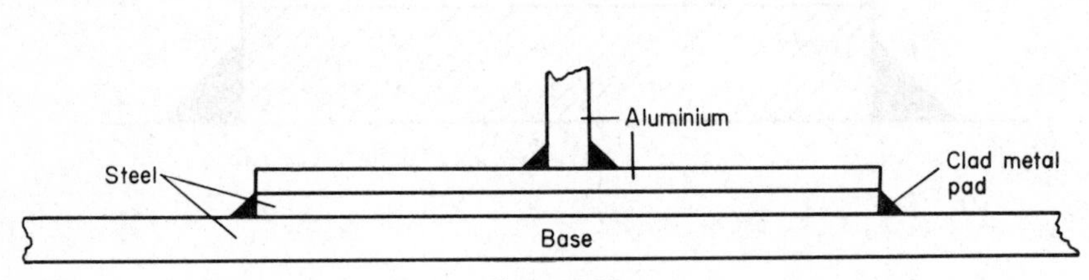

Figure 6.29

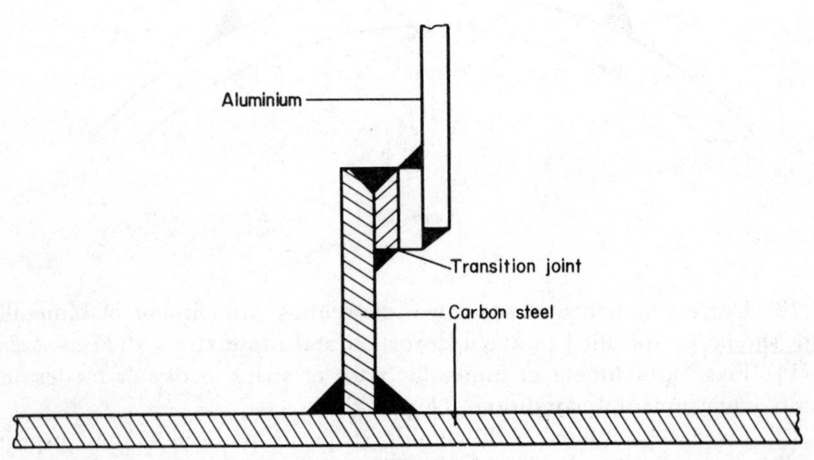

Figure 6.30

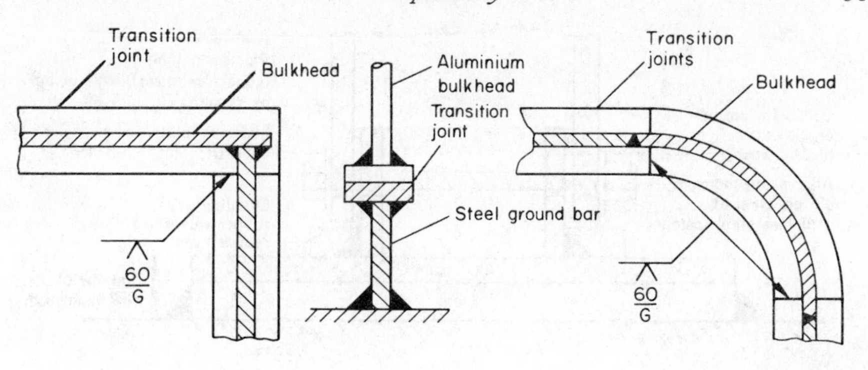

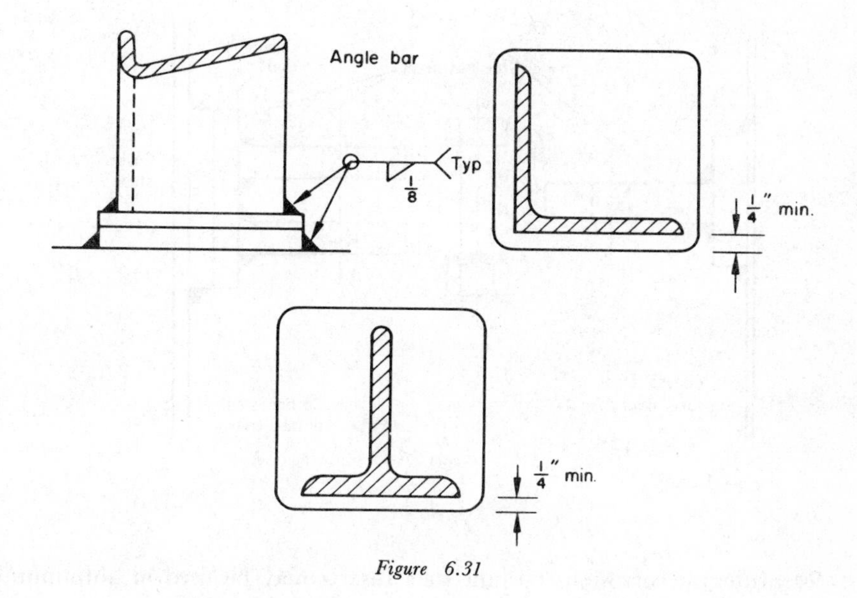

Figure 6.31

(16) Possible design variations of bimetallic pads (*Figure 6.31*)—also hollow sections, strip, 'I'-bar, etc.).

(17) Design of dismantleable bimetallic attachment (*Figure 6.32*).

(18) Typical application of stainless steel/carbon steel clad transition joint between high temperature duct and weatherside casing (*Figure 6.33*).

(19) Example of application of clad metals in design of major equipment for reduction of corrosion (*Figure 6.34*).

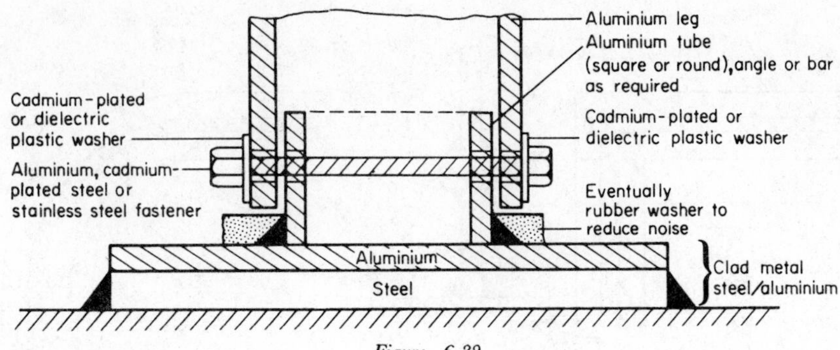

Figure 6.32

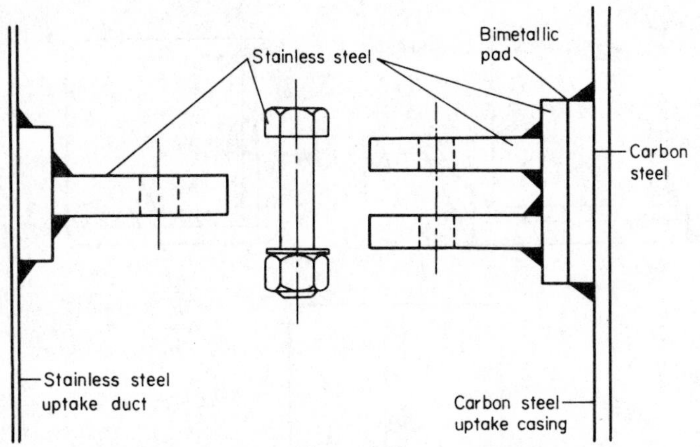

Figure 6.33

(20) Integral corrosion-resistant steel inserts may be used in aluminium casting (*Figure 6.35*).

(21) Example of dissimilar metal connection fitting requiring frequent adjustment—the bedding plate to remain undisturbed and to transfer shear loads to the structure (shear spigot could be incorporated) (*Figure 6.36*).

(22) Example of non-adjustable steel fitting secured to aluminium structure (*Figure 6.37*).

(23) Example of non-adjustable aluminium alloy fitting secured to aluminium structure with steel fastener (*Figure 6.38*).

(24) Try to conduct moisture away from galvanic couples in design (*Figure 6.39*).

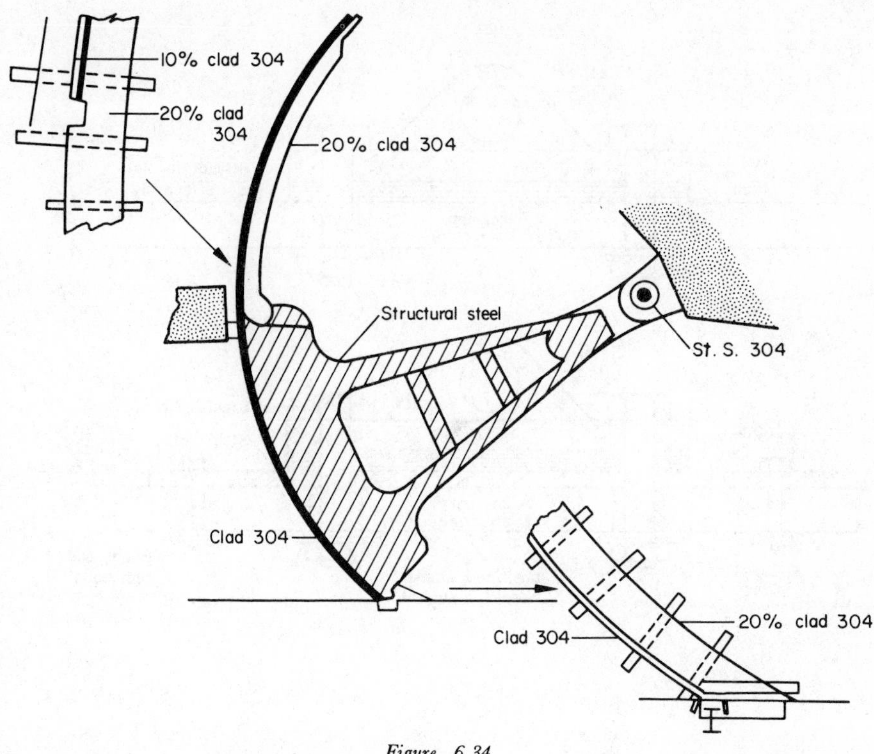

Figure 6.34

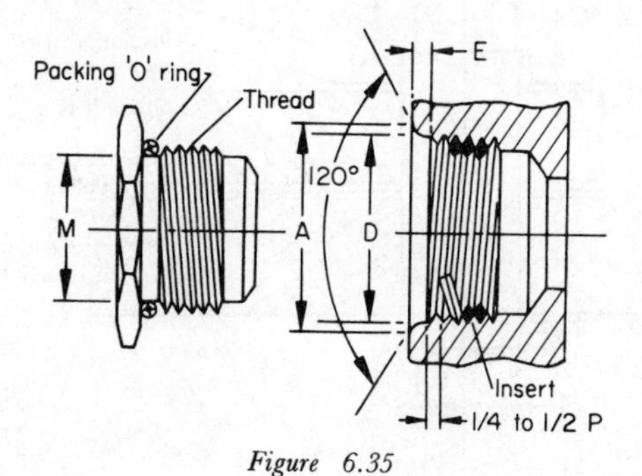

Figure 6.35

(25) Do not attach new steel to larger size old marine structures submerged in sea water—new steel will become anodic and will corrode at an accelerated rate.

(26) Plastic bearing materials should be preferred in corrosive environments and locations that are inaccessible or liable to be overlooked, for

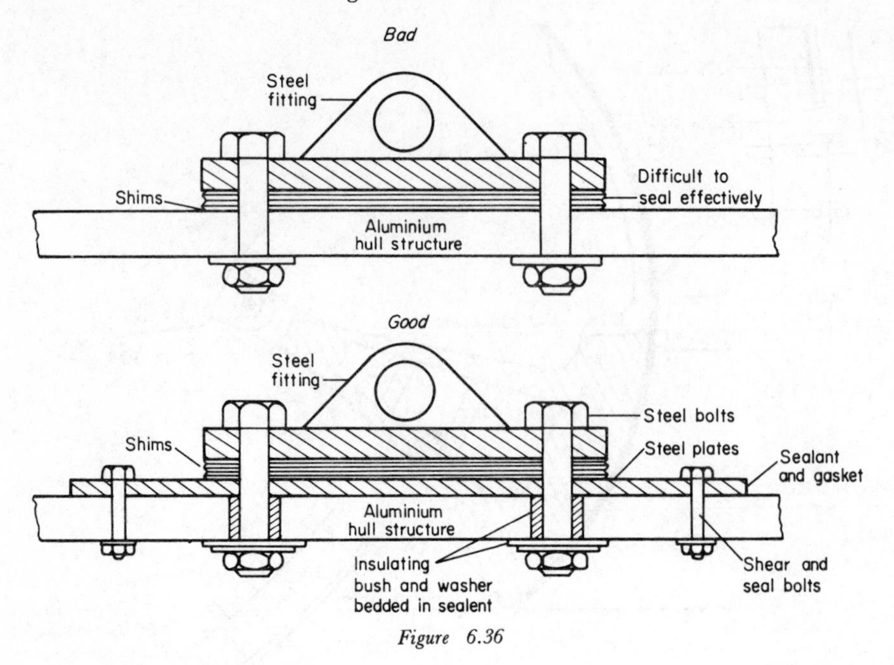

Figure 6.36

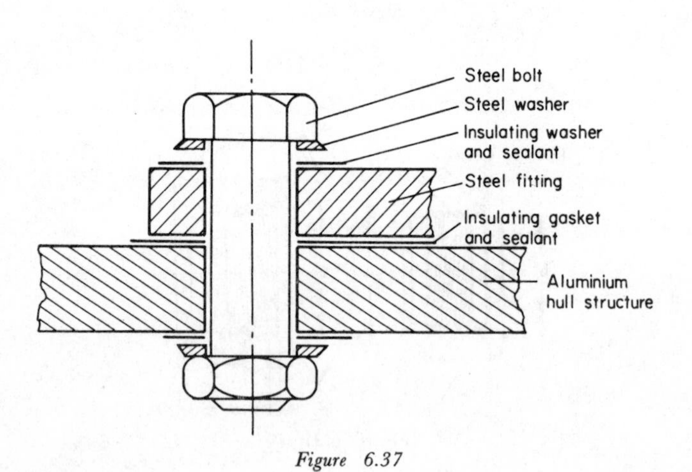

Figure 6.37

low speed services to metal bearing materials forming galvanic couples with shafts, pins or sliding surfaces—especially if the mating surfaces are equivalent to steel or harder (*Figure 6.40*). For softer shaft materials (brass or aluminium) special resins may be required.

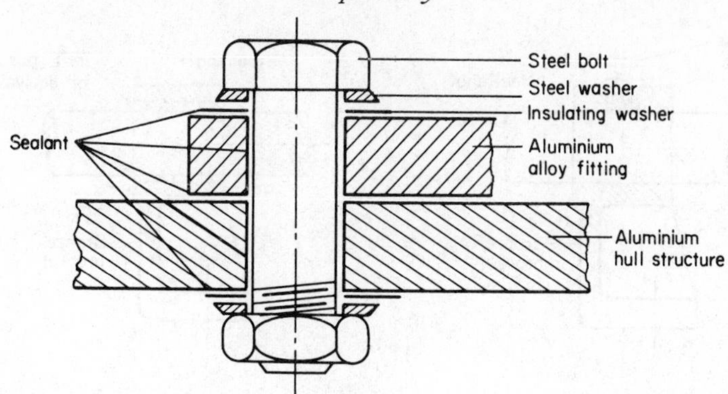

Figure 6.38

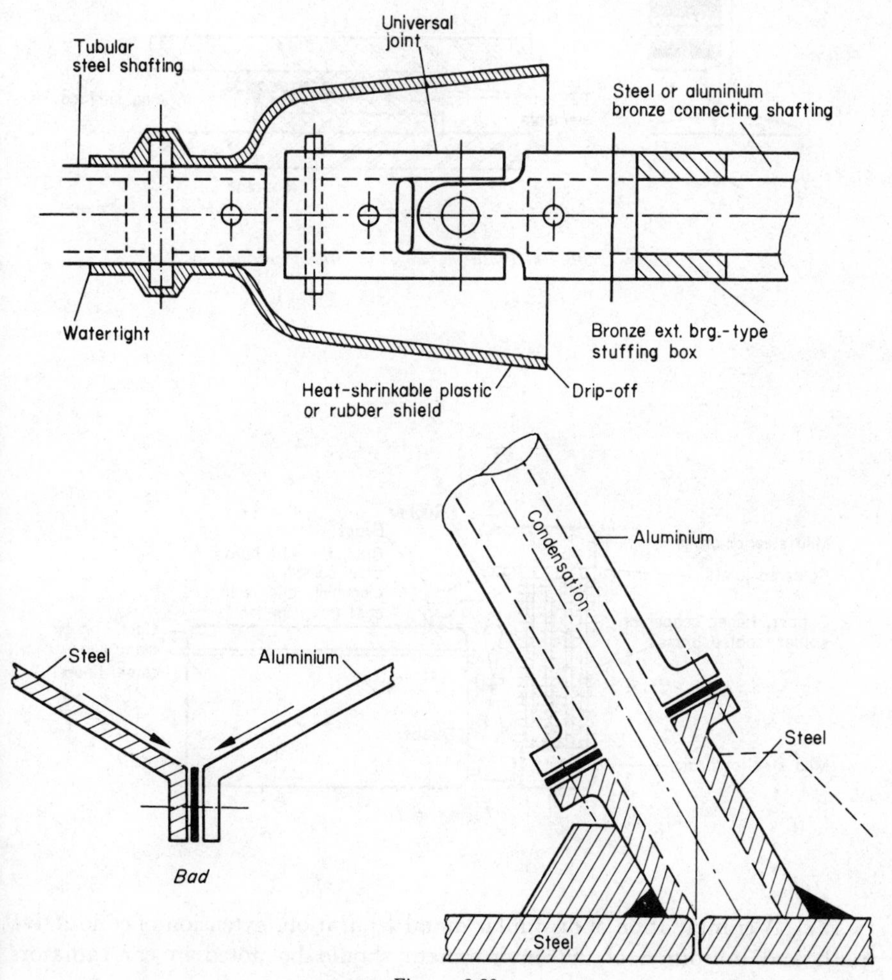

Figure 6.39

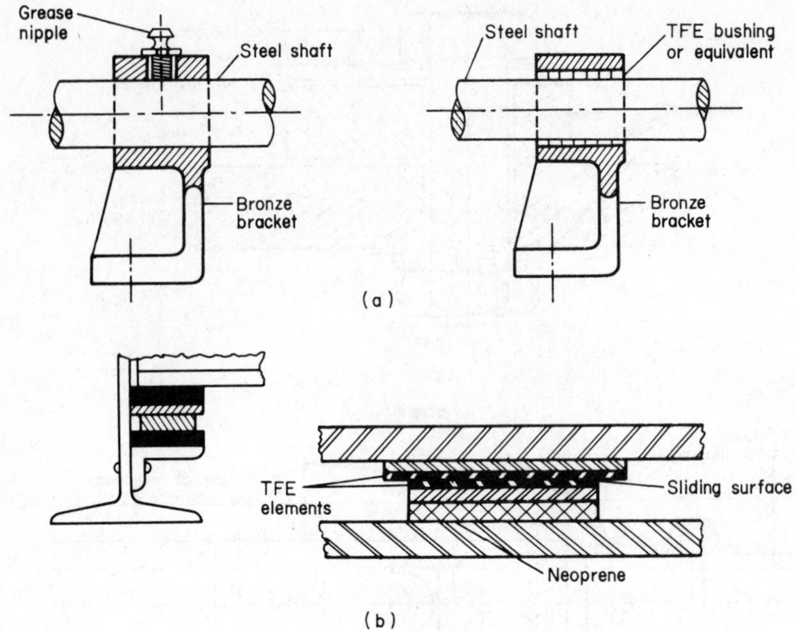

Figure 6.40. (a) Plastic bushings; (b) sliding bearings

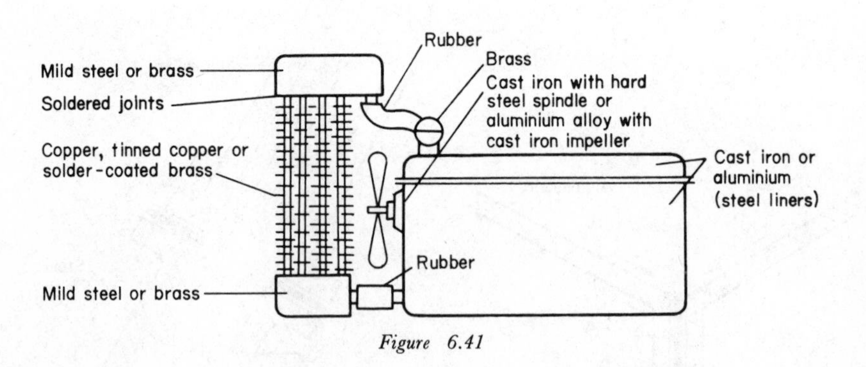

Figure 6.41

(27) A combination of dissimilar metal separation, extension of conductive path and inhibition of the environment should be noted in car radiators (*Figure 6.41*).

(28) Mercury thermometers and other sources of free mercury should be avoided in the proximity of aluminium and copper alloy structures and equipment (*Figure 6.42*).

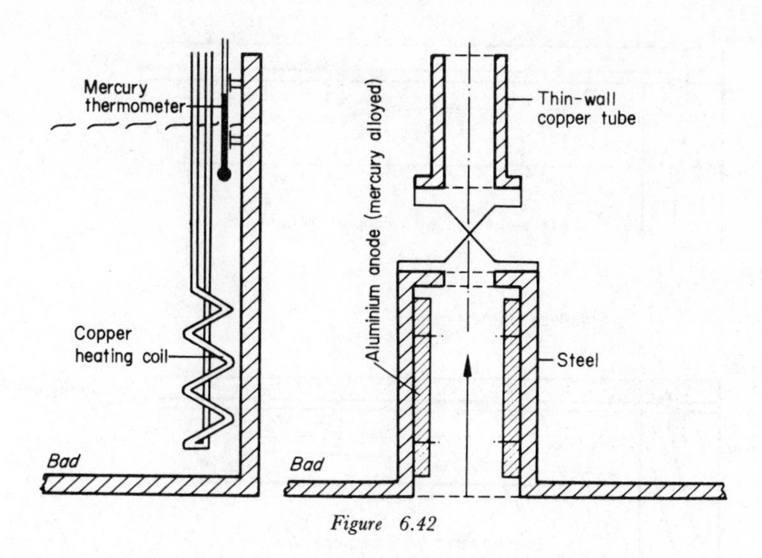

Figure 6.42

(29) The use of carbon or graphite components coupled in a conductive environment to any other metal should be avoided (*Figure 6.43*).

(30) Canvas fabric impregnated with copper salts should not be attached to steel or aluminium structures or used as a rain cover over steel or aluminium equipment.

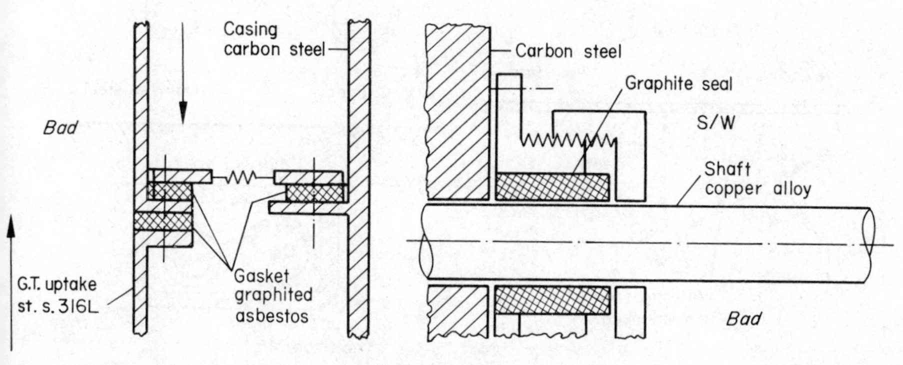

Figure 6.43

(31) Do not attach or embed metals anodic to copper into wood or plywood impregnated with copper salts (*Figure 6.44*).

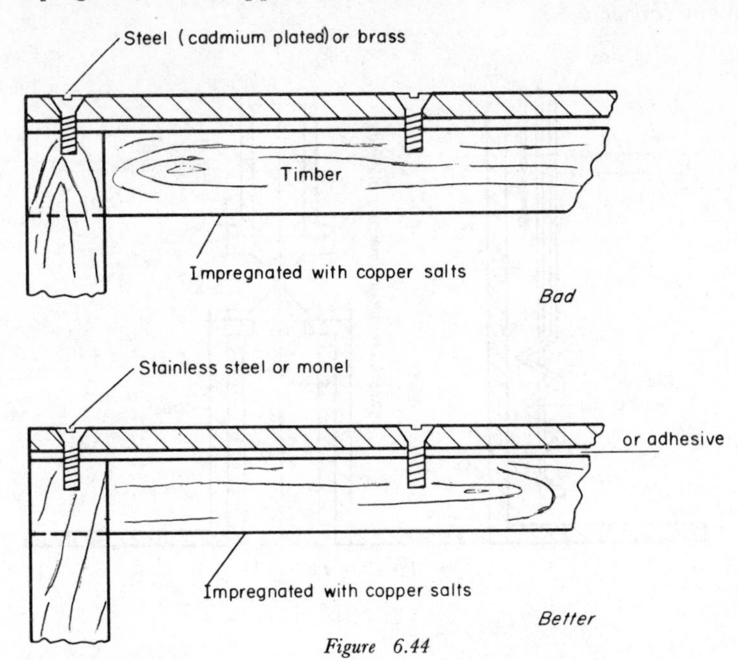

Figure 6.44

(32) Dissimilar metals embedded in porous materials in close proximity will cause galvanic corrosion (*Figure 6.45*).

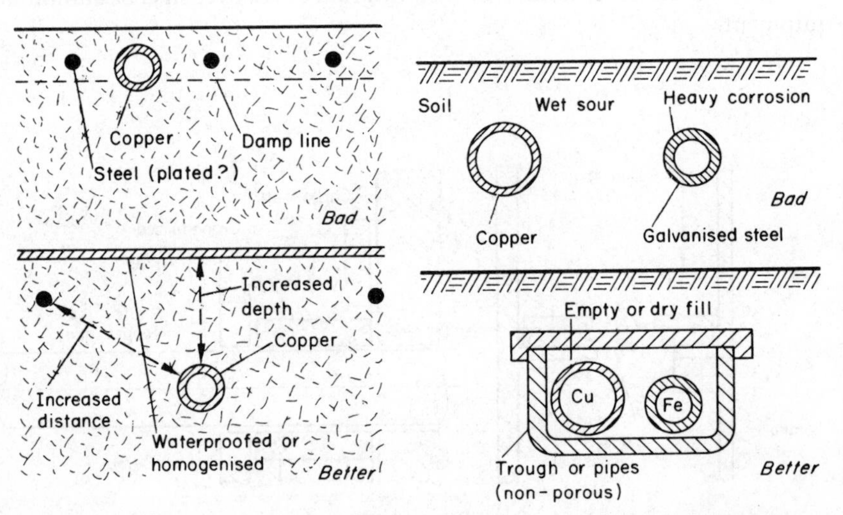

Figure 6.45

(33) Transfer of temperatures in structures or equipment which can cause adverse polarisation of metals should be avoided in design, if possible (*Figure 6.46*).

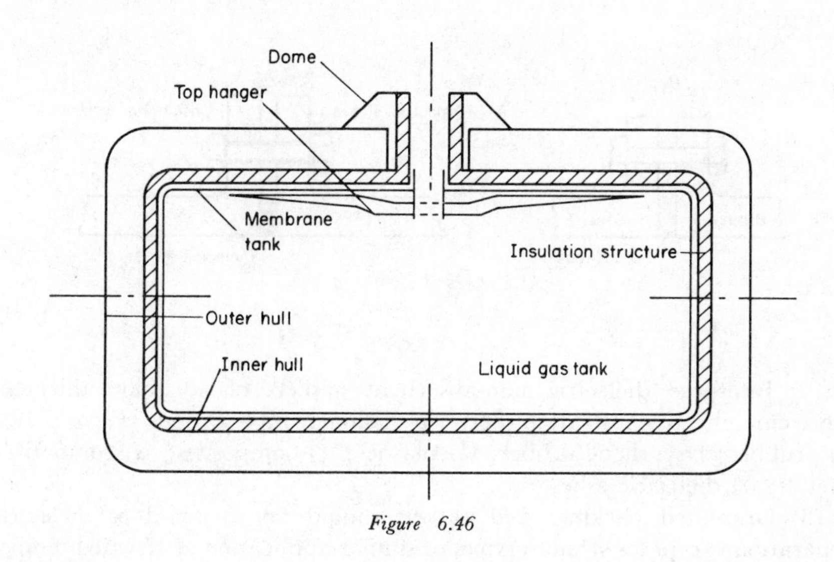

Figure 6.46

(34) Optimum selection of welding rods, joint preparation, weld configuration and sequences are vital for joining of dissimilar metals by welding (*Figure 6.47*).

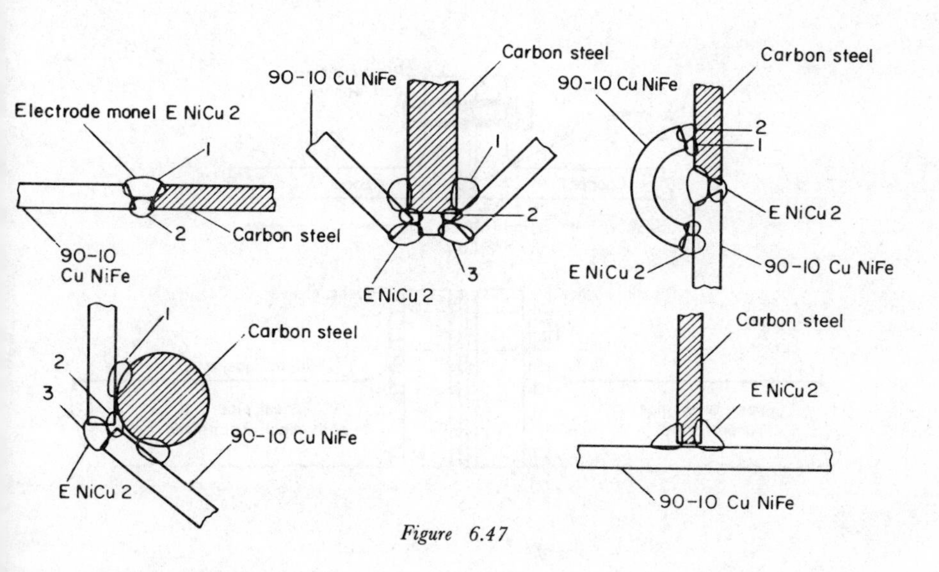

Figure 6.47

6.5 Piping Systems

(1) Secure complete and effective separation between sections of the piping of dissimilar metals (*Figure 6.48*).

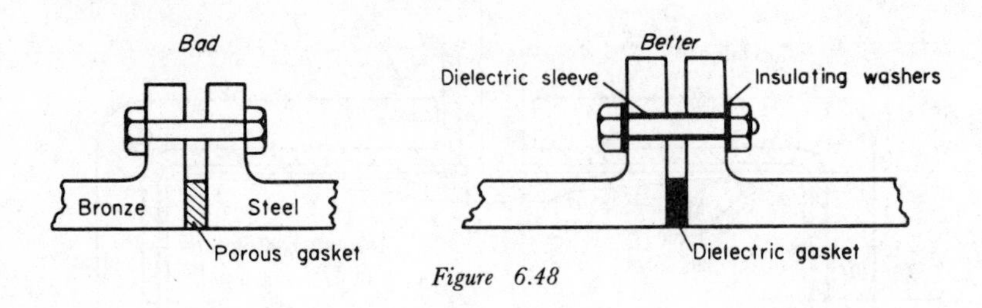

Figure 6.48

(2) Interpose dielectric non-absorbent gaskets of adequate thickness (to secure effective insulation) between dissimilar pipe sections (*Figure 6.49*): (*a*) rubber or synthetic rubber; (*b*) plastics; (*c*) composites; (*d*) compatible metals; (*e*) dielectric asbestos.

(3) Graphited packings and gaskets should not be used as dielectric separation except for steam services or similar application at elevated temperatures or in non-conductive media.

(4) Reduce galvanic corrosion of dissimilar metal pipe connections exposed to low conductivity, recirculated distilled or demineralised water (when sulphate is present), by interposing lead inserts between the faying surfaces of the two metals (*Figure 6.50*).

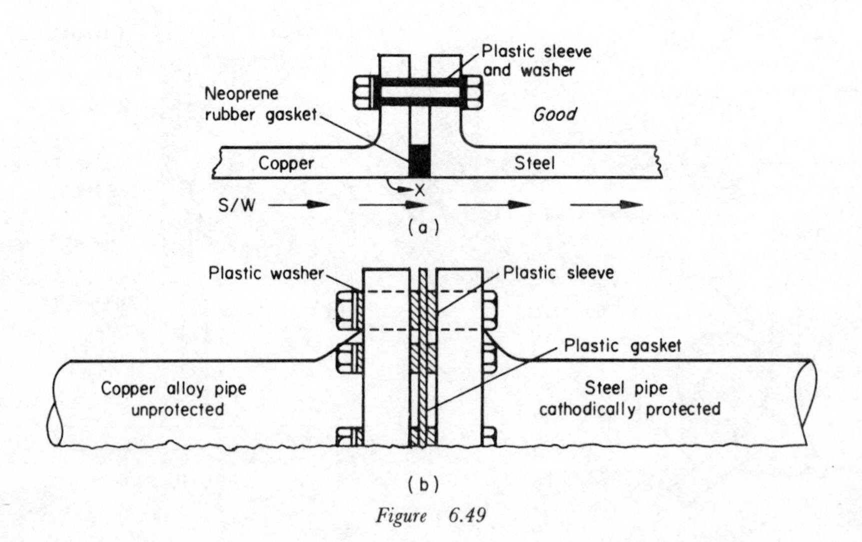

Figure 6.49

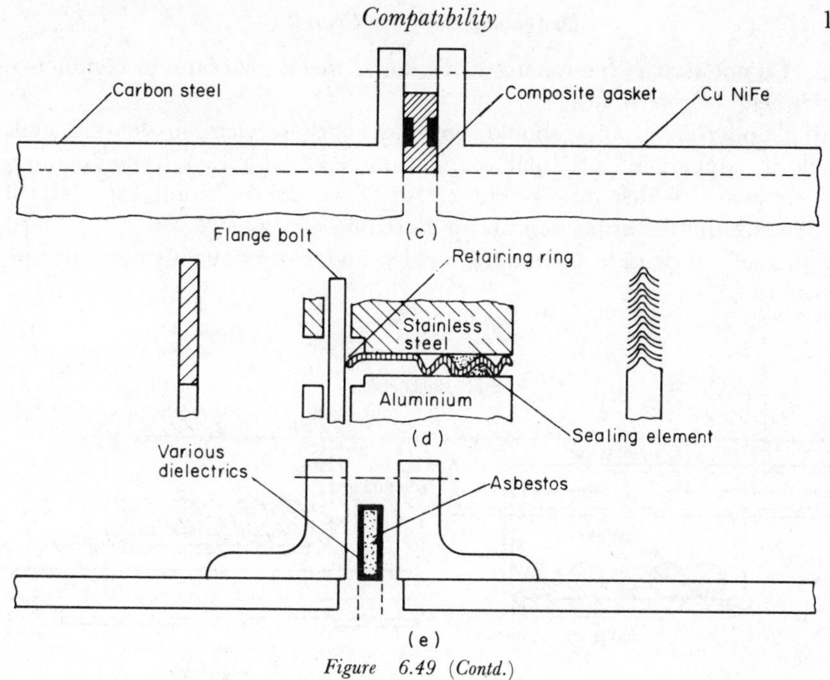

Figure 6.49 (Contd.)

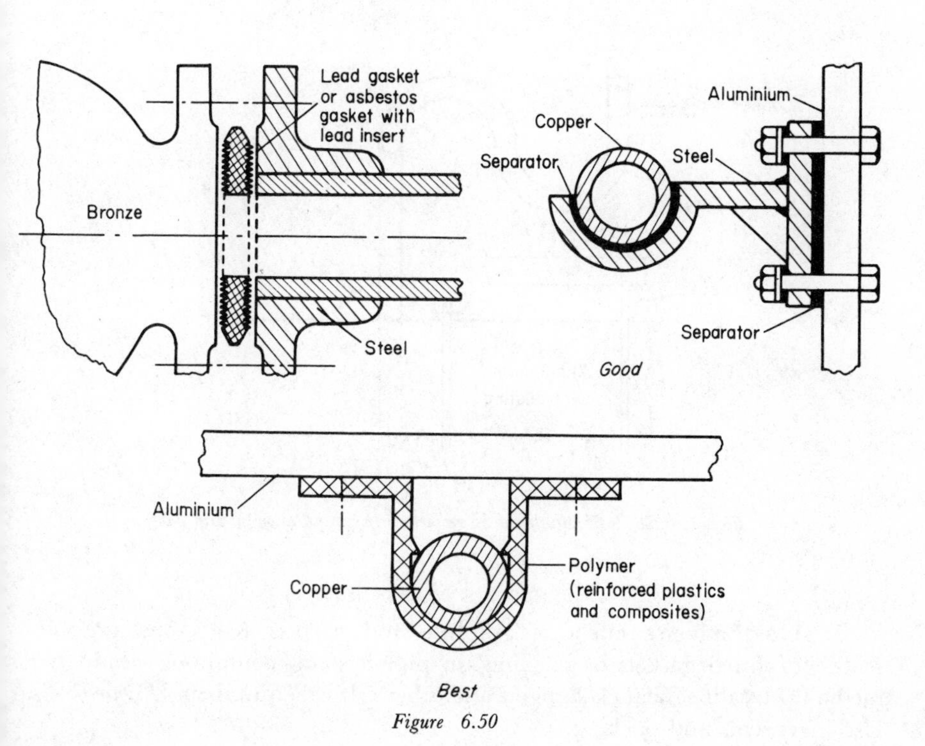

Figure 6.50

(5) Do not attach pipe systems to dissimilar metal structures by conductive attachment (*Figure 6.50*).

(6) Copper-base pipes should not pass over services made of metals anodic to copper, in particular aluminium (e.g. cableways, vent ductings, etc.), in spaces which may be subject to heavy condensation, especially if the accessibility for inspection and maintenance is poor (*Figure 6.51*). *Note:* dripping of condensate containing oxides and carbonates causes galvanic corrosion.

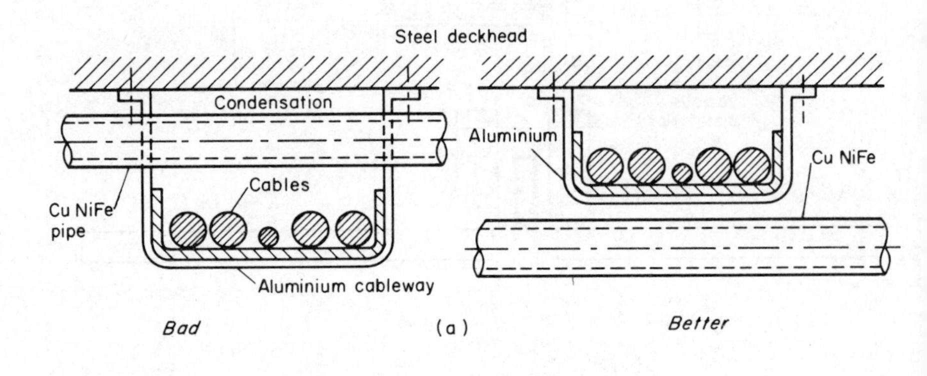

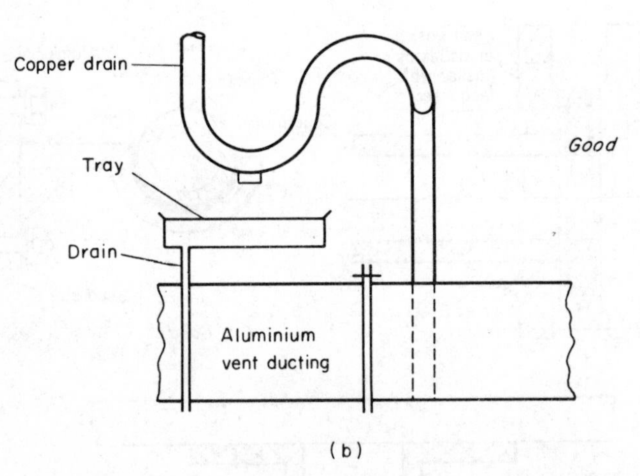

Figure 6.51. (a) Adjustment of geometry; (b) installation of drip pans

(7) Avoid adverse effect of graphite and carbon (e.g. solid graphite seals, graphited gaskets or packing) in pipe systems containing conductive media upstream of heat exchangers and other critical equipment (*Figure 6.52*). Use inert seals and packing.

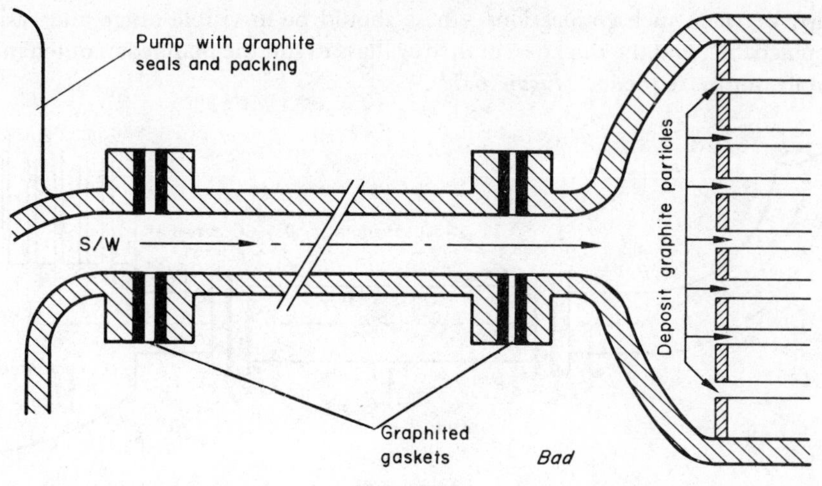

Figure 6.52

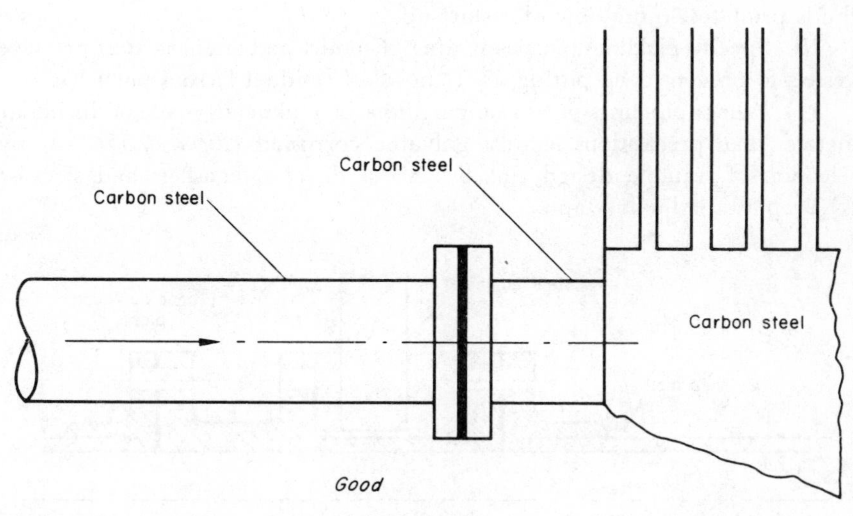

Figure 6.53

(8) Salts of copper emanating from copper-base pipes carried in solution are dangerous to carbon steel components and tanks downstream (*Figure 6.53*). If possible avoid fitting copper alloy pipes upstream of carbon steel equipment; if such fitting is necessary, interpose sacrificial pieces of mild steel

pipe between such connections—these should be in visible range and easily replaceable, and the thickness of their walls is to suit the planned maintenance programme frequency (*Figure 6.54*).

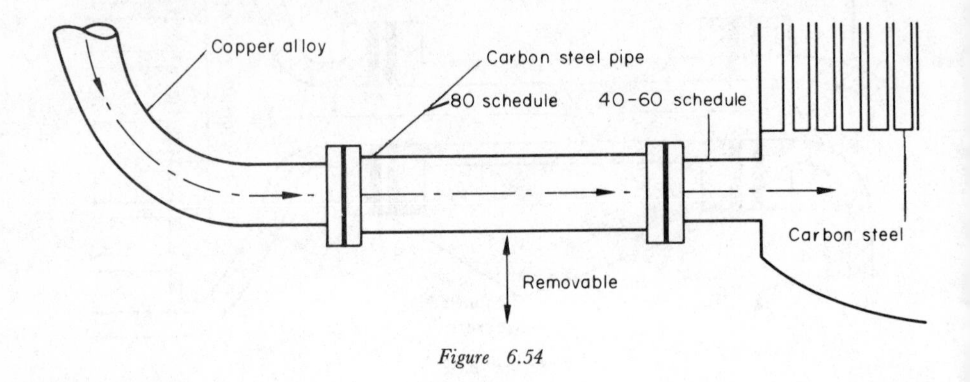

Figure 6.54

(9) Specify inspection for removal of dissimilar metals filings, tools and other objects from water tanks and other vessels containing conductive fluids prior to closing down for start-up.

(10) Specify pickling and passivating of monel and stainless steel pressure vessels to prevent deep pitting, by removal of residual ferrous particles.

(11) Where pipelines penetrate partitions or bulkheads made of dissimilar metals, take precautions against galvanic corrosion (*Figure 6.55*) : (*a*) by selection of suitable design; (*b*) by use of dielectric gaskets and sleeves; (*c*) by plastic adhesive tapes.

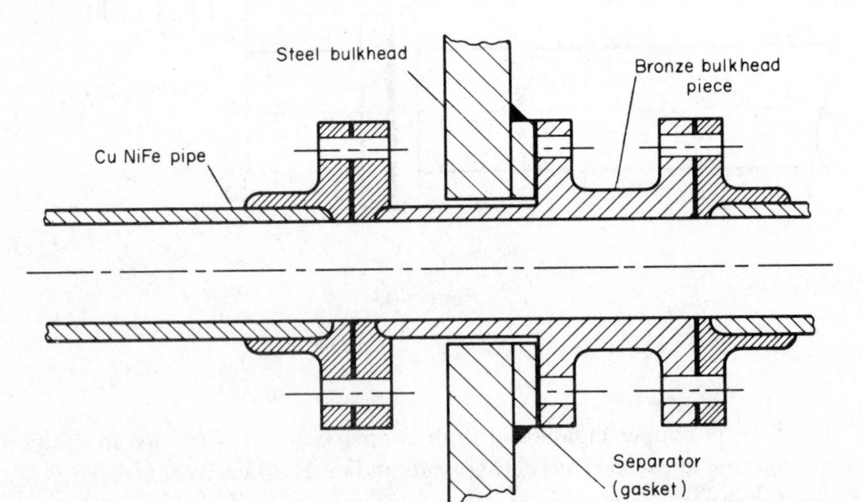

Figure 6.55

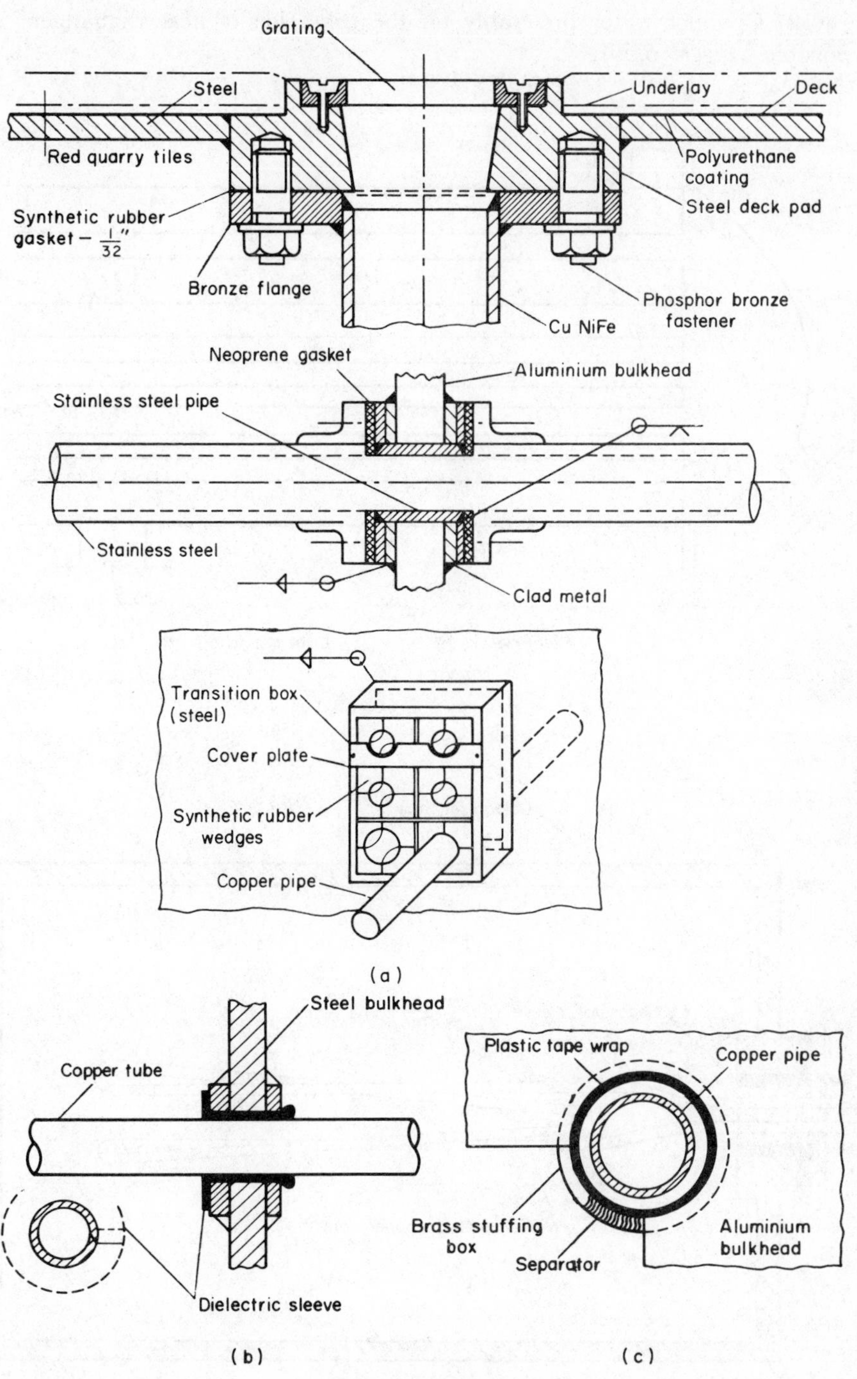

Figure 6.55 (Contd.)

(12) Conduct water preferably on the tube side of heat exchangers if possible (*Figure 6.56*).

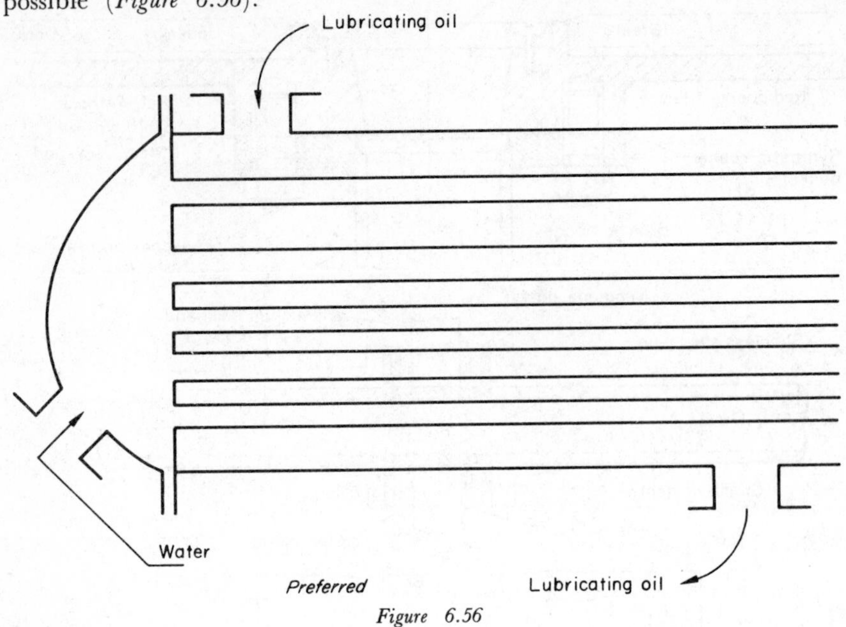

Figure 6.56

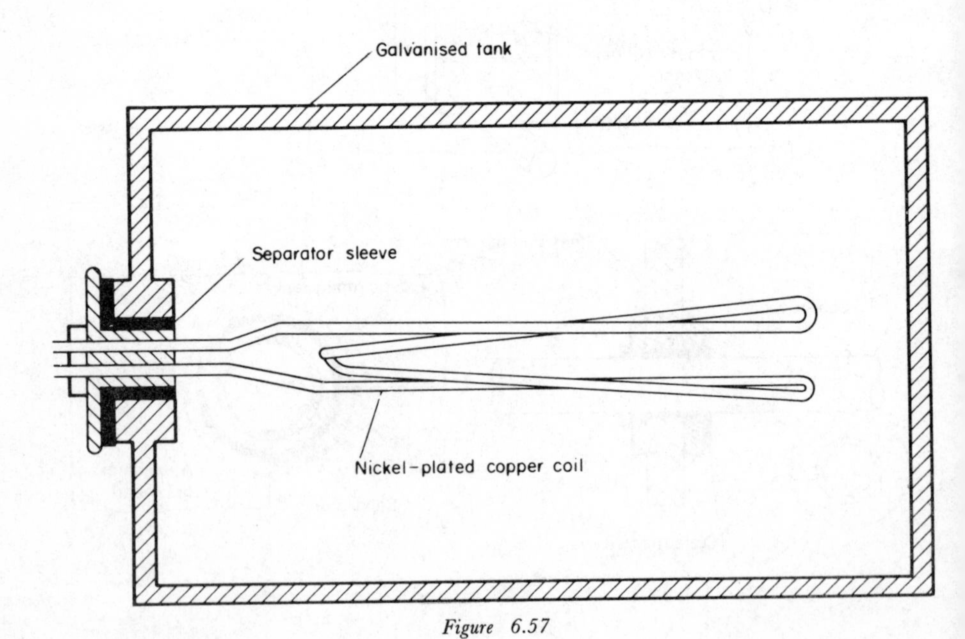

Figure 6.57

(13) In heat exchangers using copper coils the effect of copper going into solution and affecting the galvanised steel shell can be avoided by nickel plating the coils—these can then be separated by insulation from direct contact with the tank (*Figure 6.57*).

(14) Avoid accidental contact of buried pipelines with structures of dissimilar metals and other pipelines (*Figure 6.58*).

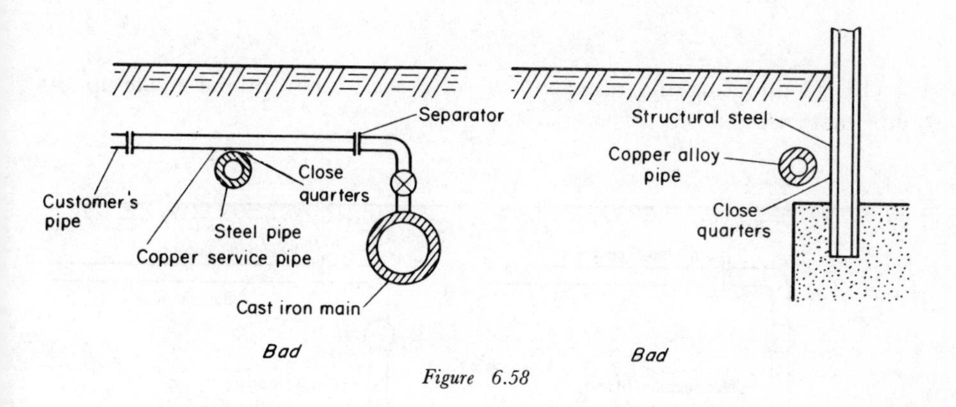

Figure 6.58

(15) Specify, where possible, uniform quality, grade and surface condition for buried pipelines—various quality sections should not be welded together (*Figure 6.59*).

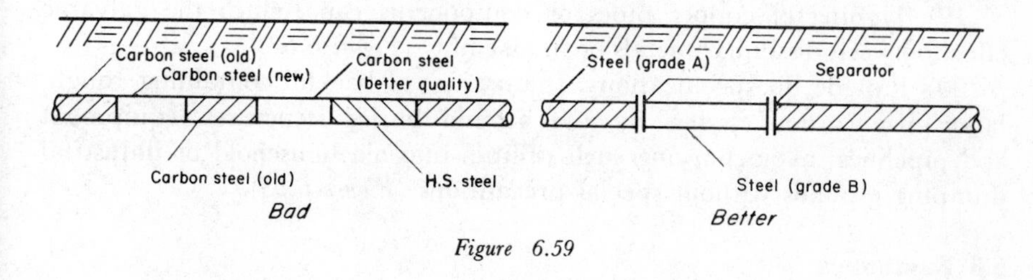

Figure 6.59

(16) Specify removal of tool scars on steel pipes submerged or buried—scars are anodic and corrode much faster than the rest of the pipe.

(17) Avoid laying radiant heating pipes into a boundary between two different environments (*Figure 6.60*).

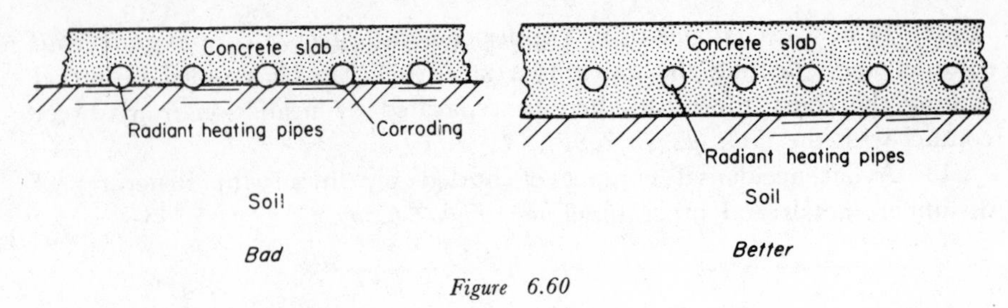

Figure 6.60

(18) Considering the resistivity of soils, select the optimal make-up of pipe system assembly (*Figure 6.61*).

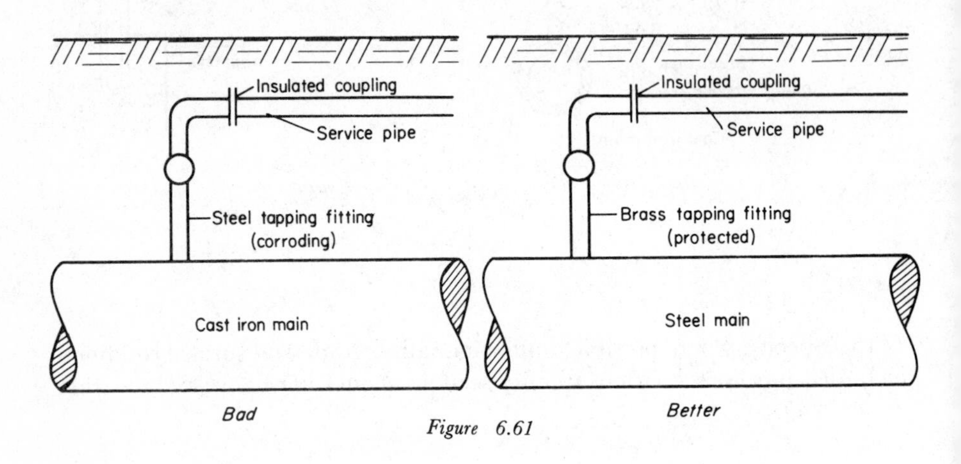

Figure 6.61

(19) Tinning of copper pipes or components can reduce the galvanic effect between dissimilar metals of an assembly (*Figure 6.62*).

(20) Provide, in specifications, against use of backfill containing rough, larger size carbon cinders by and around buried structures, equipment and pipelines; avoid burying such utilities into old household or industrial dumping grounds without special precautions (*Figure 6.63*).

6.6 Fasteners

(1) For guidelines for selection of fasteners, see Chapter 5.

(2) Selection of compatible fasteners for bimetallic connections—consult a galvanic corrosion indicator.

(3) Fasteners in dissimilar metal connections which are not compatible with either both or one of the metals in the joint, should be effectively separated from the non-compatible metal or metals by dielectric sleeves and washers (*Figure 6.64*).

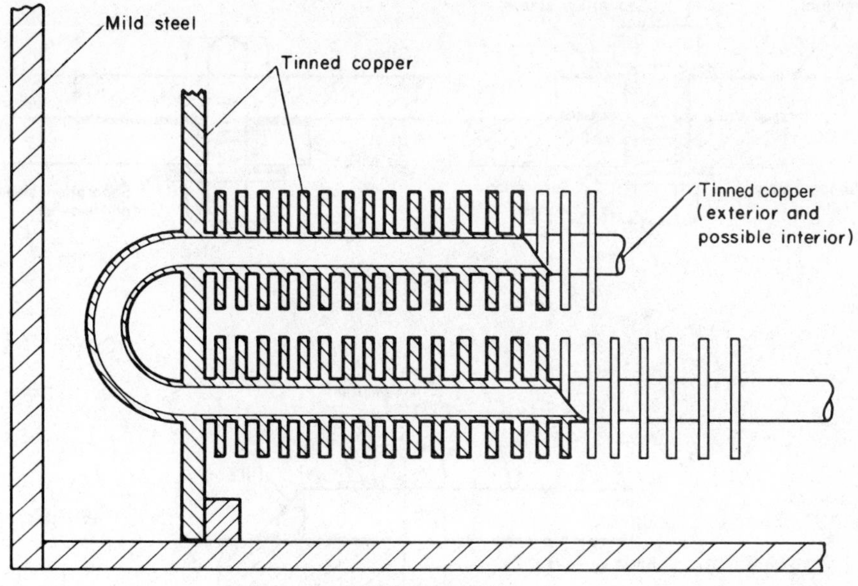

Figure 6.62

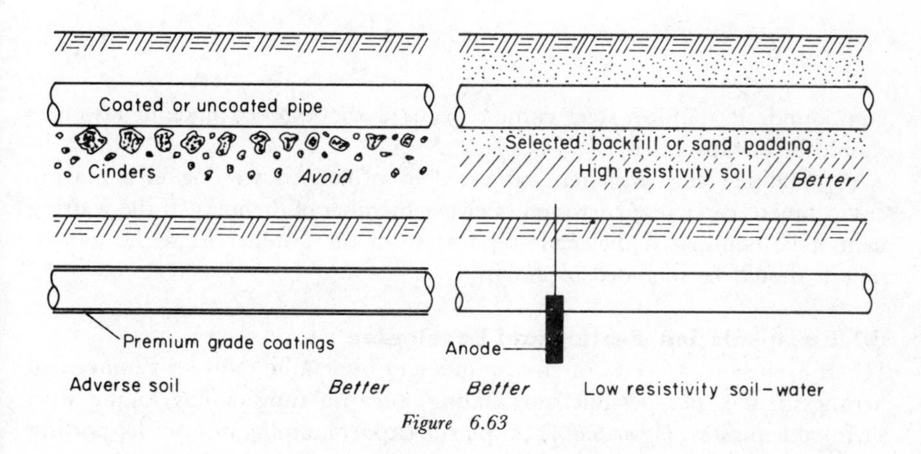

Figure 6.63

(4) If dielectric separation of fasteners in non-compatible joints cannot be implemented, the fasteners should be coated with zinc chromate primer and their exposed ends encapsulated (*Figure 6.65*).

(5) For dissimilar metal connections (aluminium to steel) in a marine environment, stainless steel fasteners installed with heads on the weatherside are preferred. Fastener to be dipped in zinc chromate primer or sealing

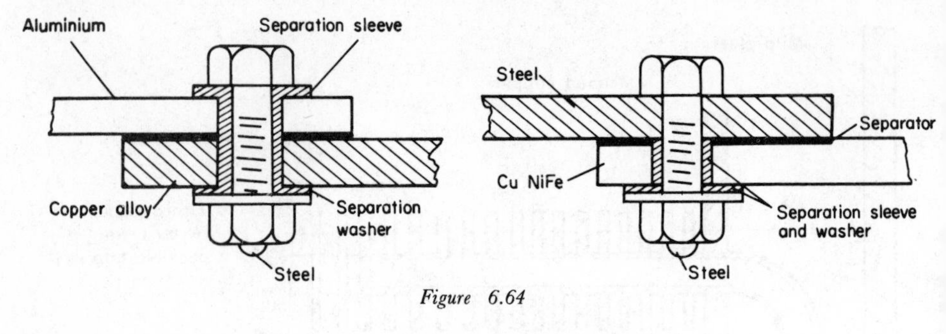

Figure 6.64

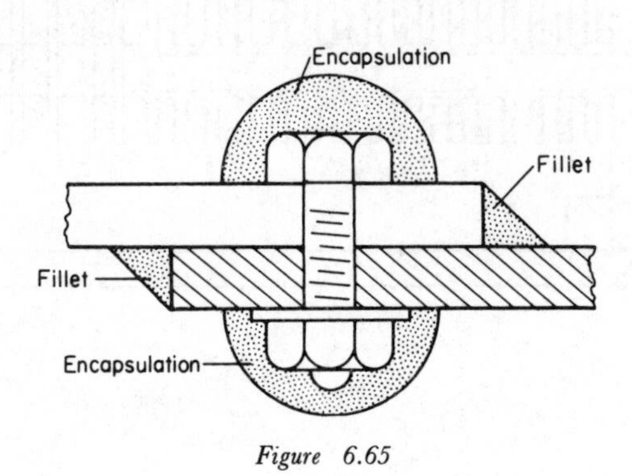

Figure 6.65

compound. If stainless steel cannot be used the exposed ends of fasteners should be encapsulated.

(6) Stainless steel fasteners subjected to prolonged wetting in a marine environment can cause corrosion of either member of the joint. If the wetting cannot be eliminated dielectric separation of the fastener from the jointed metals should be included in design.

6.7 Encapsulation, Sealing and Enveloping

(1) If exclusion of access of environment to bimetallic joint by geometrical arrangement is not possible, use sealing, encapsulating or enveloping with shrinkable plastic (*Figure 6.66*): (*a*) plastic caps containing mastic; (*b*) potting compounds (i.e. solventless epoxide) cast; (*c*) total or partial envelopment with shrinkable plastics (air and watertight) or plastic films; (*d*) application of moisture-proof coating or organic sealant.

6.8 Electrical and Electronic Equipment

(1) Restrict use of dissimilar metal connections to compatible metals (consult a galvanic corrosion indicator).

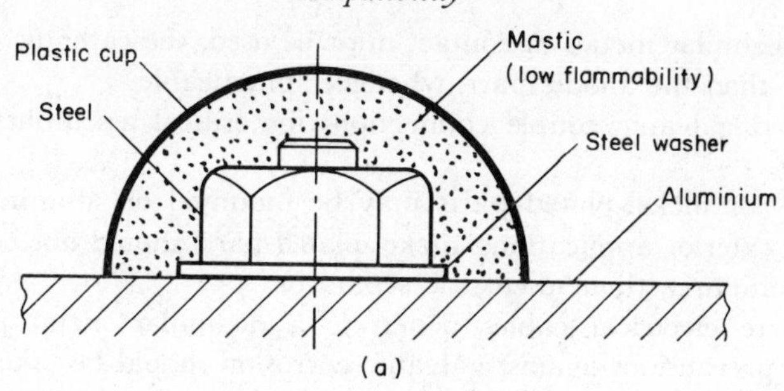
Plastic cup
Mastic
(low flammability)
Steel
Steel washer
Aluminium
(a)

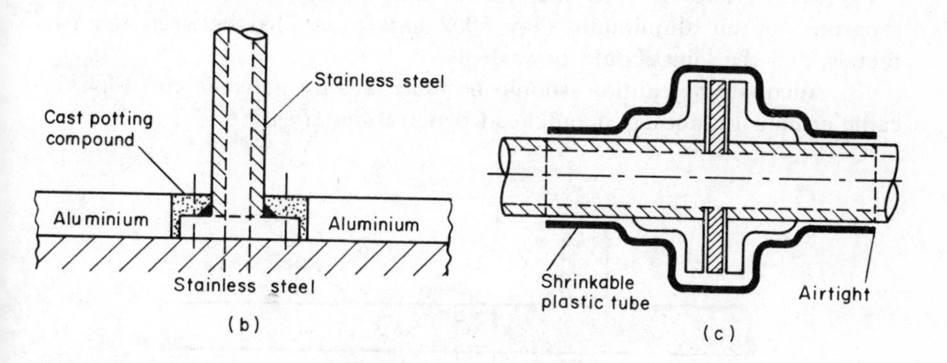
Stainless steel
Cast potting
compound
Aluminium
Aluminium
Stainless steel
(b)
Shrinkable
plastic tube
Airtight
(c)

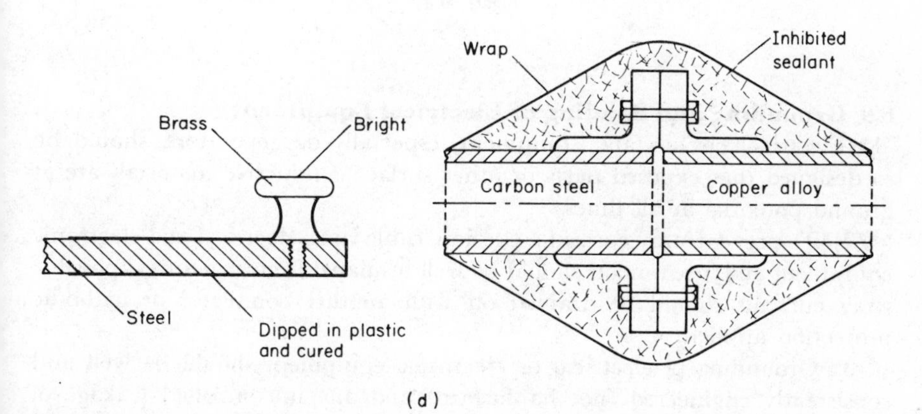
Wrap
Inhibited
sealant
Brass
Bright
Carbon steel
Copper alloy
Steel
Dipped in plastic
and cured
(d)

Figure 6.66

(2) If dissimilar metals in contact must be used, the cathodic part should be smaller than the anodic part, whenever practicable.

(3) Avoid galvanic couple connections for critical assemblies (safety or operation).

(4) Tin- or nickel-plated parts may be mounted on aluminium chassis direct; for exterior applications, nickel-plated parts should not be in contact with aluminium without dielectric separator.

(5) Where electrical cables penetrate a dissimilar metal partition or bulkhead, precautions against galvanic corrosion should be taken.

(6) Cadmium- or zinc-plated parts or zinc-base alloy parts should not be used within or in the proximity of electrical equipment subject to phenolic vapours emanating from insulating materials, varnishes or encapsulating compounds.

(7) Connections between magnesium and a dissimilar metal should be separated by an aluminium alloy 5052 gasket installed between the two metals, and the joint should be sealed.

(8) Adequate precautions should be taken against galvanic corrosion for cable armour grounding at bulkhead penetrations (*Figure 6.67*).

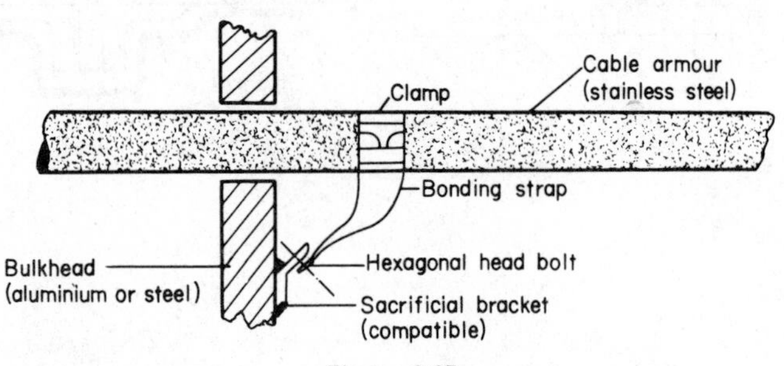

Figure 6.67

6.9 Grounding and Bonding of Electrical Equipment

(1) Electrical circuits and equipment, especially dc generators, should be so designed that exposed parts or other surface-conductive materials are at ground potential at all times.

(2) To prevent stray current corrosion, cable runs, electrical and electronic equipment and instruments should be well insulated from structures; possible stray currents should be drained off with another conductor or cathodic protection applied.

(3) Grounding of electrical or electronic equipment should be well and consistently engineered (not haphazard) and any uncontrolled leakage of electric currents into the structures or pipe systems should be prevented.

(4) When grounding cable and structure are compatible, grounding, when practicable, should be arranged by means of a bus-strap or shear-splice joint adequately insulated on the exterior (*Figure 6.68*).

(5) Copper alloy grounding conductors should not be directly attached to steel or aluminium strength structures or pipe systems but to a suitable sacrificial bracket. The material of the bracket should be compatible with the structure and a good conductor of electricity.

(6) For provision of electromagnetic compatibility for bonding by selected surface finish see Chapter 9, Section 8.

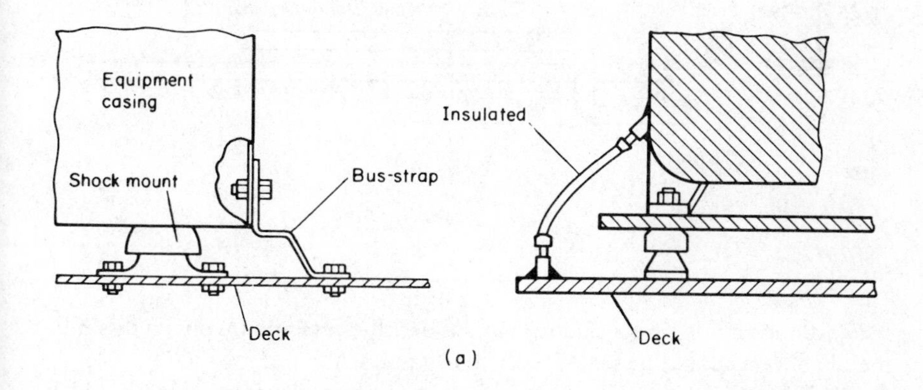

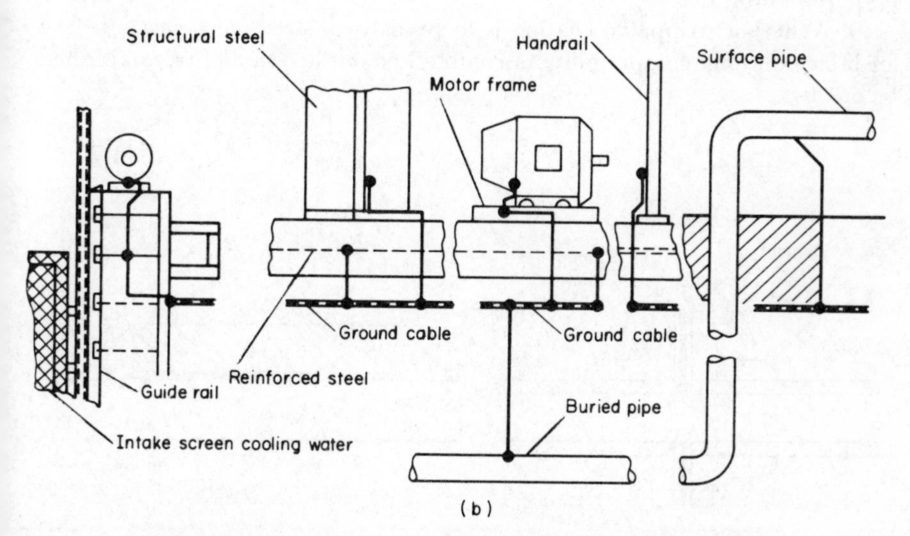

Figure 6.68. (a) Preferred; (b) typical grounding in generating stations; (c) typical interior wiring system with neutral ground at the service entrance

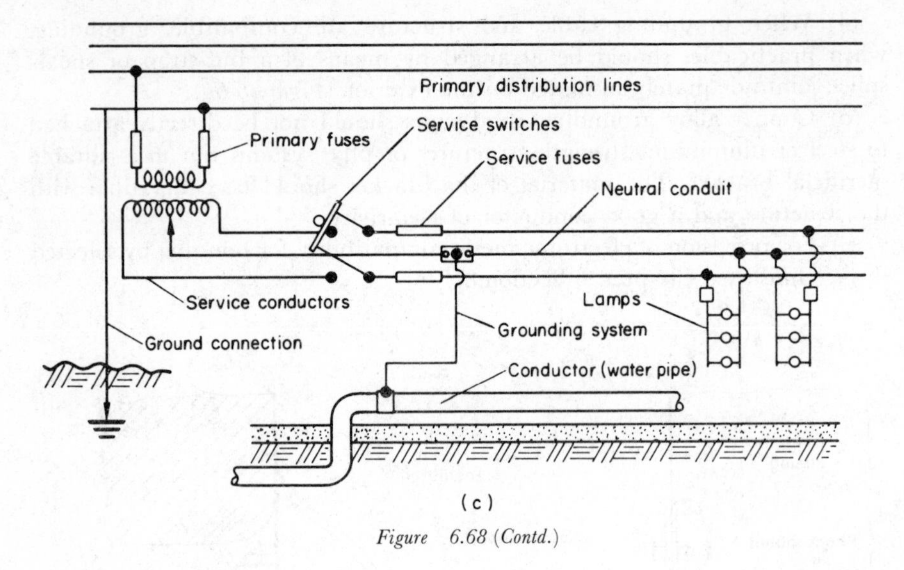

Figure 6.68 (Contd.)

(7) Bonds made by conductive gaskets or adhesives and involving dissimilar metal contact should be sealed with an organic sealant (*Figure 6.69*).

(8) If bonding dissimilar metals is unavoidable the joint area should be completely sealed after bonding with organic sealant (i.e. silicone, epoxy or polyurethane).

(9) Where a strippable coating is to be used to preserve a clean surface for later bonding, compatibility of coating and surface should be established before use.

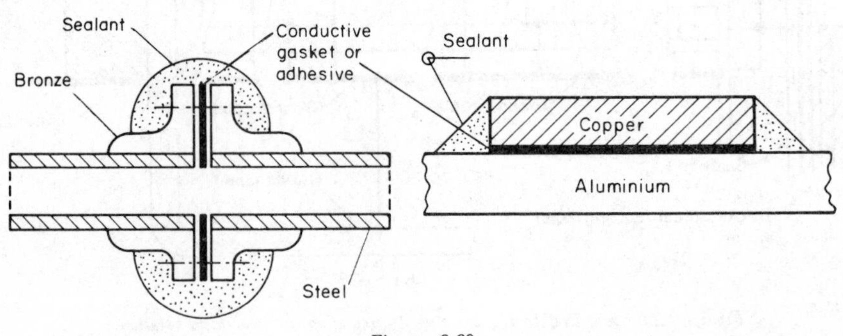

Figure 6.69

(10) When aluminium is to be electrically bonded, preference should be given to the use of clad alloys.

(11) Surfaces to be bonded should be masked prior to anodising or the insulating anodic film should be removed after anodising.

(12) When an electrical bond is to be made between dissimilar metals, the surface of one or both, where possible, should be coated with a metal compatible to both metals in the connection.

(13) Combination of conductive metal and resin or elastomer in conductive gaskets reduces contact of dissimilar metal bond with moisture; it should, however, be sealed where possible.

(14) When conductive gaskets are to be used, provision should be made in design for both environmental and electromagnetic seal—combination gaskets with conductive metal encased in resin or elastomer are preferred.

(15) Porous elastomers or resins should not be used for conductive gaskets.

(16) When wire mesh gaskets of monel or silver are used to bond aluminium or magnesium, these should be completely sealed against environment by environmental seals or external sealant bead.

(17) To avoid the adverse effect of moisture on non-metallic materials the following precautions should be taken:

(*a*) materials, unless hermetically sealed, impregnated or encapsulated, should not have greater moisture absorption than 1%;
(*b*) materials which wick or are hygroscopic should not be used;
(*c*) cut or machined edges of laminated, moulded or filled plastics should be sealed with impervious material;
(*d*) parts should have sound unbroken surfaces, free from cracks, holes or other discontinuities which would allow moisture to enter.

(18) Example of sacrificial grounding bracket (*Figure 6.70*).

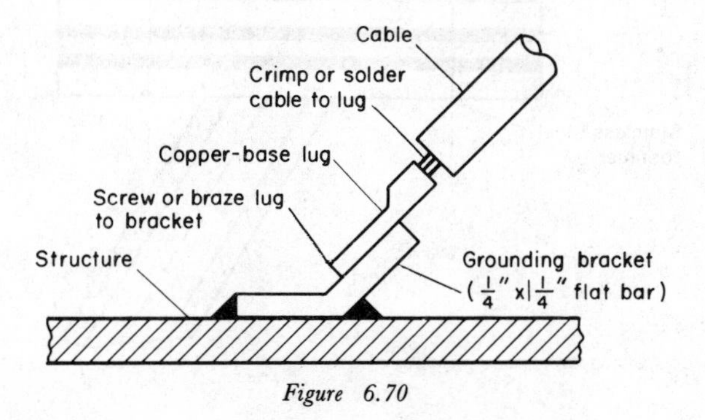

Figure 6.70

(19) Provide for complete bonding of unified piping systems containing conductive liquids between individual components by conductive fasteners, conductive gaskets or by bond straps (*Figure 6.71*).

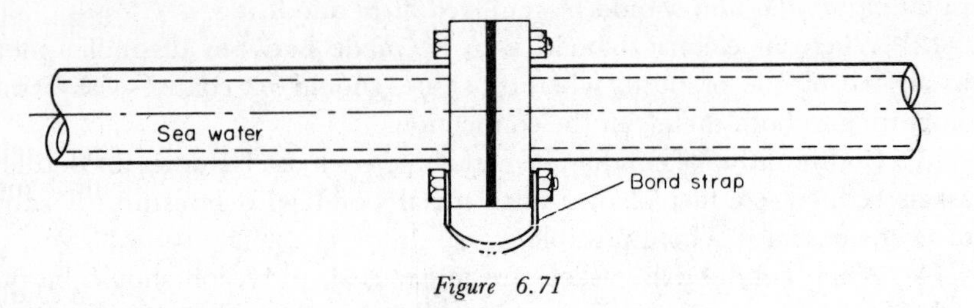

Figure 6.71

6.10 Coatings, Films and Treatments

(1) The component materials of the joint should be cleaned, pretreated and primed prior to assembly in normal conditions.

(2) Corrosion potential of bimetallic connections depends also on the relative proportion of discontinuities in protective coatings of anodic and cathodic parts.

(3) Where design or functional requirements preclude the use of dielectric separation, metallising (sherardising, galvanising, electroplating, cladding or metal spraying) with anodic metal (to one or both members of the connection) of all or at least some of the faying surfaces (components, fasteners, etc.), or coating with sufficient dry thickness (3–15 mils) of zinc-rich paint, can help to reduce or delay the galvanic reaction between the base metals (*Figure 6.72*).

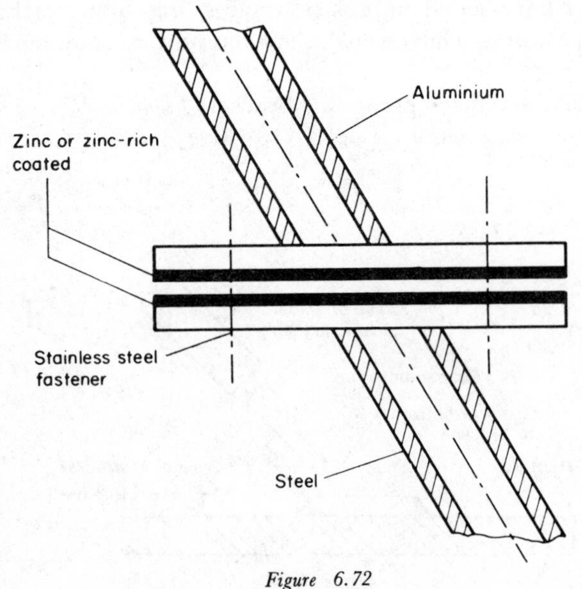

Figure 6.72

(4) Where two metals, one of which has been metallised, are coupled, the coating, not the base metal, should be regarded as contacting metal for evaluation of emf potential.

(5) Metal-to-metal contact should be considered as existing between parts that have been painted; all paint coatings are permeable to a certain degree, as far as galvanic corrosion is concerned.

(6) When using metallic coating over whole bimetallic assembly, the coating metal should be less noble than either of the component metals—or at least the cathodic one.

(7) Anodic films on aluminium base alloys should be considered a part of dielectric separation.

(8) The effect of conversion coatings (chromates, phosphates) applied to dissimilar metal couples can vary (*Figure 6.73*):

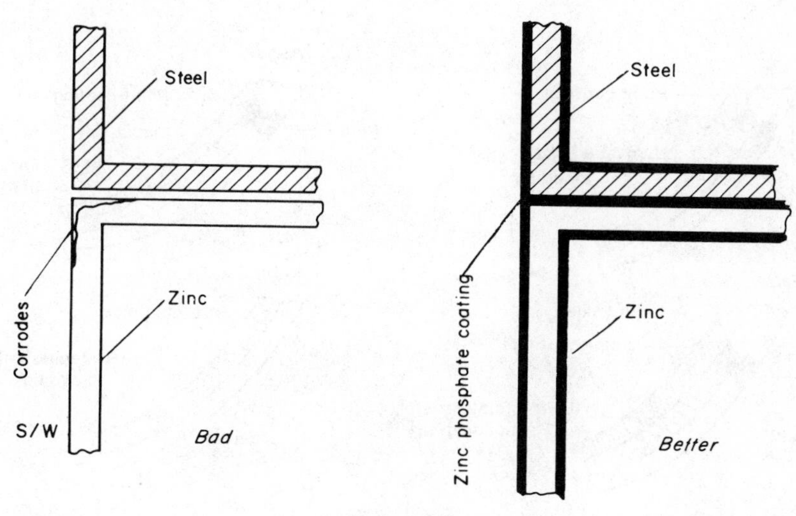

Figure 6.73

(*a*) chromate- and phosphate-treated zinc and cadmium-coated metals are not dielectrically separated when in contact;
(*b*) chromate- and phosphate-treated metals in a disimilar metal couple can sometimes obtain a reduction of galvanic corrosion which is caused by electric current transfer in a conductive medium.

(9) The individual galvanic effect of metallic coatings should be evaluated before their application in design for prevention of corrosion (*Figure 6.74*).

(10) Balance the extent and location of coating applied on galvanic couples for a most effective protection (*Figure 6.75*).

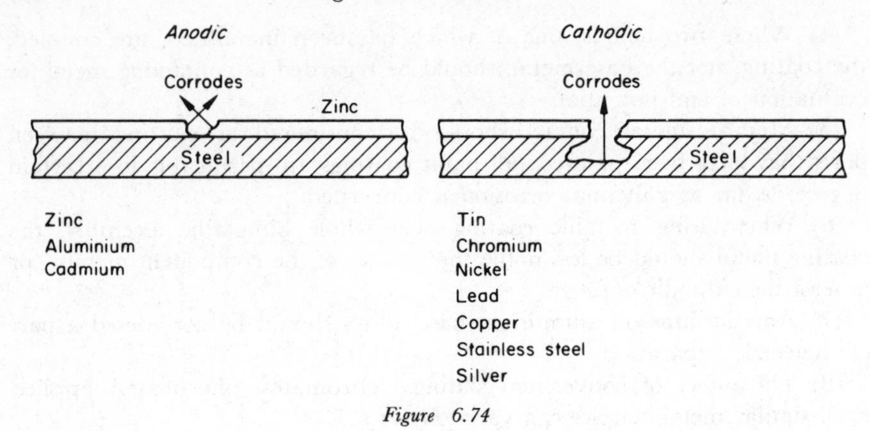

Figure 6.74

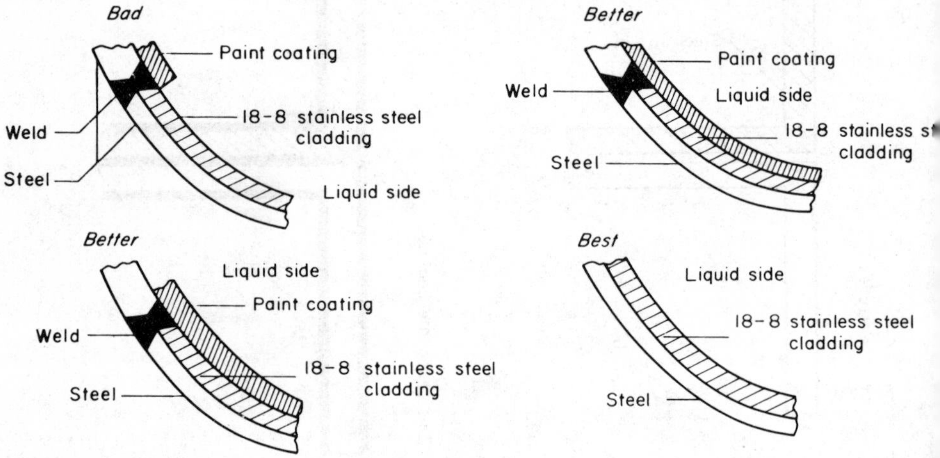

Figure 6.75

(11) In an aluminium alloy hinge fitting, the cadmium-plated steel bushes should be shrink assembled to ensure maximum electrical contact resistance (*Figure 6.76*).

(12) Zinc, as a coating of reinforcing rods and other steel embedded in concrete, helps to prevent or delay formation of rust on such reinforcement in a marine environment.

(13) Avoid application of coatings containing metals or their active compounds on top of coatings containing zinc or aluminium, on structures submerged in conductive environment (sea water).

(14) For mobile joints various combinations of metallising and plastic coating can be used (e.g. nylon, ptfe, etc.) instead of dielectric separation between dissimilar metals.

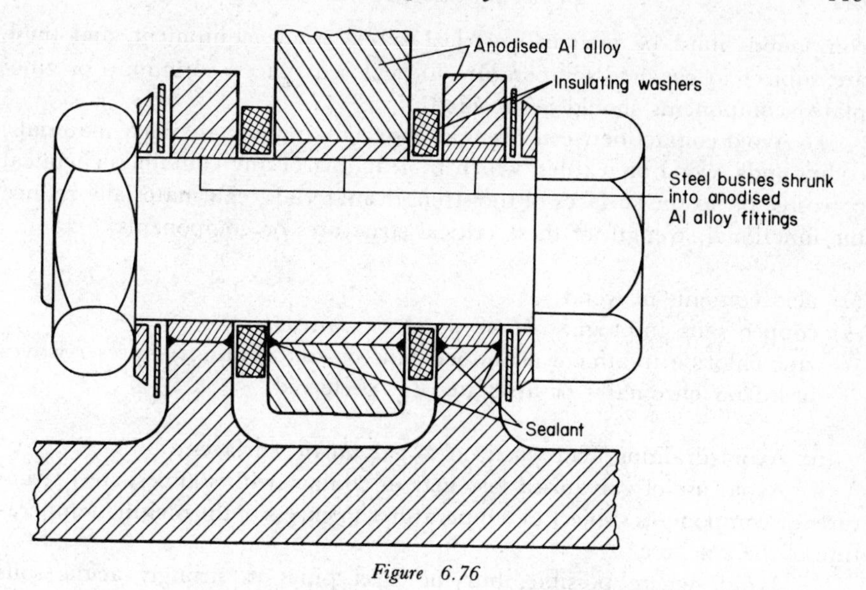

Figure 6.76

(15) Avoid specifying application of any coatings containing metals or their active compounds, which are cathodic to the substrate on structures and equipment destined for a conductive environment.

(16) Consider the possibility of the applied metallic coatings changing their polarity versus their substrate in any probable environmental medium.

6.11 Chemical Compatibility

(1) Avoid the use of materials for design of connections which are mutually incompatible by reason of their chemical contents under particular environmental conditions, e.g. vulcanised rubber which contains sulphur, affecting metal in contact, etc.

(2) Avoid materials which, under ambient conditions or when under fire or in high temperature conditions, out-gas or liberate corrosive fumes in the vulnerable proximity of materials which can be adversely affected by such fumes and their functional stability impaired:

(*a*) partially cured or under-cured organic materials;
(*b*) insulating materials emitting phenolic vapours, varnishes or encapsulating compounds, within totally unventilated spaces of electronic equipment containing cadmium- or zinc-plated or zinc-base alloy parts;
(*c*) vinyl paints emitting hydrochloric acid vapours at temperatures over 150°F (66°C).

(3) Where phenolic insulating materials, varnishes or encapsulating

compounds must be used in electrical or electronic equipment, and these are subject to elevated temperature in enclosed spaces, cadmium- or zinc-plated components should be avoided.

(4) Avoid contact between strength materials and any auxiliary materials, compounds, wood or textiles, which by leaching of any contained chemical corrosive on to the surfaces of the strength materials, can materially reduce the functional strength of these critical structures or components:

(*a*) acid contents in wood;
(*b*) copper salts impregnation of wood or canvas;
(*c*) zinc chloride treatment of timber (zinc or zinc coatings)—preservatives based on chromates or arsenates are preferred.

(5) Avoid draining acid wood on to galvanised surfaces.

(6) Avoid use of galvanised fasteners in contact with stainless steel structures or components subject to temperatures in excess of the melting temperature of the zinc, etc.

(7) Avoid, where possible, burying steel pipes in strongly acidic soils (lack of polarisation); lead or aluminium should not be used for buried structures, equipment and pipes in highly alkaline soils. Provide, if necessary, for change of surrounding media (backfill, sand pads); use insulating coatings, cathodic protection and these separate or combined.

6.12 Environment

(1) Galvanic corrosion of dissimilar metals can be eliminated, delayed or at least reduced by induction of environmental changes at bimetallic connections:

(*a*) change of temperature;
(*b*) reduction or increase of aeration to suit the metals;
(*c*) reduction or increase of movement of fluids to suit the metals;
(*d*) adjustment of chemistry.

(2) Increase concentration of the inhibitor for reduction of galvanic corrosion in comparison with the one used for reduction of corrosion of a single metal.

(3) Specify corrosion inhibitor (zinc chromate, zinc chromate paste, etc.) for galvanic connections when possible.

6.13 Stray Currents

(1) Avoid passage of electric current between metal and its environment; e.g. buried or submerged pipelines, tank bottoms and structures, electric traction, welding plants, power undertakings and cathodic protection schemes.

(2) Structural weldments should be so designed that the welded sections can be grounded *in situ* in the close proximity of the weld.

(3) Use insulating couplings to separate metallic structures for control of stray current corrosion (*Figure 6.77*).

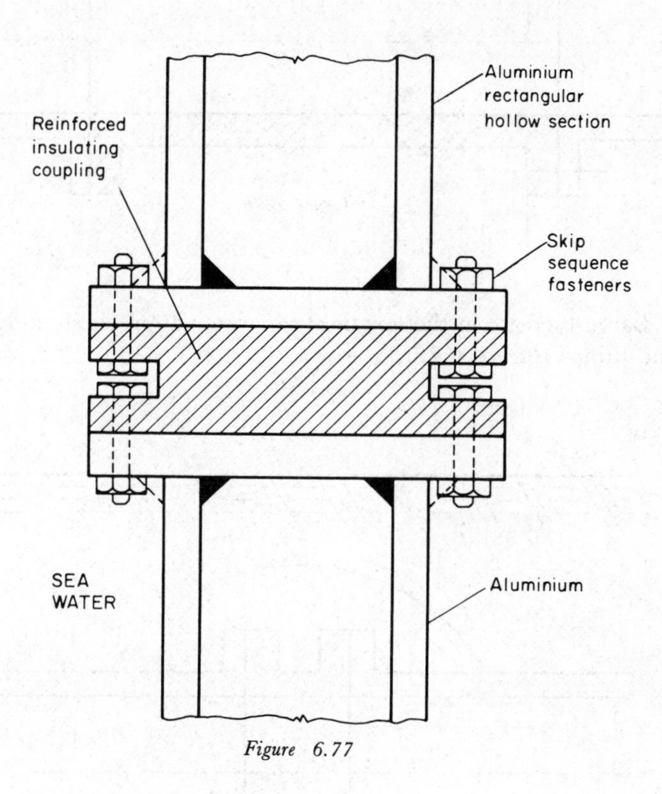

Figure 6.77

(4) The current jump depends on magnitude of potential difference, electrical conductivity of liquid in the pipe, soil or surrounding medium, geometric configuration of the pipe or structure and insulator, the temperature and surface films.

(5) Non-metallic nipples, pipe materials or sections of the structures can be used, instead of narrow washers or sleeves, to increase the length of insulating couplings (*Figure 6.78*).

(6) Minor increase in the length of insulator does not increase appreciably the effectiveness against current jump.

(7) Major increase in the length of separator (e.g. short length of non-metallic pipe) has no great effect on control of external current jump (e.g. in soil or other conductive media).

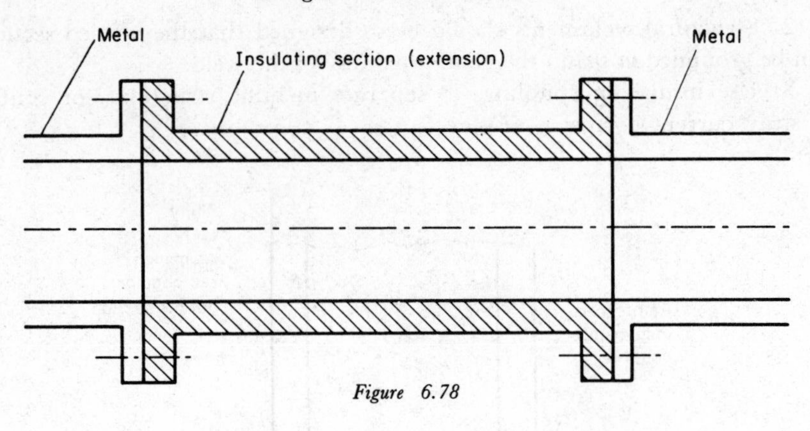

Figure 6.78

(8) Large increase in the length of separator appreciably reduces internal current jump (*Figure 6.79*).

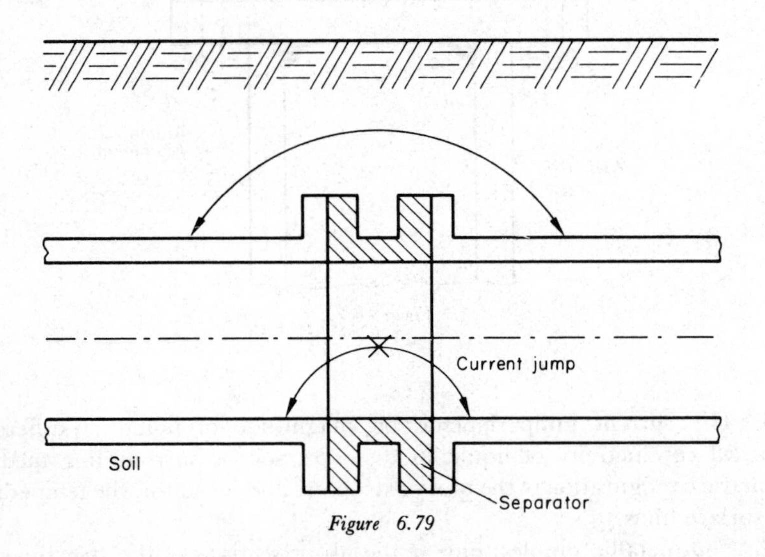

Figure 6.79

(9) Surface films on involved metallic structures influence the effect of the separator length (*Figure 6.80*).

(10) Consider the possibility of a change of polarity by heat or cold transfer in basic structures or surface coatings, by the effect of water currents or by variable oxygenation, and specify appropriate remedial actions in design (see Chapter 3).

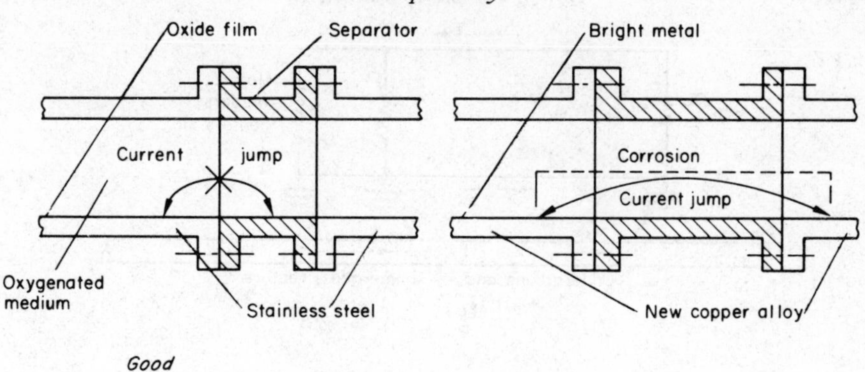

Figure 6.80

(11) Determine and evaluate the local sources of stray currents for their effect on the designed utility (underground and submerged) (*Figure 6.81*).

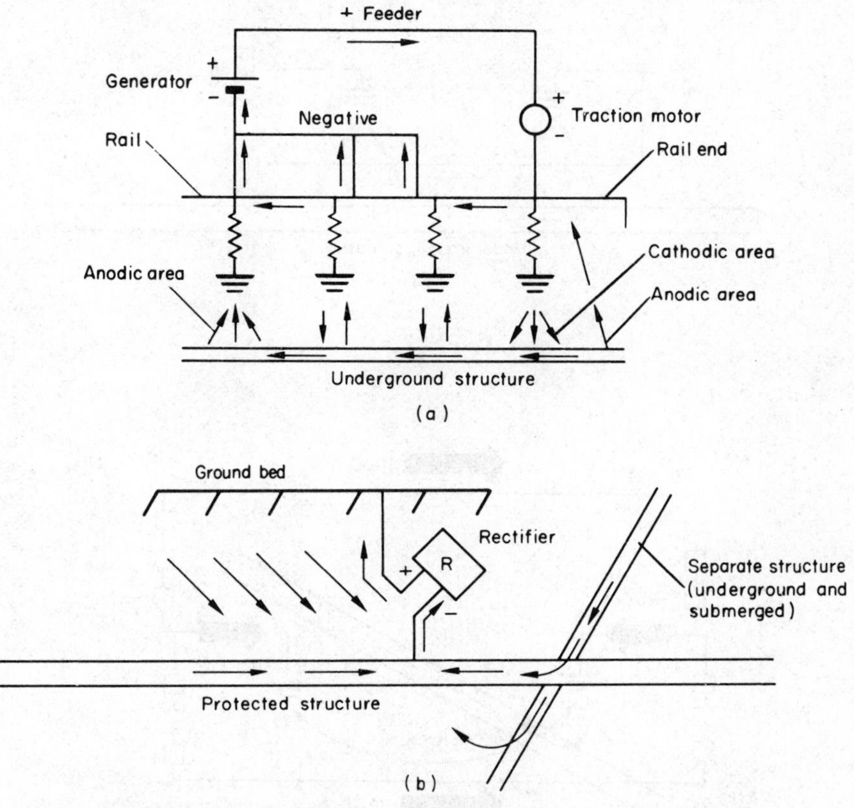

Figure 6.81. (a) Railway traction systems; (b) cathodic protection systems; (c) direct current distribution or industrial plant; (d) bipolar direct current transmission lines; (e) monopolar and homopolar high voltage direct current (hvdc)

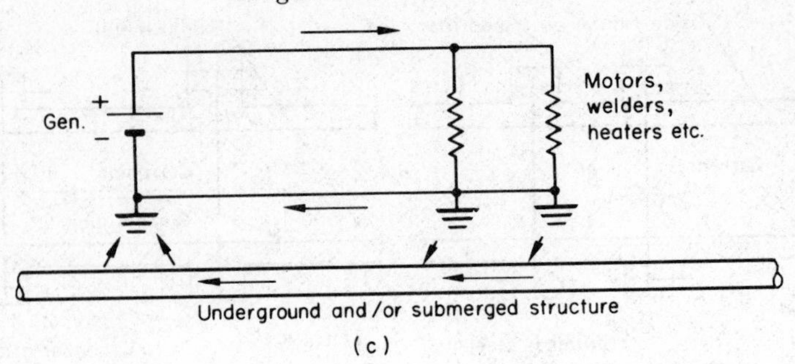

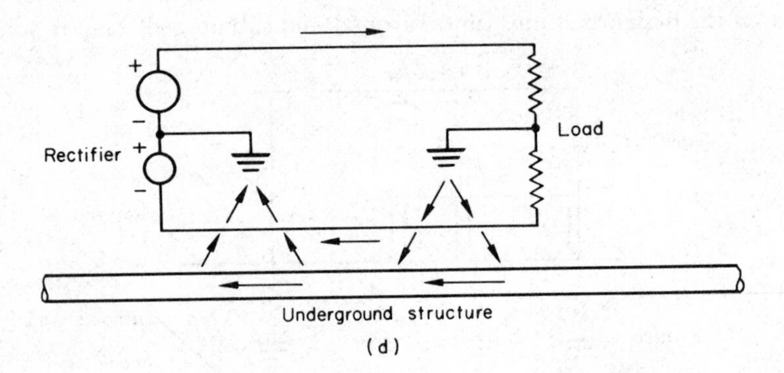

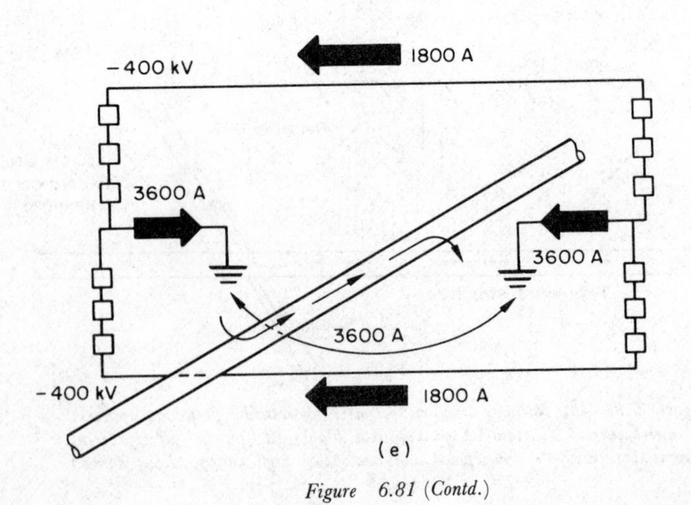

Figure 6.81 (Contd.)

(12) Evaluate the possibilities of generating stray currents within the designed complex and take preventive design actions.

(13) Provide, by all available design means, for reduction of possibilities to form optimal conditions for creating critical leakages of current.

(14) Reduce the leakage current by increasing the resistance between the source and earth, by rail bonding and rail-to-negative ties, by increasing the conductivity of the conductor (rail, lead), by proper scheduling of substation operation or by welding across each rail section.

(15) Reduce the pick-up or discharge of leakage current on the critical structure by increasing the earth contact resistance, by providing isolation from the earth, by using insulating coating (organic liquid or tapes), by changing over to non-metallic materials, by placing the structure in conduits and by flushing the ducts with water in highly salted areas.

(16) Where possible interrupt the continuity of the leakage current path back to the substation by introduction of insulating couplings in the critical structure. Where continuity of plant for protection or interferencc reasons is necessary, bridge the insulating couplings with resistors or capacitors.

(17) Provide drain wires between anodic locations of critical structure and the negative feeder, rail or substation ground bus of sufficient conductivity to return the leakage without creating adverse conditions on the drained structure. These can include reverse current switches or diode arrangements in case of current reversal.

(18) Use cathodic protection, preferably with automatic control.

(19) To avoid electrolysis damage in the vicinity of the supply point (higher current density) bond metallic structures normally to the negative bus-bar.

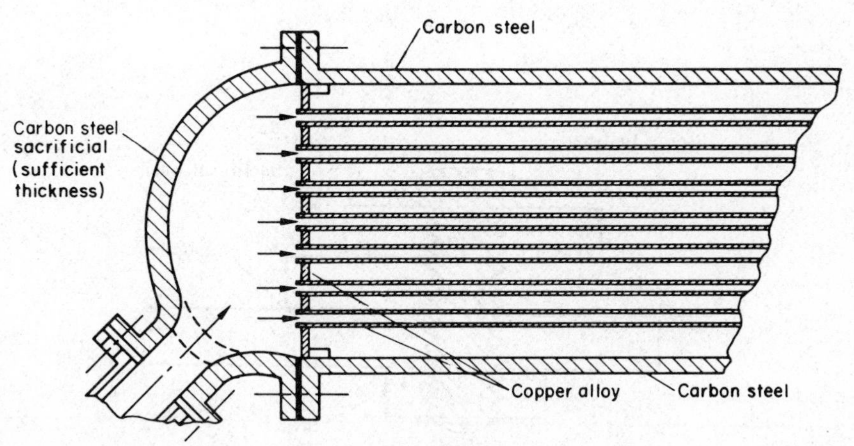

Figure 6.82

6.14 Beneficial Results

(1) Use of steel casing in heat exchangers with copper alloy tubes and tube sheets reduces corrosion of copper metals (*Figure 6.82*).

(2) Use of galvanised or metallised steel washers in contact with the anodic member of the connection reduces galvanic attack on this metal (*Figure 6.83*).

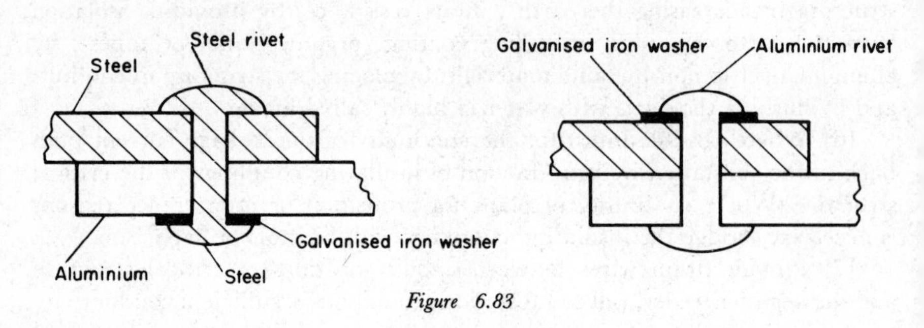

Figure 6.83

(3) Use of solid sacrificial washers for protection of anodic metal of the couple (*Figure 6.84*).

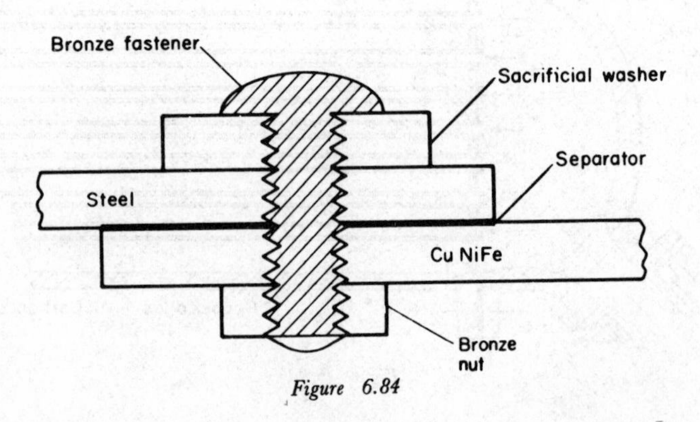

Figure 6.84

(4) Use of sacrificial pieces for cathodic protection (*Figure 6.85*).

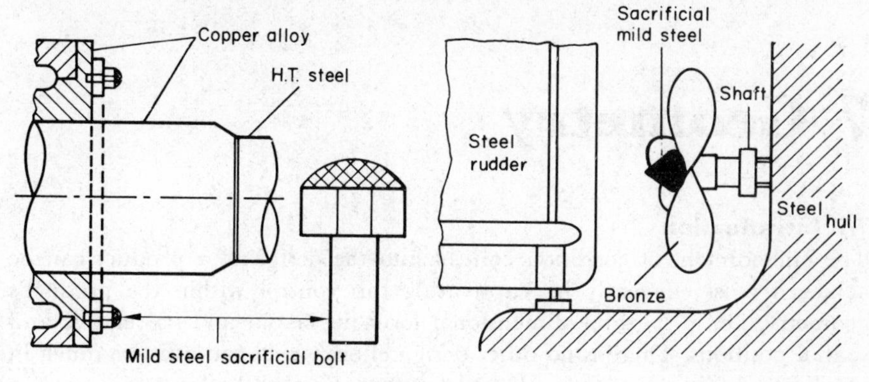

Figure 6.85

(5) Use of sacrificial metals for prevention of stress corrosion cracking (*Figure 6.86*).

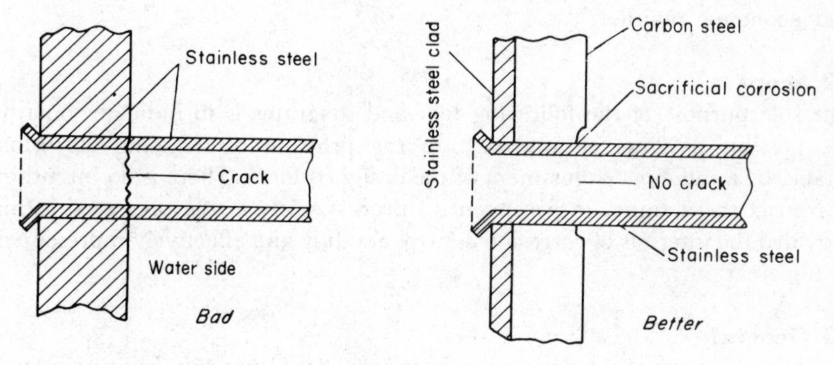

Figure 6.86

(6) Use of sacrificial metals for protective coatings (*Figure 6.87*).

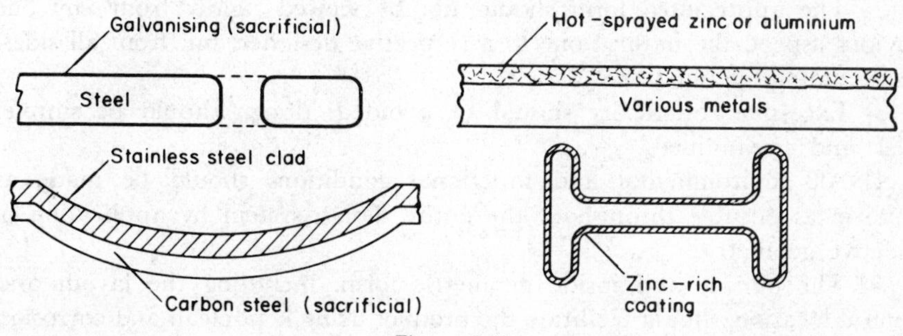

Figure 6.87

7 Geometry

7.1 Introduction

The embodiment of corrosion control into the design of a product can be achieved most efficiently by captivating this control within the product's geometry, i.e. in its three-dimensional form, its layout and its relative and spatial positions. There is no other design effort which can assist so much in prevention of corrosion for such a comparatively small outlay.

Whereas basically the pattern of a utility depends on its functional, material and fabrication requirements, it is within the scope of a good designer to select from the available possibilities only such geometric shapes or combinations of forms that help to reduce corrosion attack in the most efficient and economic manner.

7.2 Scope

The sole purpose of the following text and diagrams is to indicate some of the possible avenues of approach to the problem of reducing corrosion attack, by a judicious adjustment of the designed form. There is no intention to restrict the designer in his inventive process solely to the presented form, provided the interests of corrosion control are duly and effectively represented in his creation.

7.3 General

(1) The geometry of the designed component should not only be appreciated within the narrowly defined lines of the component itself, in its own splendid isolation; its interdependence with other components within the system, within the utility and the space generally should also be considered.

(2) The appreciated form should not be viewed rigidly from any one obvious aspect, the natural one to a respective designer, but from all sides, i.e. including the blind one.

(3) Excessive complexity should be avoided; design should be simple, sleek and streamlined.

(4) All environmental and functional conditions should be made as uniform as possible throughout the entire design system by application of selective geometry.

(5) The outside and inside geometric form, including the layout and general location, should facilitate the product being kept clean and corrosion free at all stages of fabrication, assembly and during service—based on

normal operation and breakdown conditions—without excessive effort.

(6) The design should prevent the adverse corrosive influence of one component of the utility on another in various media, due to spillage, emission of fumes or vapour, thermal and chemical effects, transfer of corrosive matter, formation of hot spots, etc., within the selected pattern.

(7) Where water can be deposited by rain, spray or condensation, all reasonable design precautions should be taken to provide free access of drying air to the wetted surfaces. Fast drying of such surfaces should be secured primarily by an appropriate selection of individual shapes, as well as by their proper combination and attachment.

(8) Shapes which, contrary to their function, retain corrosive combinations of air and electrolyte should be avoided. The designed product should neither collect nor retain unwelcome compound corrosive media within their form and frame.

(9) Access and retention of unwelcome solid contaminants or wastes, which may act within the designed form by their absorbence and retention of moisture or their abrasive action, should be avoided by selected form and arrangement.

(10) The geometry of the product should be designed for exclusion or inclusion of oxygen, as relevant to the requirements of the particular construction material (e.g. active/passive metals require oxygen for the build-up of protective film; corrosion of other metals or alloys is aggravated by the presence of oxygen).

(11) Design forms should be chosen which lessen the effect or reduce the occurrence of such types of corrosion which depend directly or indirectly on the geometry of the product for their occurrence and degree of aggressiveness (see Chapter 3).

(12) Such shapes, forms, combinations of forms and their style of attachment should be selected, whose fabrication, joining technique and treatment will not aggravate corrosion.

(13) Those geometric forms should be chosen which can assist in securement of the optimal results from the selected corrosion preventive measures, at their initial application and at any future repetitive application.

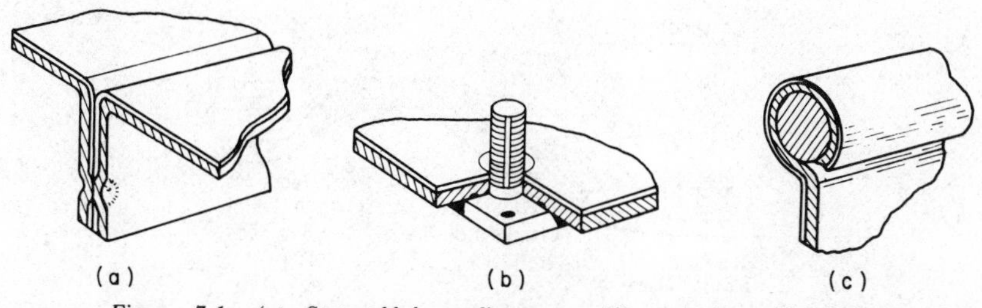

Figure 7.1. (a) Spot-welded standing seam; (b) projection-welded bolt; (c) reinforced rolled edge

(14) Where materials which are pretreated prior to fabrication or assembly are used, geometric form allowing fabrication and assembly without major damage to the pretreatment should be chosen (*Figure 7.1*).

(15) Access to corrosion-prone areas should be considered of the prime importance.

(16) The effect of corrosion on operability and performance of the product at the given geometry should be considered, particularly in areas not subject to periodic examination.

(17) Size and shape structural members and components to avoid double dipping or progressive galvanising—single immersion is preferred.

7.4 Structures and Equipment

(1) Locate utility where it cannot be adversely affected by natural and climatic conditions or by a corrosive pollution (gaseous, liquid or solid) borne by prevalent winds or sea and river currents from near or distant sources (*Figure 7.2*).

(2) Select optimum arrangement and layout within the utility to prevent adverse effect of one part of assembly on another (based on normal operation and breakdown conditions) (*Figure 7.3*).

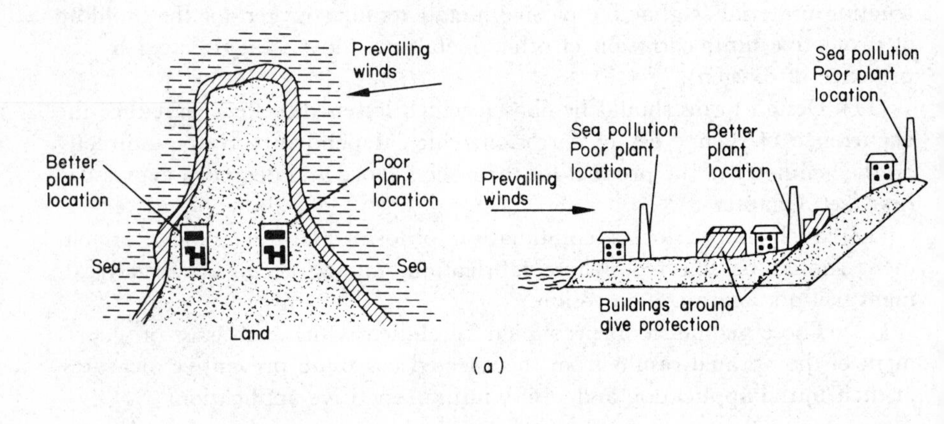

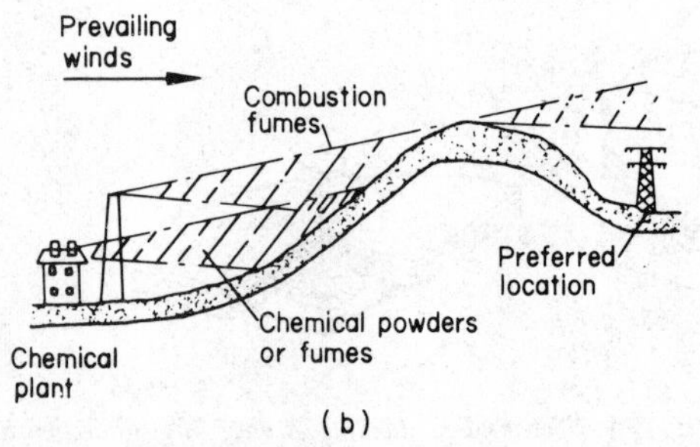

Figure 7.2. (a) Sea air and spray; (b) industrial pollution; (c) sea currents

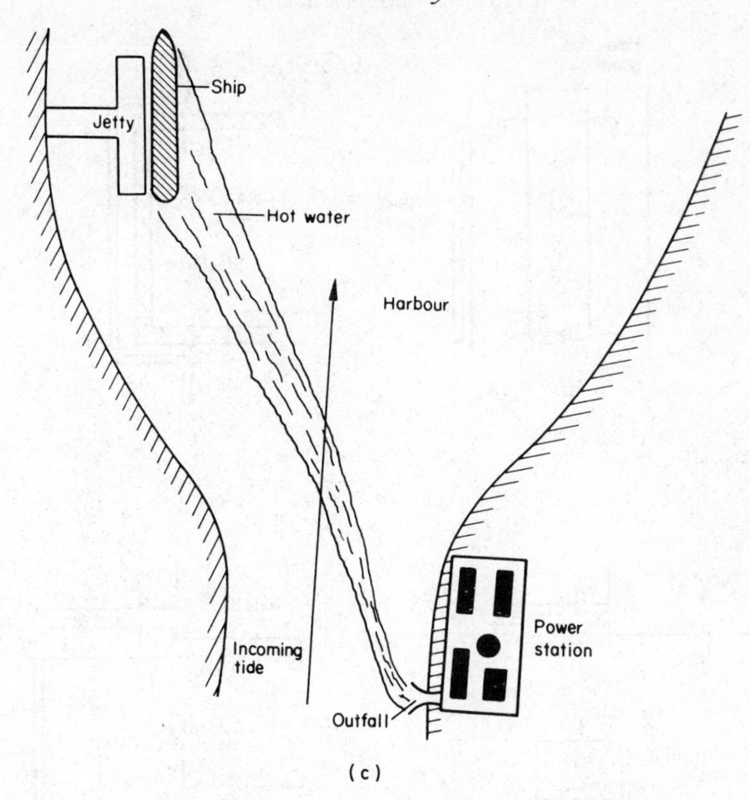

Figure 7.2 (*Contd.*)

(3) Avoid undrainable traps accumulating liquids and absorbent solid wastes (*Figure 7.4*).

(4) Provide adequate drainage, scuppers and limberholes (*Figure 7.5*). Scuppers should be fitted at the lowest possible position in a space to ensure full drainage of space. The ship's movement should be taken into consideration when choosing optimum position for a scupper.

(5) Design self-draining structures (*Figure 7.6*).

(6) Avoid small enclosures within structural frame (*Figure 7.7*).

(7) Provide for removal of moisture or other corrosive media from critical spaces and for their speedy drying (*Figure 7.8*).

(8) Prevent access of abrasives and other solid contaminants to critical spaces (*Figure 7.9*).

(9) Prevent condensation in critical spaces by selected geometry (*Figure 7.10*).

(10) Provide for dry connections and avoid accumulation of moisture and other corrosive media in the close proximity of joints (*Figure 7.11*).

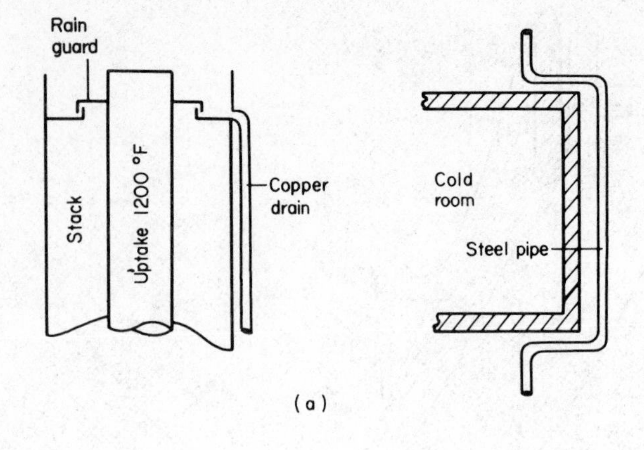

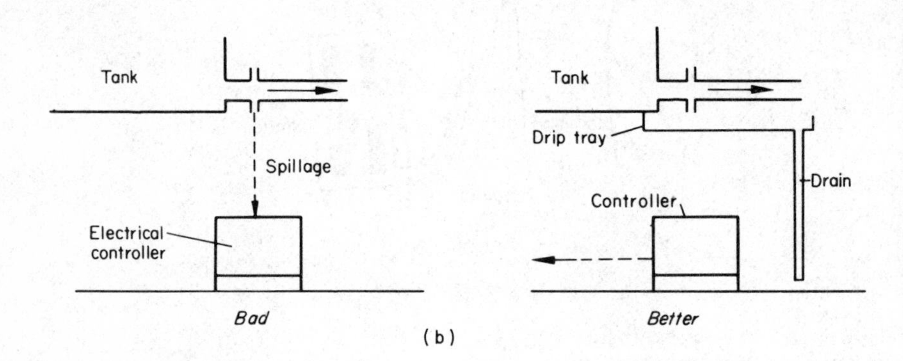

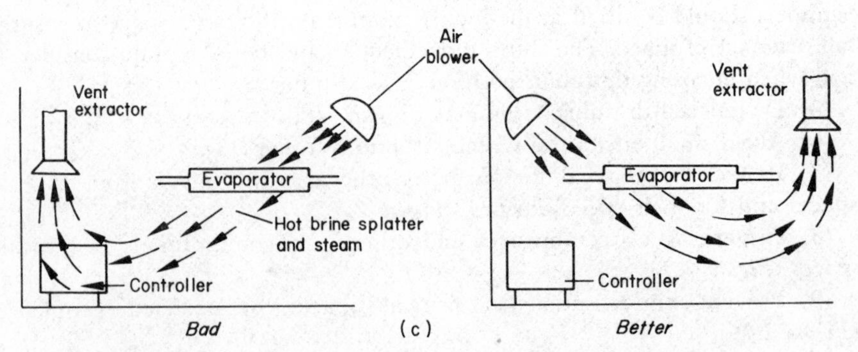

Figure 7.3. (a) Heat and cold emission and transfer; (b) spillage; (c) contamination; (d) re-ingestion of solids, contaminated liquids and gases; (e) vibration transfer

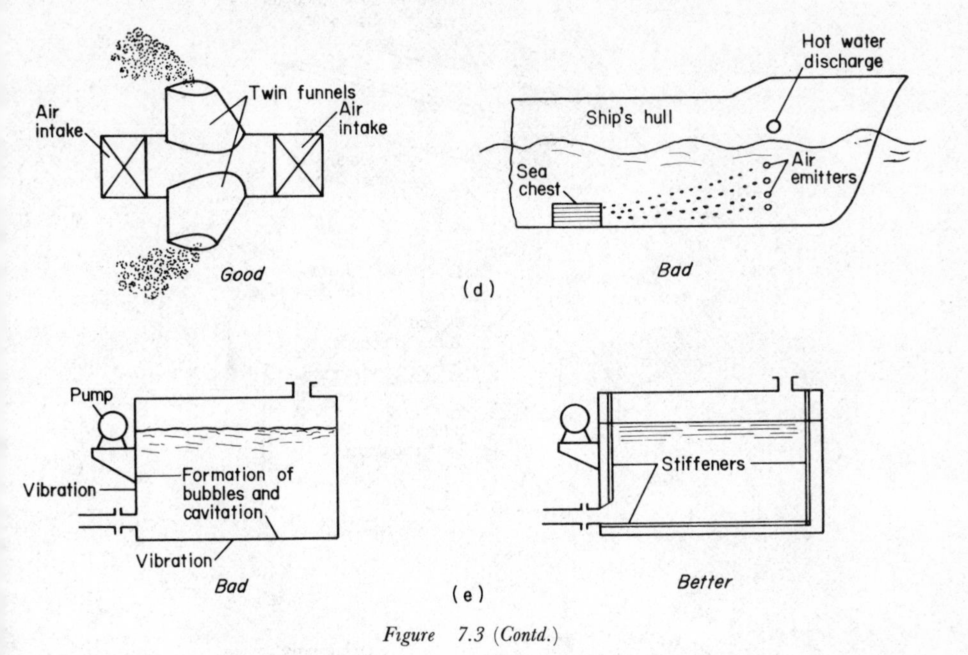

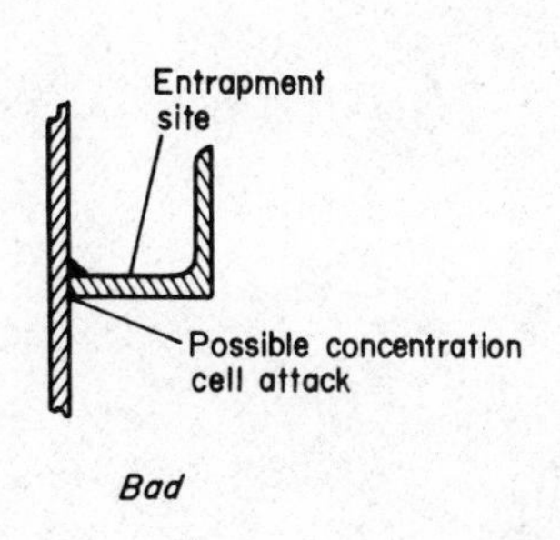

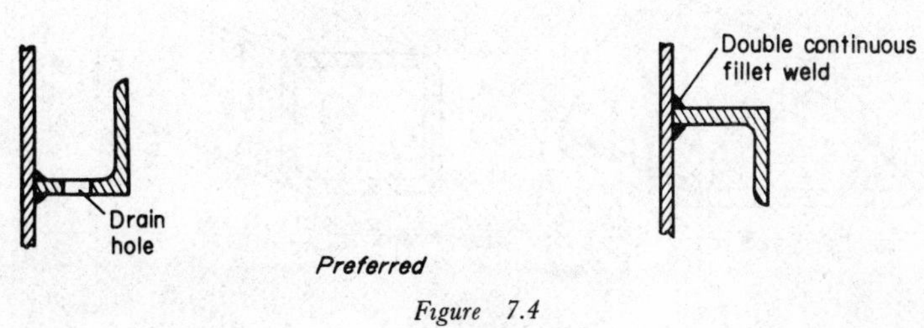

Figure 7.3 (Contd.)

Figure 7.4

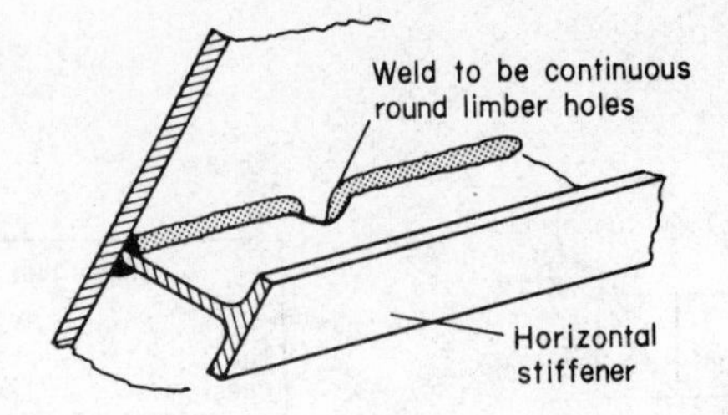

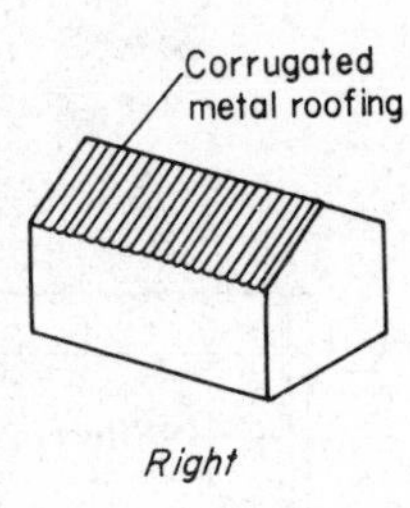

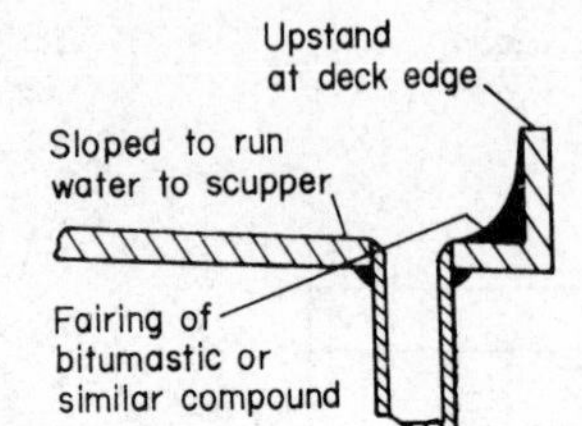

Right

Figure 7.5

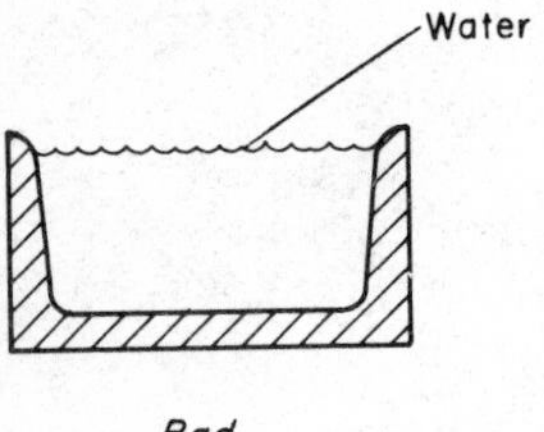

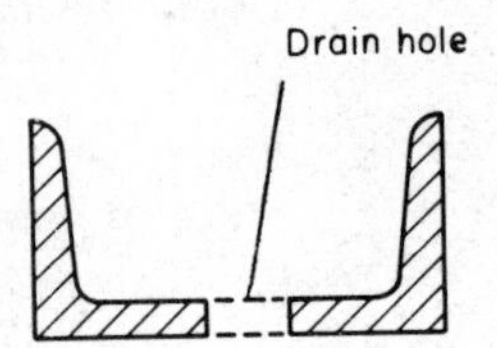

Better

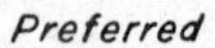

Preferred

Figure 7.6

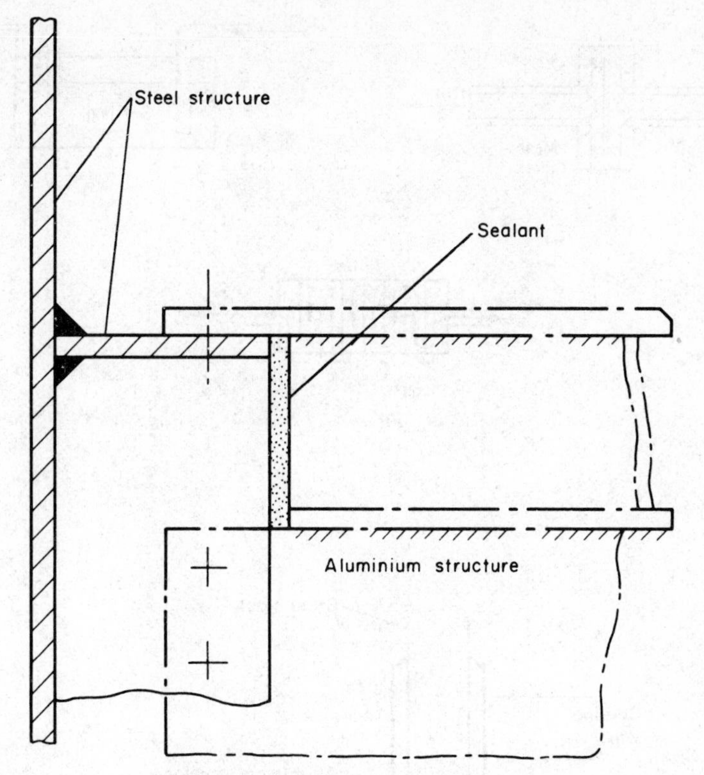

Figure 7.7

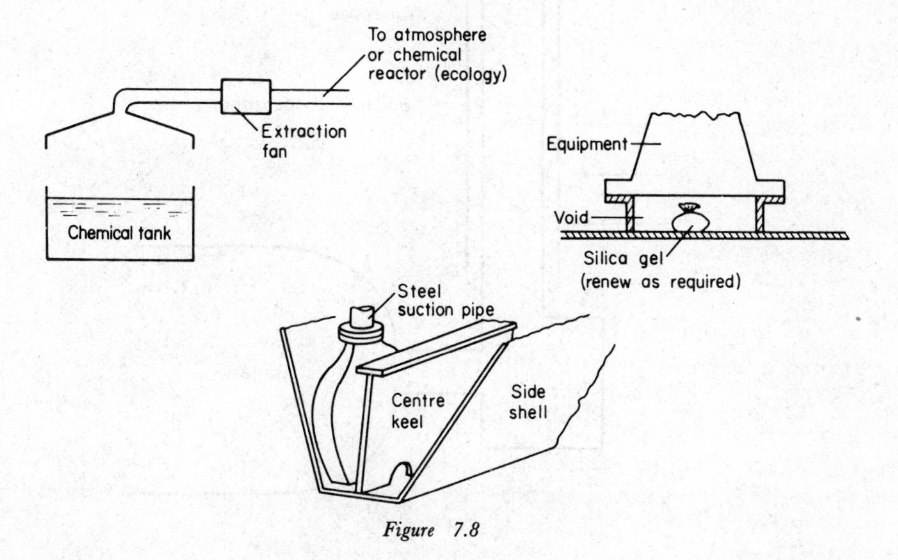

Figure 7.8

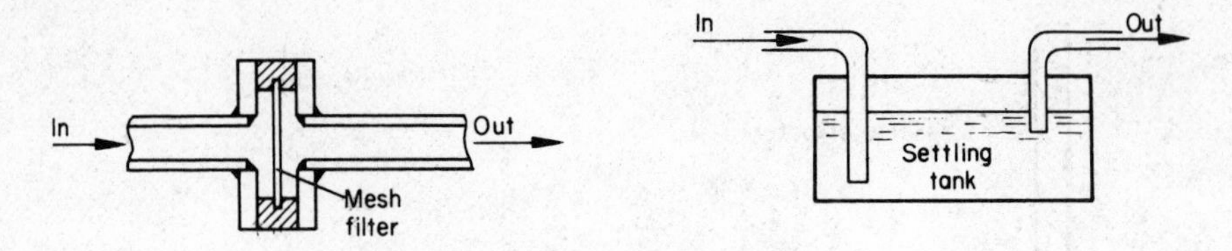

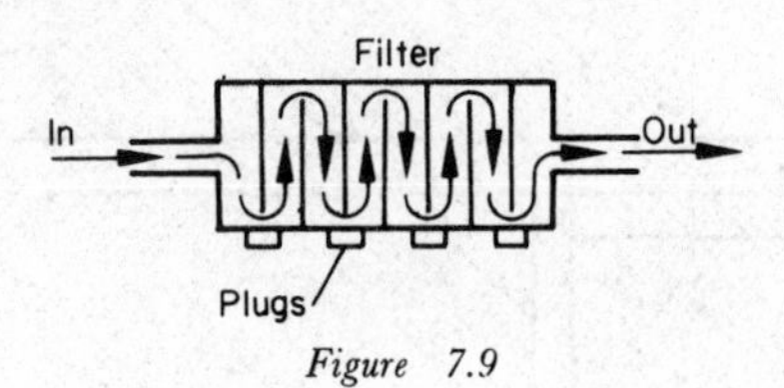

Figure 7.9

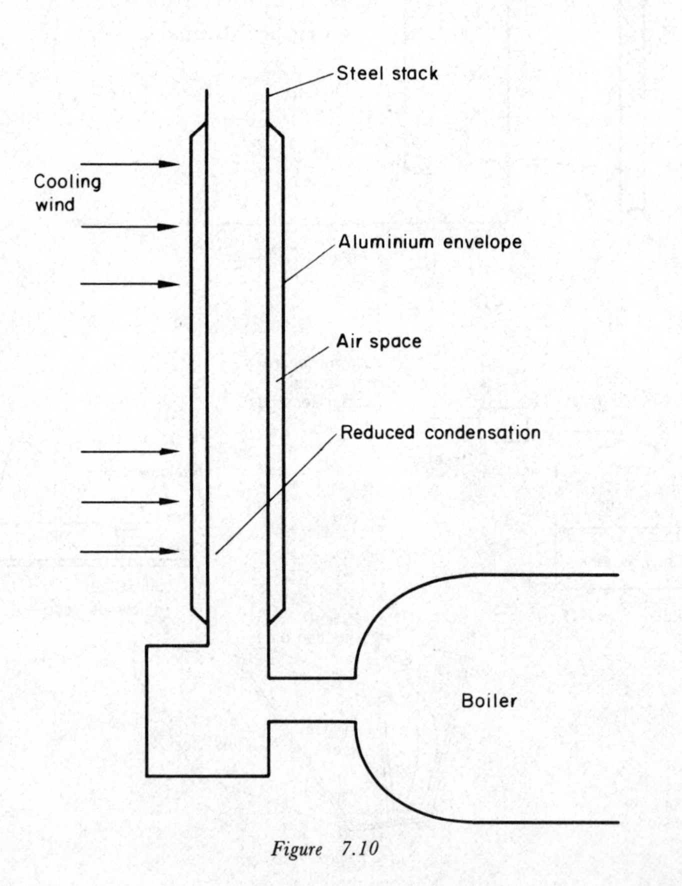

Figure 7.10

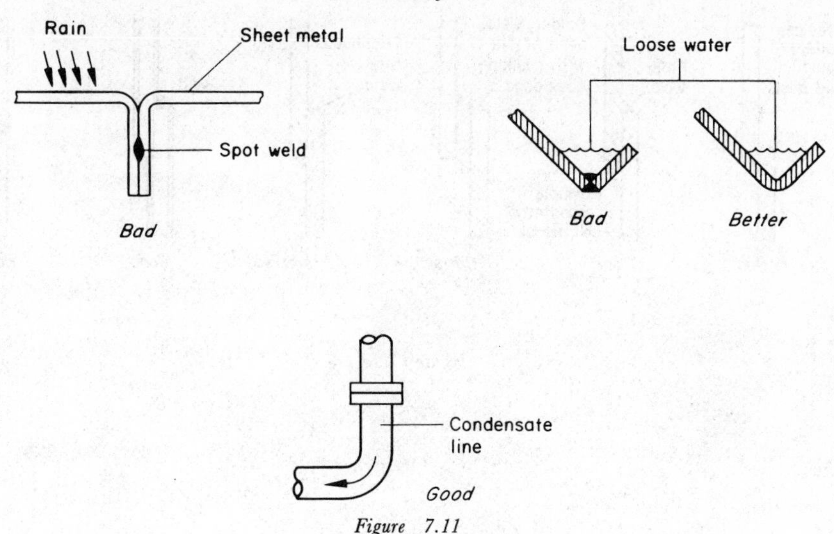

Figure 7.11

(11) Where crevices cannot be avoided take precautions to prevent ingress of corrodant by improving the geometry, fit or surface texture (*Figure 7.12*).

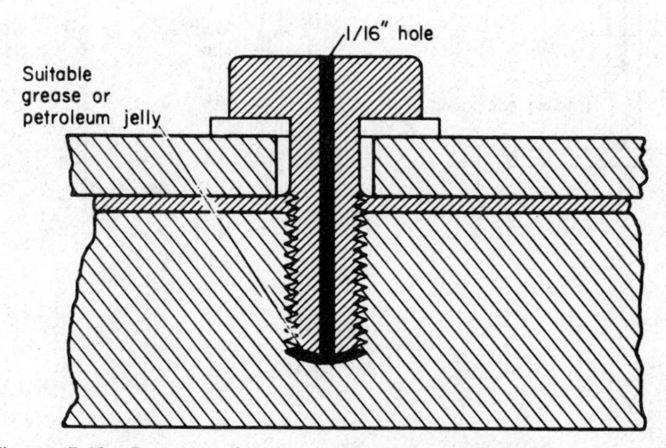

Figure 7.12. Prevention of crevice corrosion—submerged stainless steel fasteners

(12) Where possible avoid laps and crevices or seal these effectively, especially in areas of heat transfer, between metal and a porous material or where aqueous environment contains inorganic chemicals or dissolved oxygen (*Figure 7.13*).

(13) Face laps downwards on exposed surfaces (*Figure 7.14*).

(14) Make every effort to give design a shape or form which will reduce

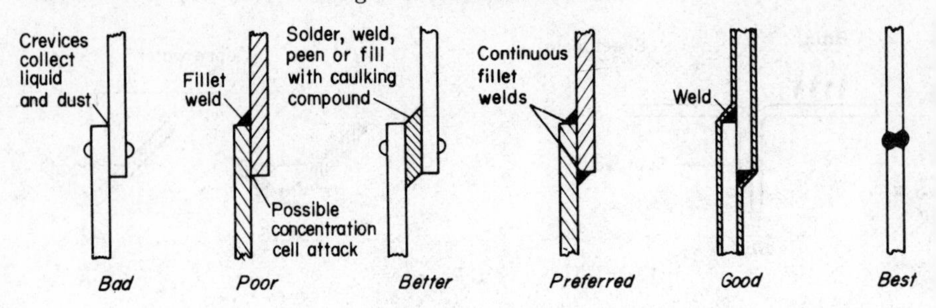

Figure 7.13

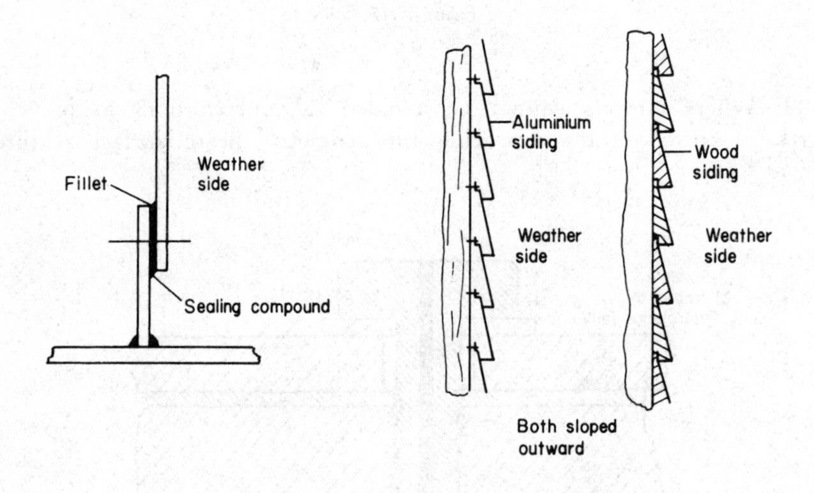

Figure 7.14

the effect of excessive velocity, turbulence of flow and formation of gas bubbles (*Figure 7.15*).

(15) Avoid forms which aggravate rapid surging of flow (*Figure 7.16*).

(16) Avoid gaskets overlapping on the inside and outside of the equipment or structures (*Figure 7.17*).

(17) Sponson plates could be preferable to a group of brackets (*Figure 7.18*).

(18) Select a combination of shapes whose form of attachment will not aggravate corrosion (*Figure 7.19*).

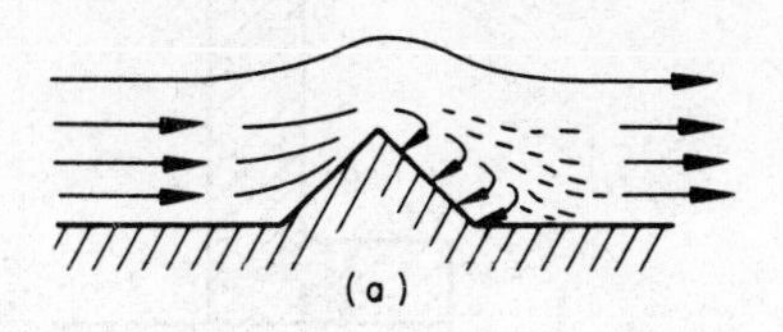

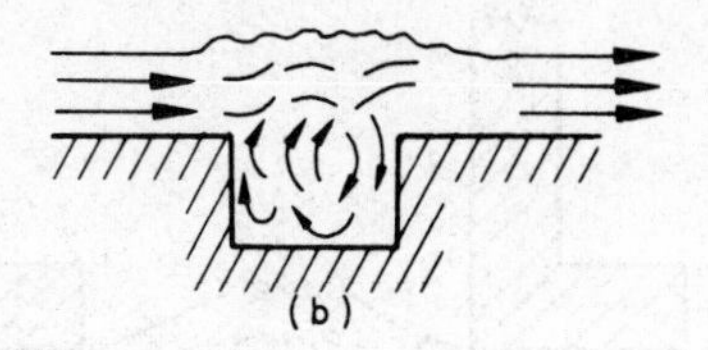

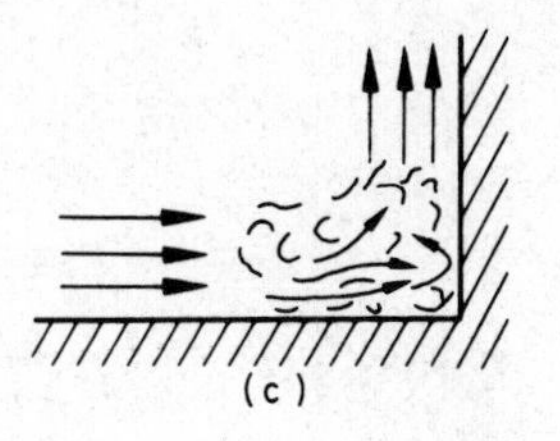

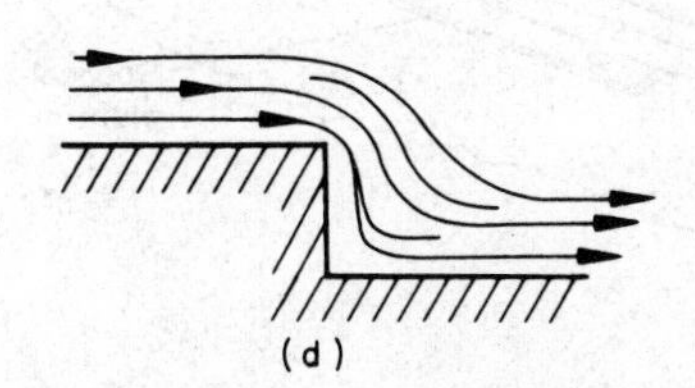

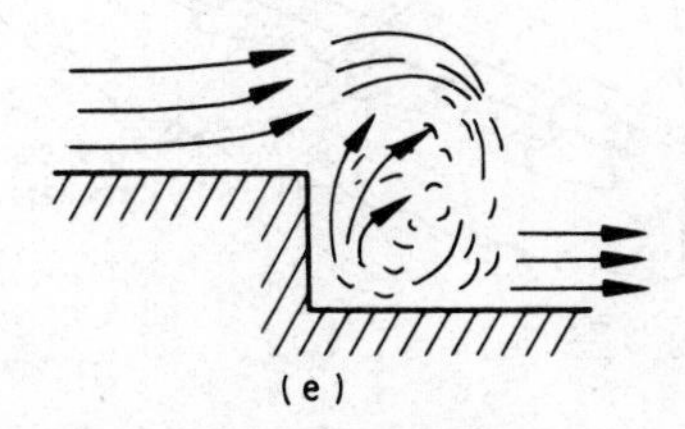

Figure 7.15. (a) Effect of projection; (b) effect of groove or crevice; (c) effect of corner; (d) effect of weir (low flow velocity); (e) effect of weir (high flow velocity)

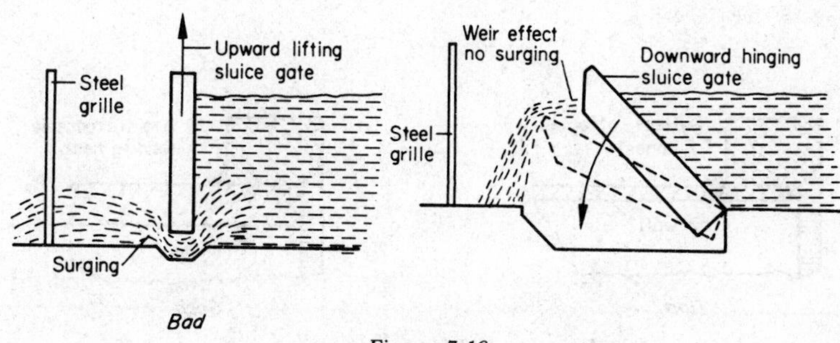

Figure 7.16

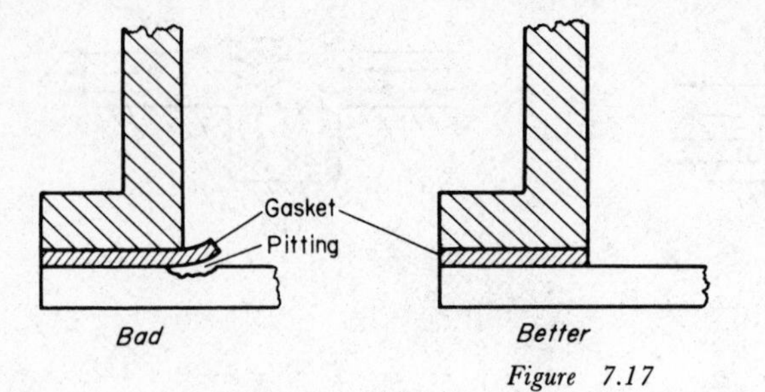

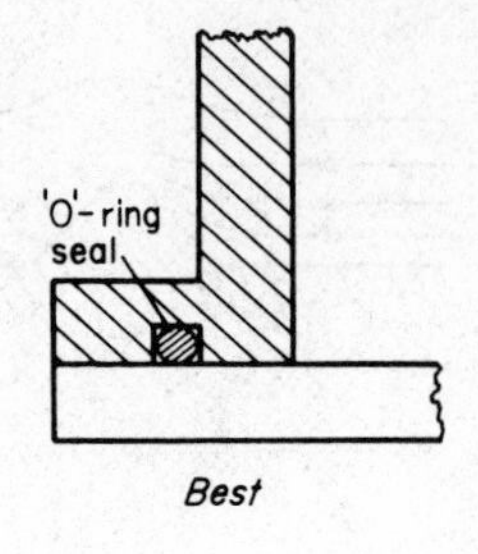

Figure 7.17

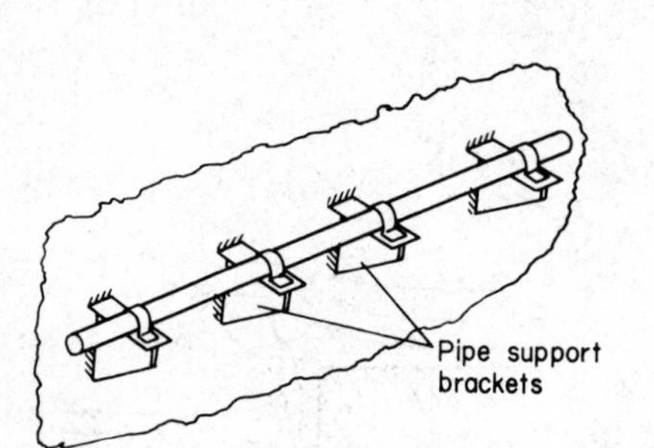

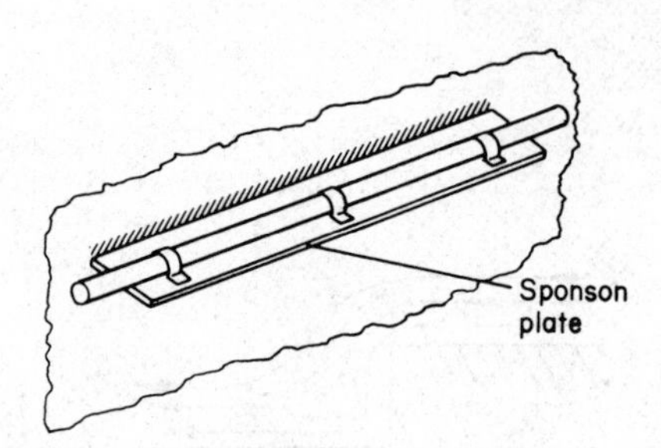

Figure 7.18

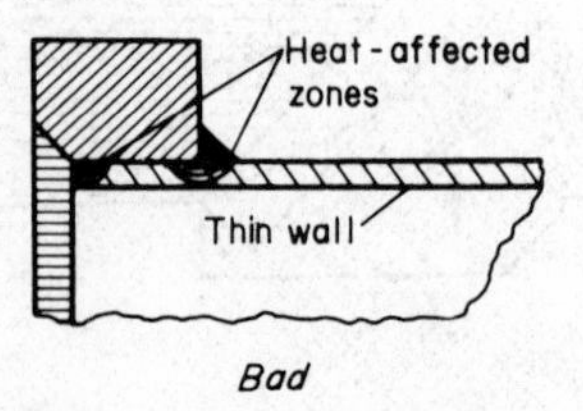

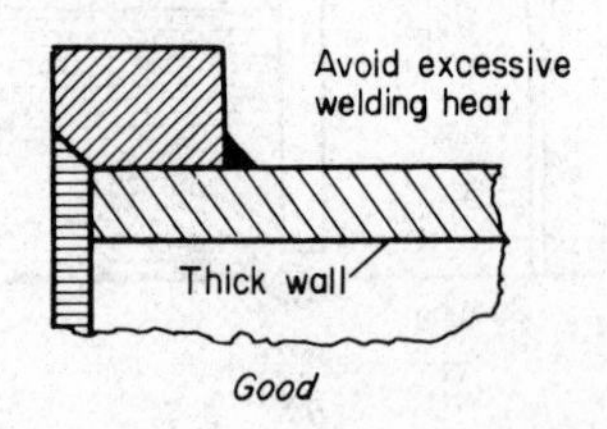

Figure 7.19

(19) Avoid asymmetrical shapes of unequal thickness for galvanising, where possible. Avoid extremes in weight and cross-section of design members. Consult galvaniser for reinforcing and bracing to prevent warpage and distortion (*Figure 7.20*).

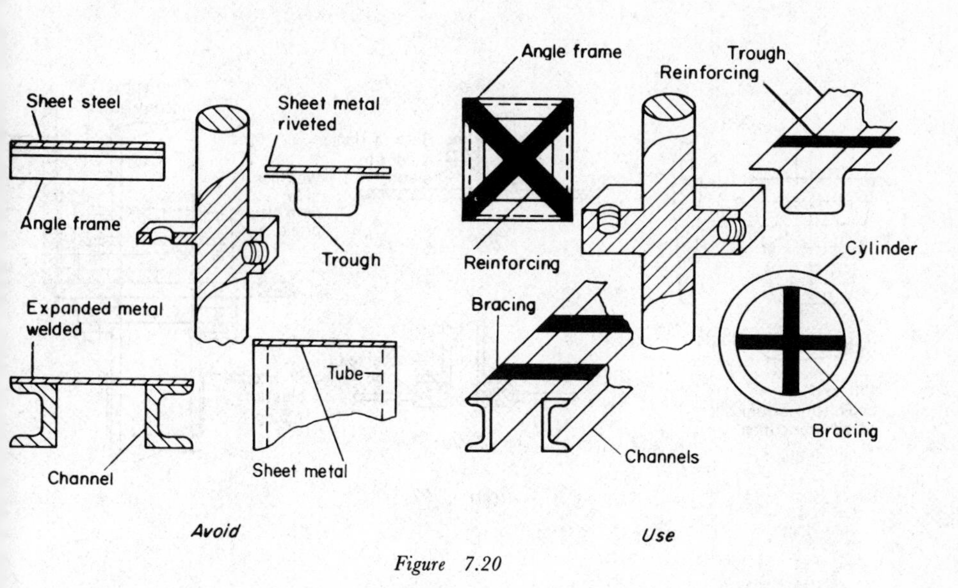

Figure 7.20

(20) Arrange the layout for an easy initial preservation and repainting (*Figure 7.21*).

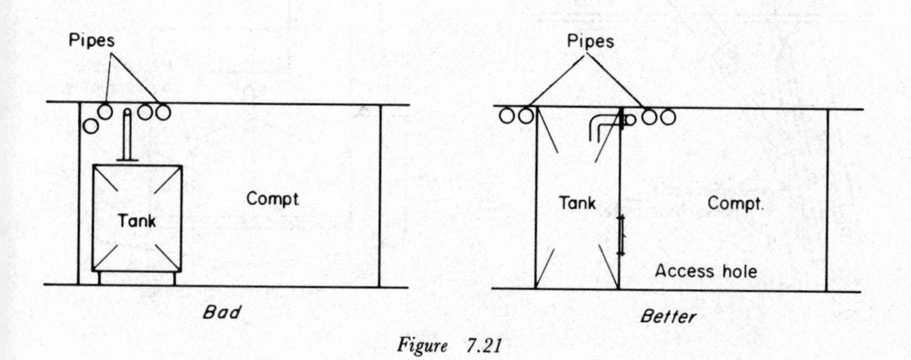

Figure 7.21

(21) Make surfaces of connections accessible for easy cleaning or maintenance. If access has to be considerably reduced or denied, an effective overall sealing, encapsulation or envelopment should be provided (*Figure 7.22*).

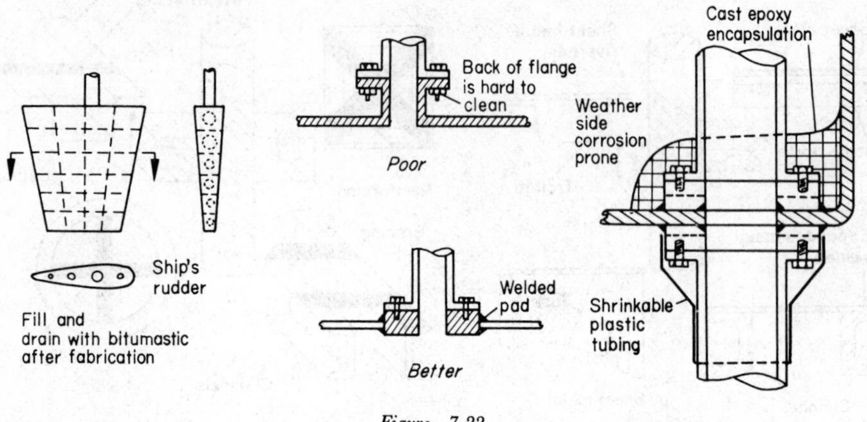

Figure 7.22

(22) Provide replaceable impingement plates and baffles (*Figure 7.23*).

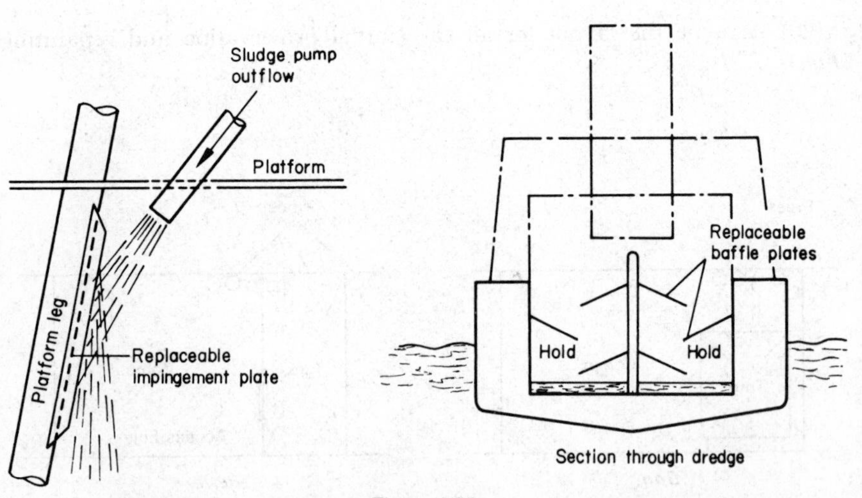

Figure 7.23

(23) Secure access of oxygen to structures made of active/passive metals by selective geometry (*Figure 7.24*).

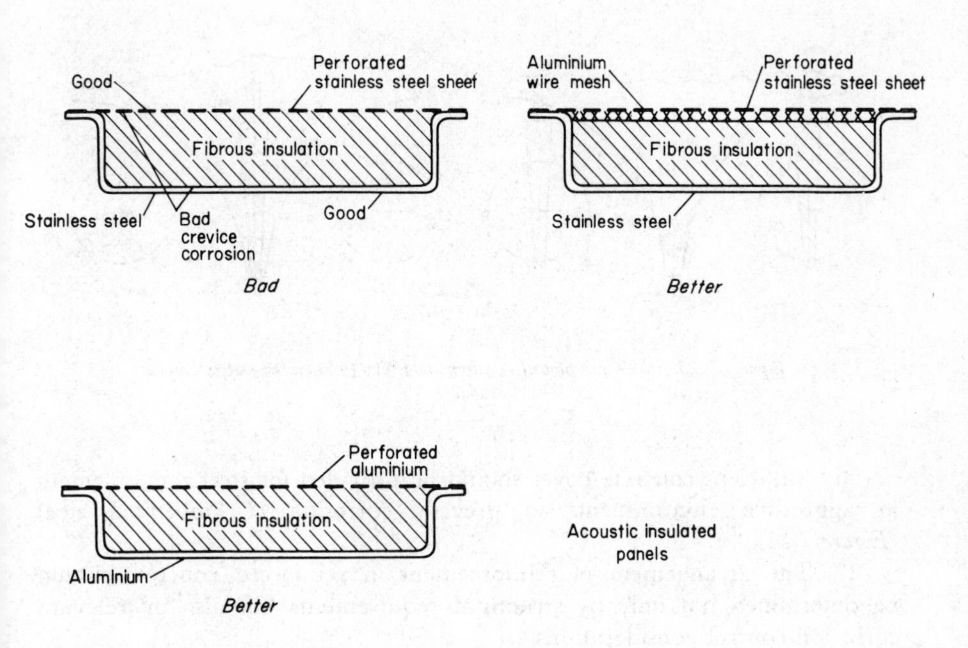

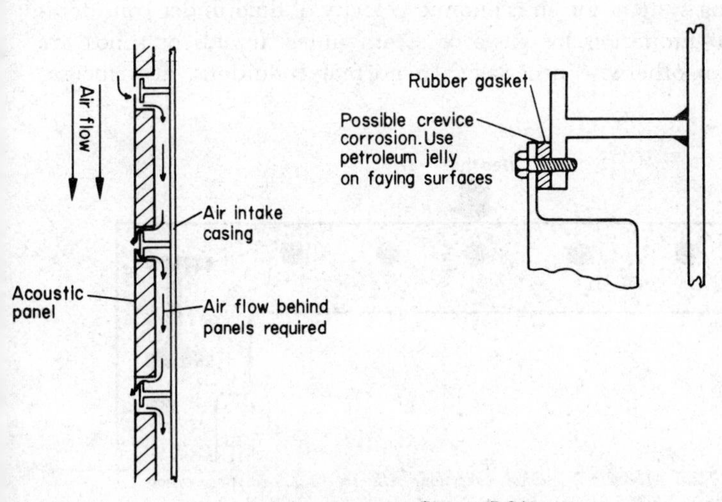

Figure 7.24

(24) Shape propellers or impellers to avoid high turbulence formation and reduce low pressure build-up at the tips of propeller blades (test design in cavitation tunnel).

(25) Avoid adverse position of rudder in the way of turbulence caused by propeller (cavitation) (*Figure 7.25*).

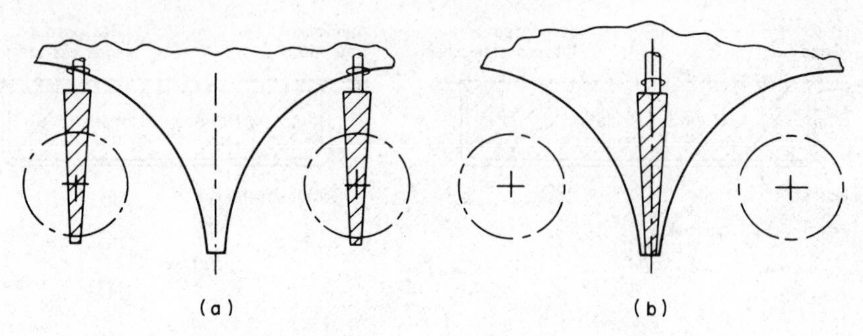

Figure 7.25. (a) Twin propellers (twin rudders); (b) twin propellers (single rudder)

(26) Sufficient concrete cover should be provided for steel reinforcement in aggressive environments to prevent corrosion of embedded steel (*Figure 7.26*).

(27) The arrangement of reinforcement in reinforced concrete should be determined not only by structural requirements but also by relevant corrosion-control considerations.

7.5 Piping Systems

(1) Design piping systems for an economic velocity of fluid under consideration (there is no limitation for gases or steam unless liquids or solids are entrained), unless otherwise necessary. In normal conditions, at velocities

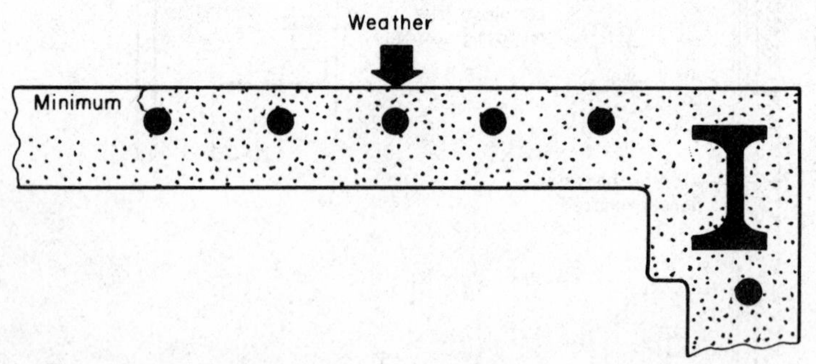

Figure 7.26. Minimum: optimal conditions 1/2 in (1.27 cm); corrosive conditions 2 in (5.08 cm); hydraulic structures 2–3 in (5.08–7.62 cm)

of 2–10 ft/s(61–305 cm/s) there should be no severe corrosion in absence of other factors. Relative to piping bores maximum fluid speeds may vary from a mean velocity of 3 ft/s (91 cm/s) for a 3/8 in (0.95 cm) bore to 10 ft/s (305 cm/s) for an 8 in (20.32 cm) bore. *Note*: economic velocity is also governed by the material used.

(2) Higher velocities than those mentioned above may, however, be required to provide a uniform and constant oxygen content in fluids, which is needed for formation of protective films on active/passive metals and those metals which are subject to pitting, e.g. stainless steel (austenitic—minimum 5 ft/s (152 cm/s) required), monel, aluminium alloys, etc.

(3) Provide for removal of rust, debris and other solid contaminants (entrained or formed on stream) from the system (*Figure 7.27*).

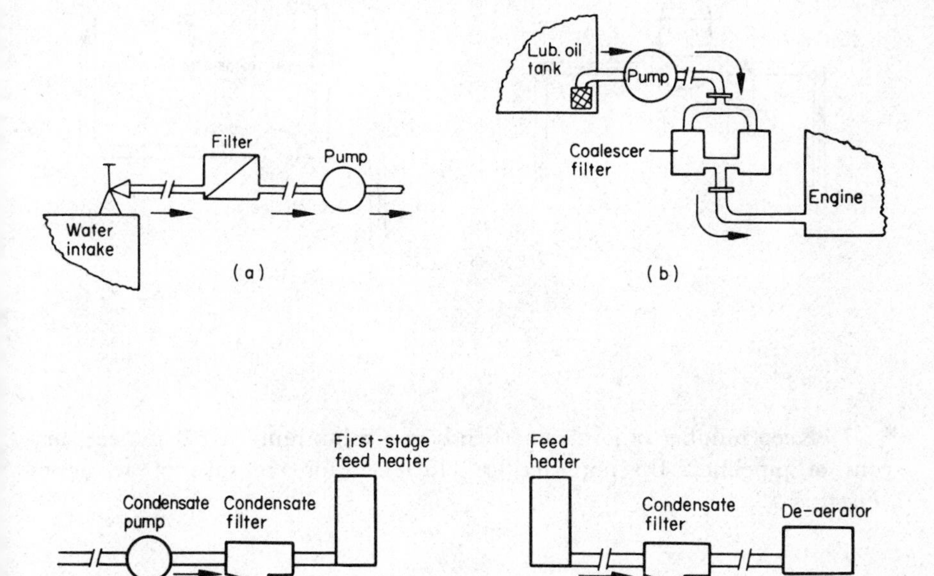

Figure 7.27. (a) Water; (b) lubricating oil; (c) steam

(4) Provide for removal of liquids from compressed air, gas and steam systems (*Figure 7.28*).

(5) Provide for removal of entrained air and gases from the liquids in piping systems (*Figure 7.29*).

(6) Provide for complete sealing of equipment against uncontrolled leakage of liquids, blow of air or steam and dissipation of fumes both from inside out and from outside in.

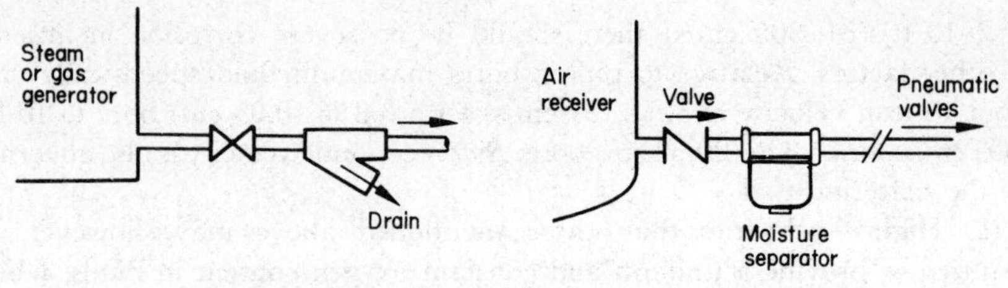

Figure 7.28

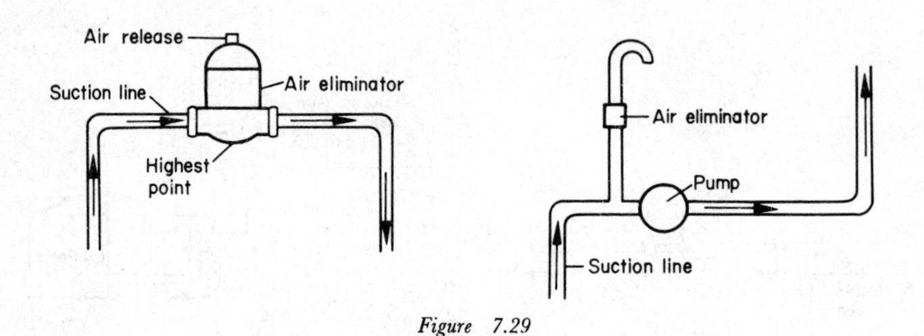

Figure 7.29

(7) Keep number of joints and bends to a minimum—avoid unnecessary runs of pipelines. Do not sacrifice efficiency for the sake of aesthetics (*Figure 7.30*).

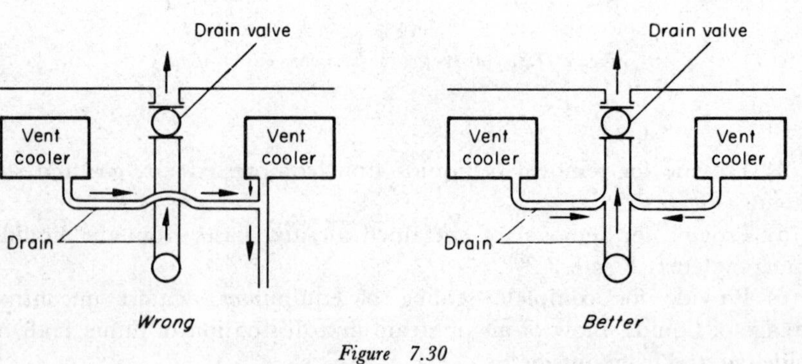

Figure 7.30

(8) Streamline the interior of piping systems for an easy drainage (*Figure 7.31*): (*a*) avoid stagnancy producing stubs and dead ends; (*b*) slope all pipelines continuously downstream to their outlets or other terminals, if possible (except rising vents), for complete emptying; (*c*) provide drainage in dipped sections of pipes; (*d*) slope elbows for drainage if possible.

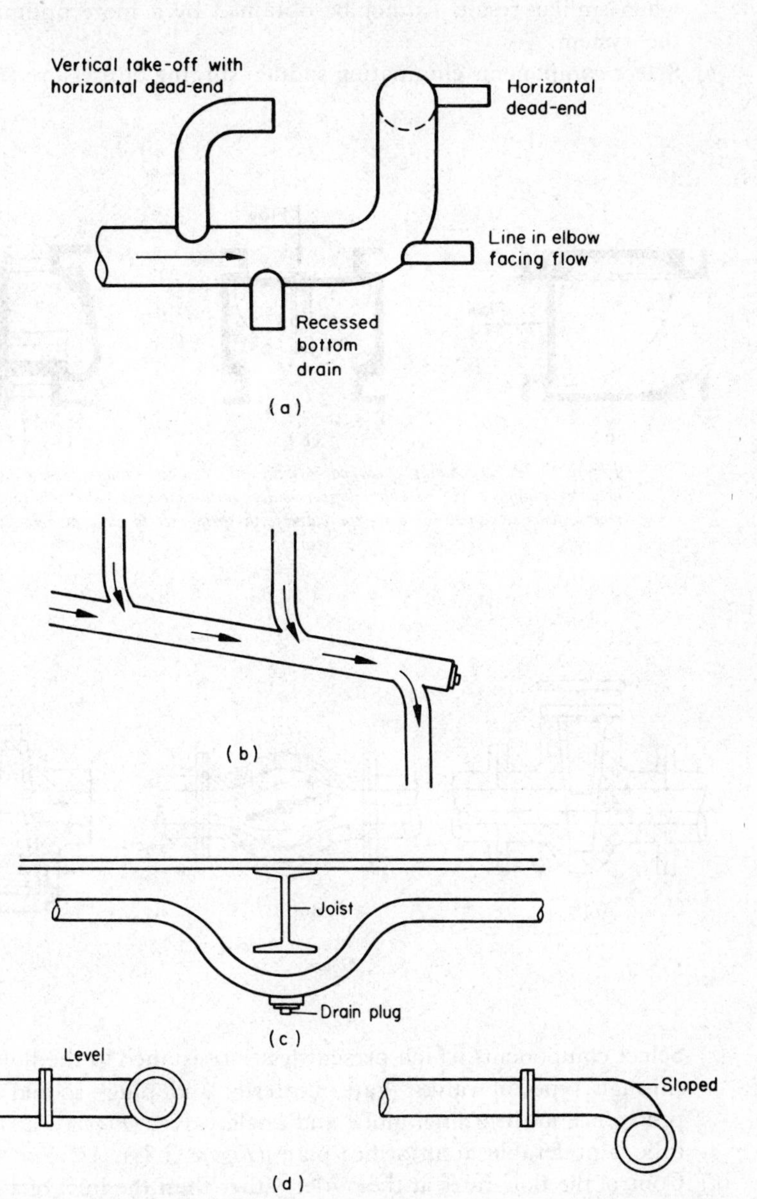

Figure 7.31

(9) Avoid turbulence, rapid surging, excessive agitation and impingement of fluids in the system:

(*a*) Use turbulence-forming components (i.e. throttle valves, orifices and similar flow-regulating devices) only when absolutely necessary or when similar results cannot be obtained by a more optimal design of the system.

(*b*) Select components eliminating sudden surging of pressure (*Figure 7.32*).

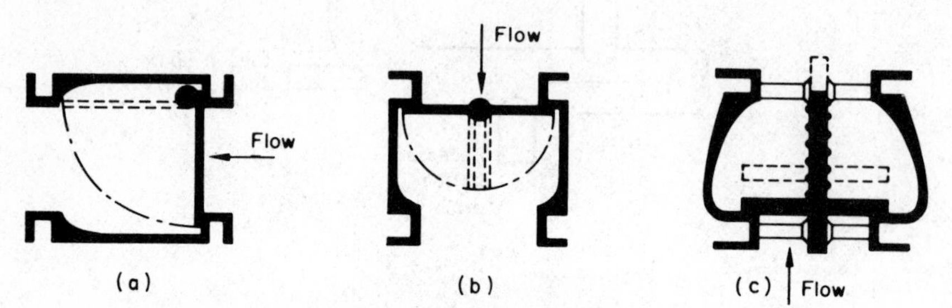

Figure 7.32. (a) Swing check valve with 90° clapper swing—reverse flow actuates clapper; (b) double clapper valve with spring-assisted closing—reverse flow actuated; (c) disc-type valve with spring set to close at zero flow velocity

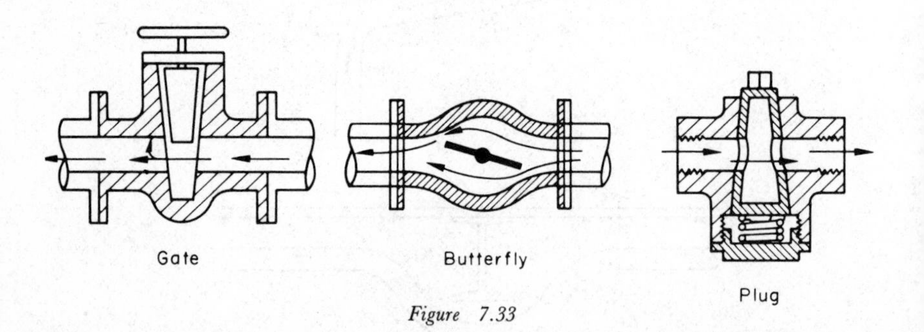

Figure 7.33

(*c*) Select components which present lesser resistance to the flow. Straight-through types of valves (gate, butterfly and plug) should be used in preference to throttling, globe and angle valves (*Figure 7.33*). A venturi tube is preferable to an orifice plate (*Figure 7.34*).

(*d*) Control the flow rates at the outlet rather than the inlet of a component prone to impingement (*Figure 7.35*).

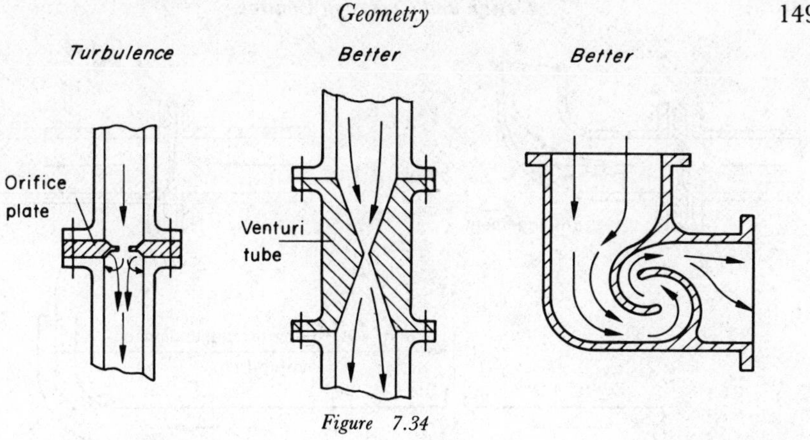

Figure 7.34

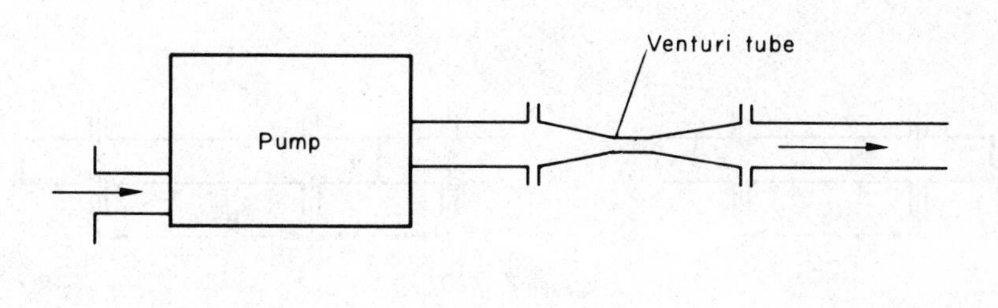

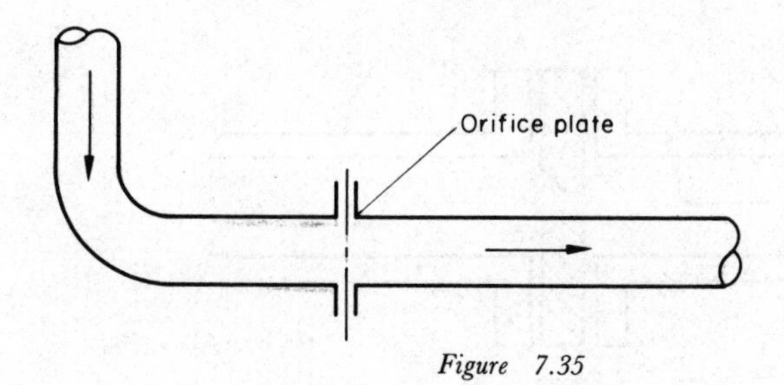

Figure 7.35

(*e*) Avoid using orifices or other flow-reducing devices in the close proximity of bends or changes of direction downstream (*Figure 7.36*).

(*f*) If the orifice plates are unavoidable use at least two plates; the distance between multiple-type plates should be mismatched (*Figure 7.37*). There should be a minimum of two pipe diameters between orifice plates.

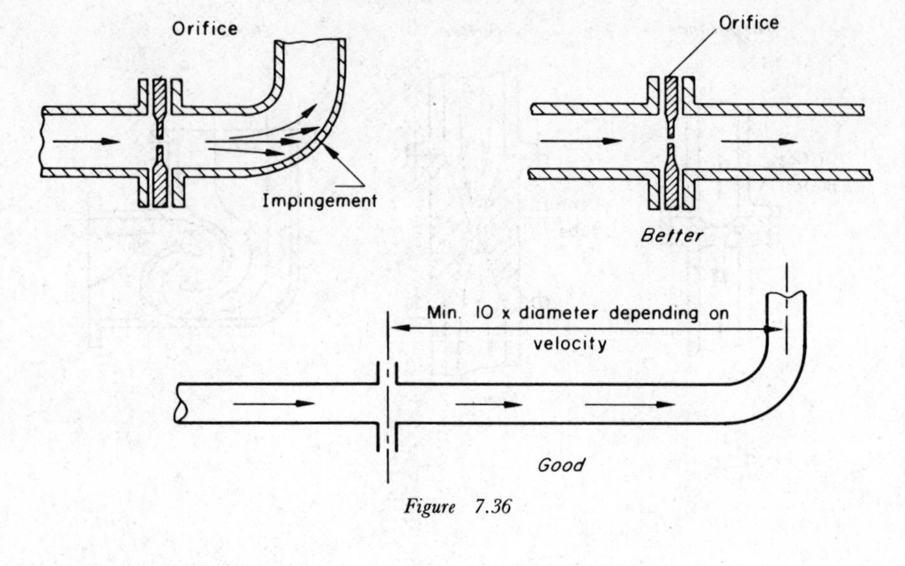

Figure 7.36

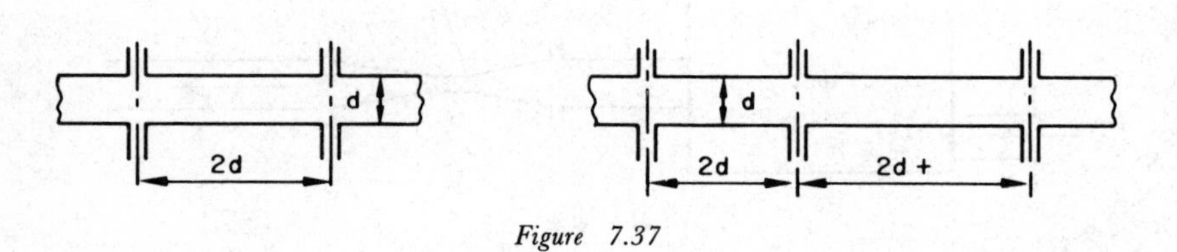

Figure 7.37

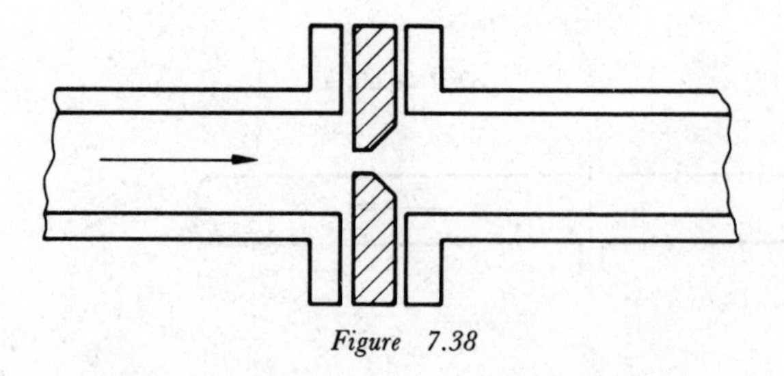

Figure 7.38

(*g*) Chamfer the upstream side of the orifice (*Figure 7.38*).
(*h*) If possible use sintered metal plugs as liquid or gaseous flow throttle controls (*Figure 7.39*).
(*i*) Avoid any sudden changes (sharp bends) in the direction of fluids in pipelines and fittings, especially in those made of lead, copper and their alloys (*Figure 7.40*).

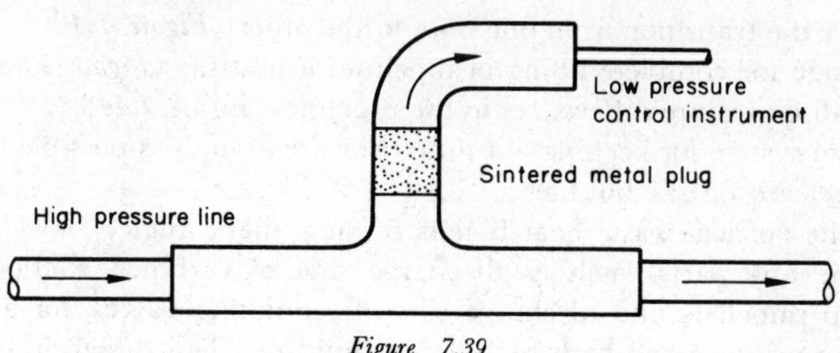

Figure 7.39

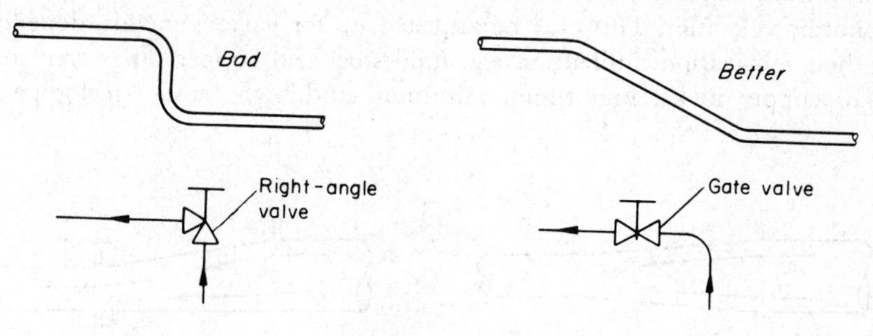

Figure 7.40

(*j*) Avoid surging and other sudden changes in velocity of fluids in pipelines and fittings, by the careful selection of component fittings within the system which do not cause surging (e.g. porous metal snubber and filter) and by introducing replaceable baffles (e.g. opposing orifice) (*Figure 7.41*).

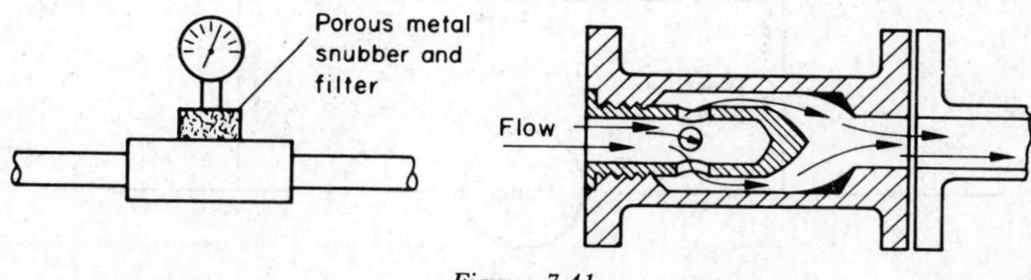

Figure 7.41

(*k*) Taper the transition from one bore to the other (*Figure 7.42*).
(*l*) Arrange for complete filling of pipelines if possible (*Figure 7.43*).
(*m*) Equalise pressure differences in the pipelines (*Figure 7.44*).
(*n*) Design system for keeping absolute pressure as high as possible to restrict the release of gas bubbles.
(*o*) Set the vertical waste heat boilers off at a slight angle.
(*p*) Shape any parts, such as discharge side of turbines, suction side of pump impellers and discharge side of regulating valves, for avoidance of low pressure and high turbulence build-up. Test design in cavitation tunnel.

(10) The bend radii of pipes should be as large as possible. Normally, a minimum of three times the diameter of the pipe should be enforced for economic velocities. This may be adjusted up for various metals, depending on their fabrication difficulties, e.g. mild steel and copper pipe three times, 90/10 copper nickel four times, minimum and high tensile steel pipe five

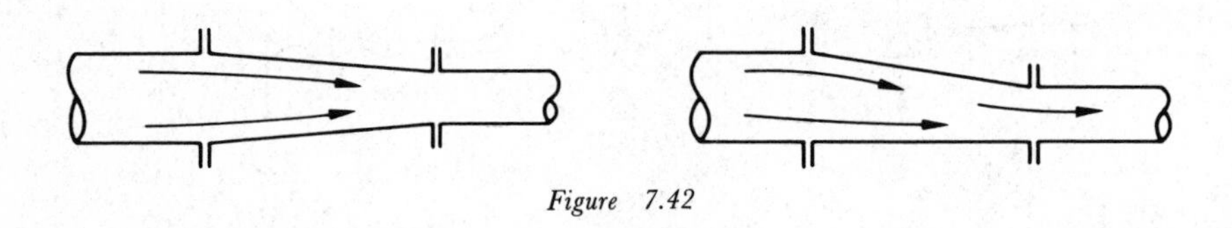

Figure 7.42

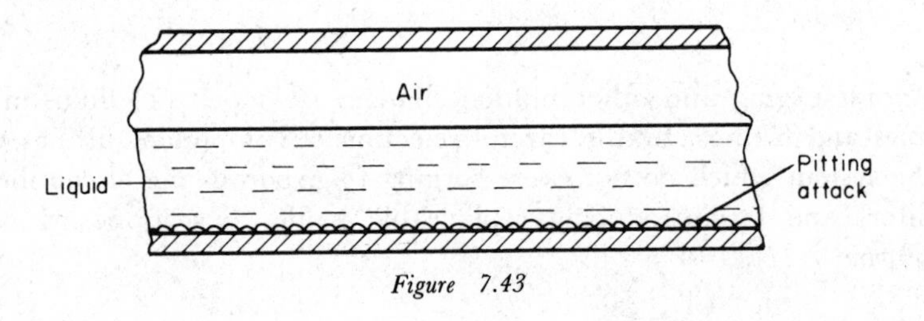

Figure 7.43

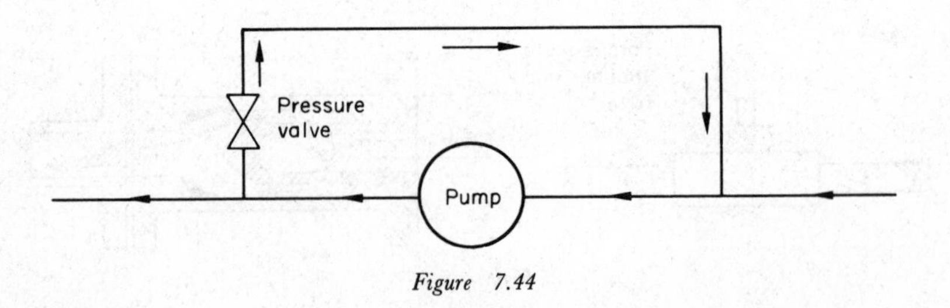

Figure 7.44

times the diameter of the pipe minimum. Adjustment for high velocities is, of course, also required—the higher the velocity the larger the radius of the pipe (*Figure 7.45*).

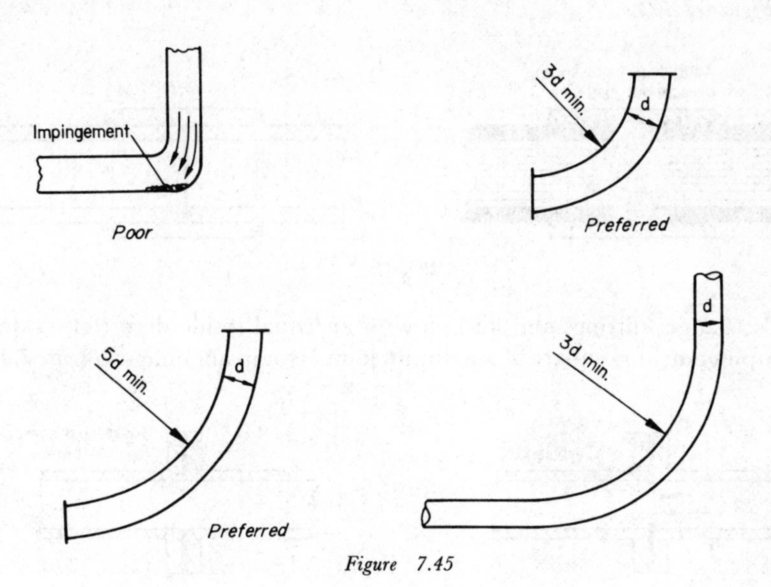

Figure 7.45

(11) Elbows of similar radii, i.e. minimum three diameters, would be advantageous if these are commercially available.

(12) Avoid branching off in tees on high velocity connections—laterals are preferred (*Figure 7.46*).

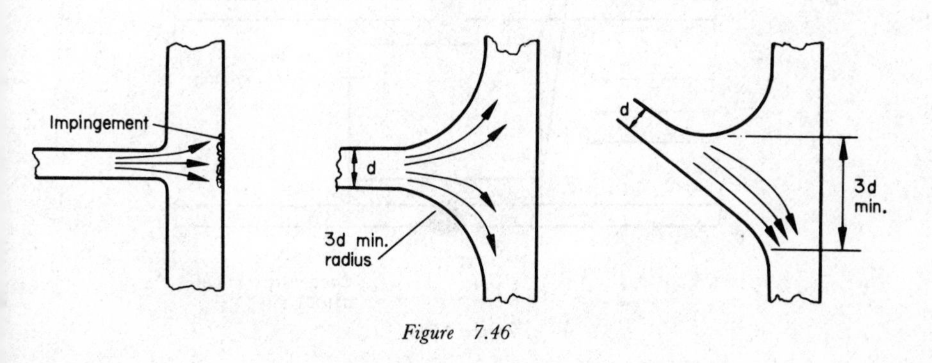

Figure 7.46

(13) Select optimal form of take-down joints which does not cause turbulence:

(*a*) Avoid selection with a possibility of inaccurate and incomplete fitting (*Figure 7.47*).

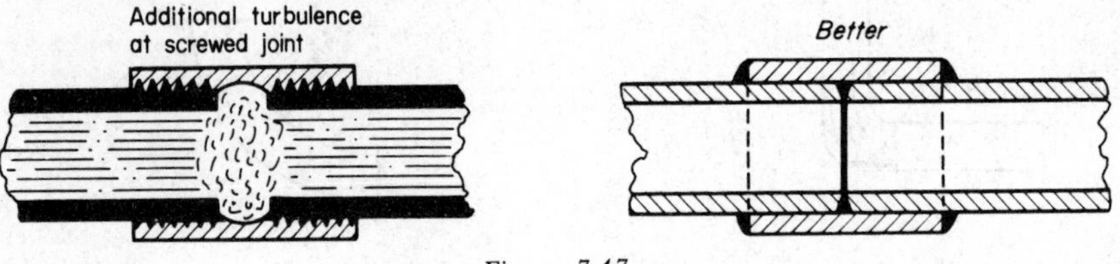

Figure 7.47

(*b*) Use flanges, fittings and gaskets with an equal inside diameter—rate of impingement = square of maximum joint error in alignment (*Figure 7.48*).

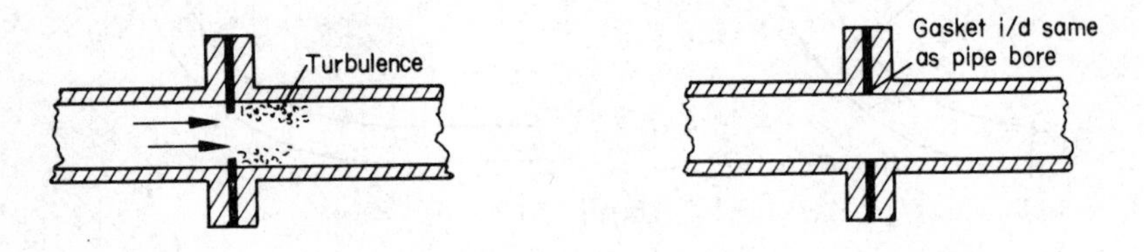

Figure 7.48

(*c*) Avoid excessive bead reinforcement and backing rings in welded pipelines (*Figure 7.49*).

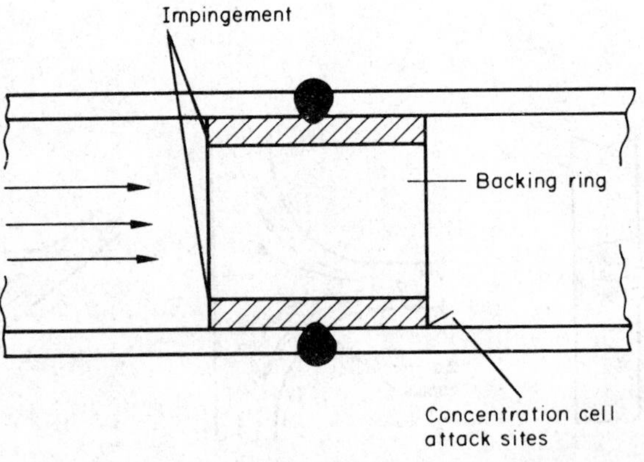

Figure 7.49

(*d*) Select take-down joints with a minimum probability of misalignment—if possible self-aligning—optimal maximum misalignment approximately 0.006 in (152 μm) (*Figures 7.50* and *7.51*).

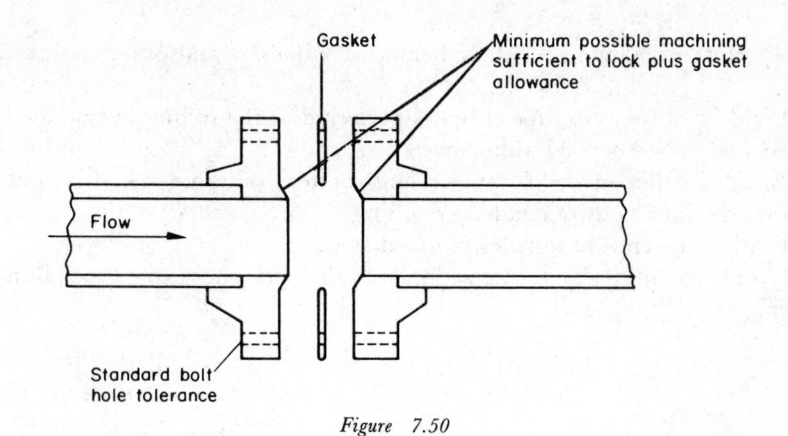

Figure 7.50

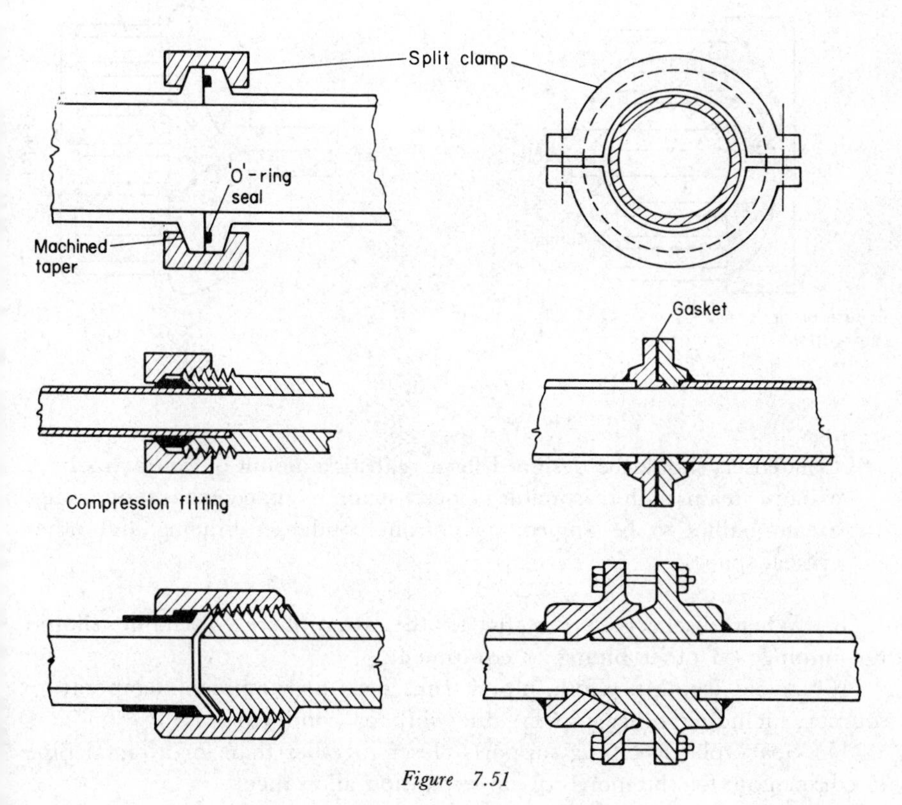

Figure 7.51

(*e*) Select 'O'-ring joints for alignment of pipes; not optimum, however, for stainless steel pipes.
(*f*) Restrict gasket bore to not more than 1/64 in (395 μm) larger than the pipe bore.

(14) Heat exchangers, coolers, heaters, condensers and other equipment.

(*a*) Welding of tubes in tube sheets is preferred to the rolling-in system.
(*b*) Extend tubes beyond tube sheets.
(*c*) Insert ferrules made of same metal, better resistance metal or plastic into the inlet ends of condenser tubing.
(*d*) Feather the ends of ferrules to avoid step.
(*e*) Avoid cooling water starvation at the periphery of tube bundle (*Figure 7.52*).

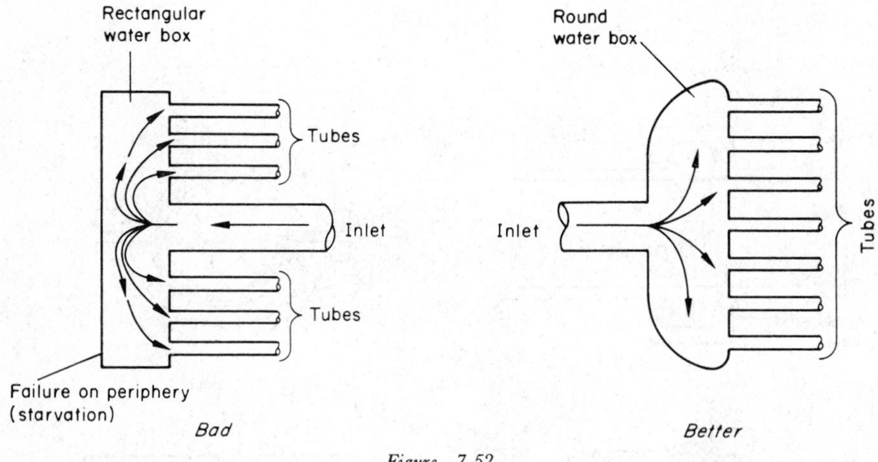

Figure 7.52

(*f*) Condensers should be designed for a realistic amount of excess auxiliary exhaust steam, with reasonable velocity steam inlet and exhaust openings. Steam baffles to be angled away from condenser bracing and other critical spaces.

(15) When discharging directly to the atmosphere, discharge should not impinge on other piping or equipment.

(16) Avoid locating plastic piping runs near high ambient temperature sources, including other piping, ductwork or conductors.

(17) Space plastic piping supports closer together than for a metal pipe to compensate for the more critical expansion allowance.

(18) Avoid formation of hot spots by attachment (*Figure 7.53*).

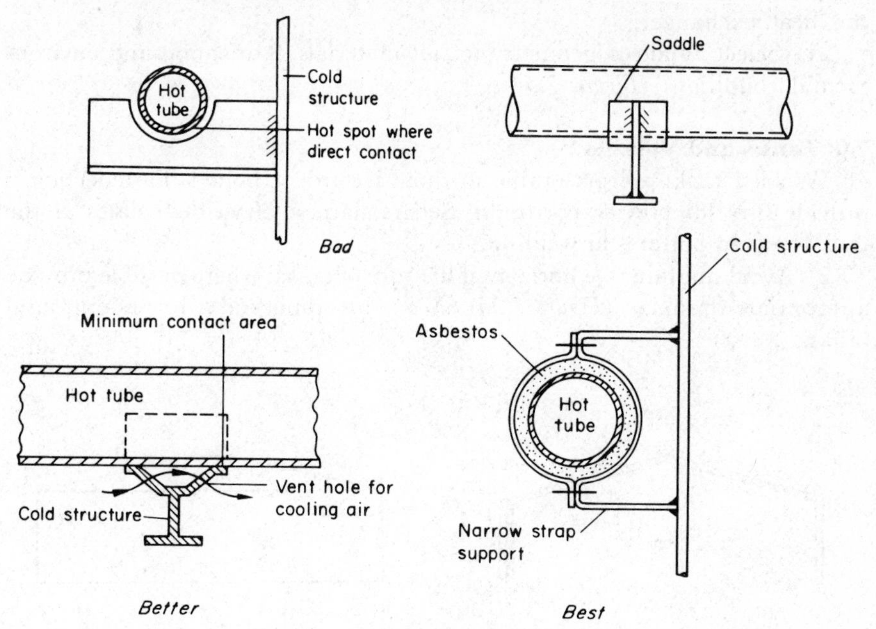

Figure 7.53

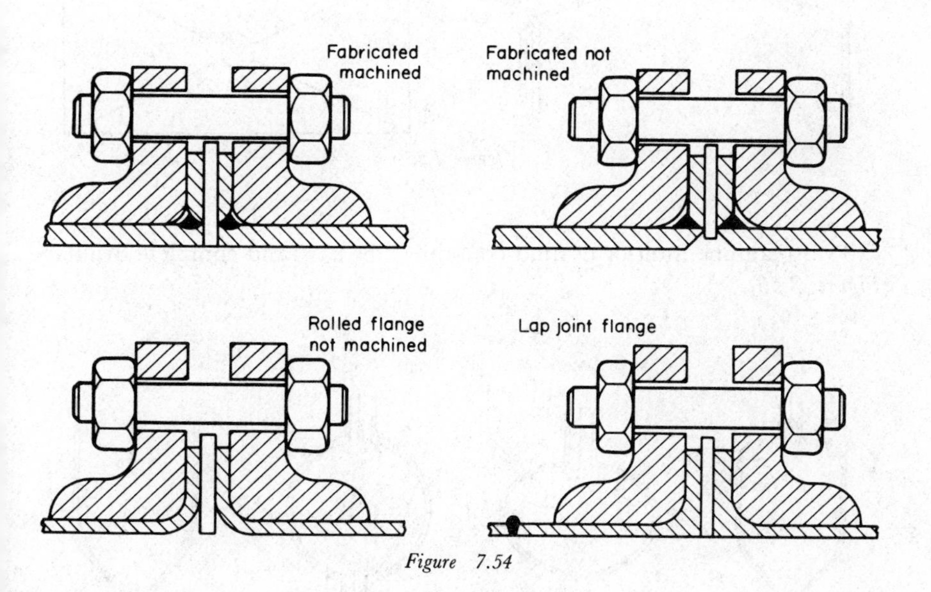

Figure 7.54

(19) Slant slightly the heat exchanger tubes, so that they will drain properly.

(20) Secure approximately equal water velocity through all the tubes in the heat exchanger.

(21) Select balanced geometry to suit materials, fabrication and environmental conditions (*Figure 7.54*).

7.6 Tanks and Vessels

(1) Welded tanks are preferable to those riveted or bolted. Fastener joints provide sites for crevice corrosion. Secure flatness of welded plates of the tank tops and bottoms in welding.

(2) Avoid undrainable horizontal flat tops of tanks; where possible provide appropriate drainage (*Figure 7.55*). *Note*: this applies also to underground tanks.

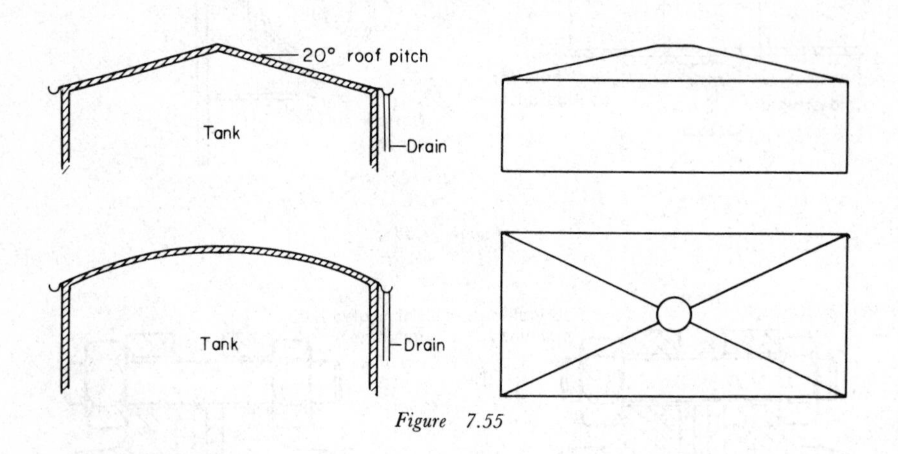

Figure 7.55

(3) Streamline interior of fluid containers for easy and complete drainage (*Figure 7.56*).

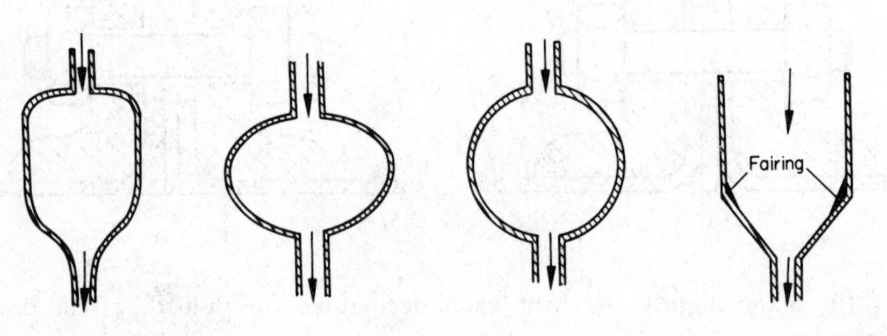

Figure 7.56

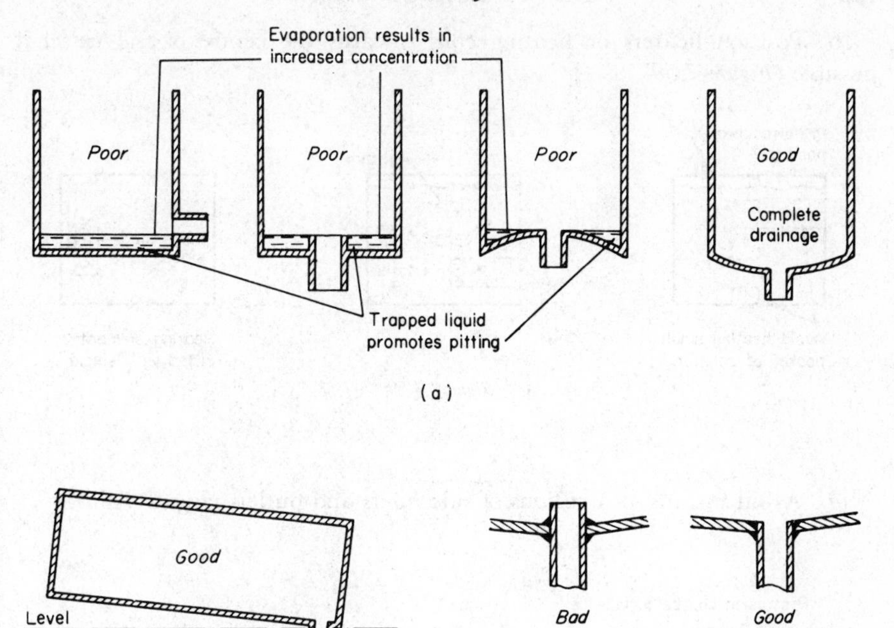

Figure 7.57. (a) Shape of bottom; (b) sloping of tanks; (c) tank drains—detail

(4) Slope tank bottoms towards drain holes to prevent collection of liquids after emptying of tank (*Figure 7.57*).

(5) Direct inlet pipes towards the centre of the vessel (*Figure 7.58*).

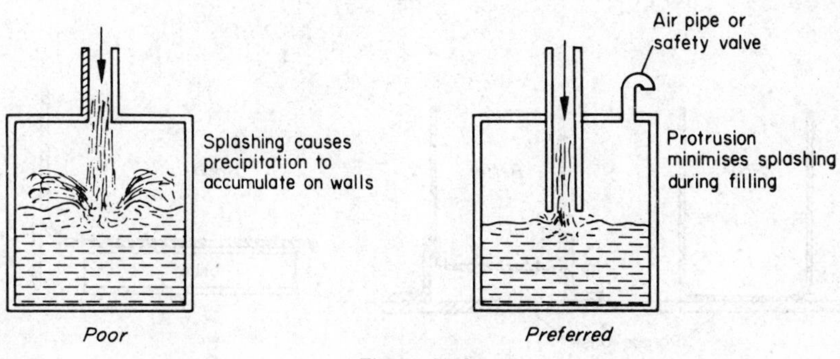

Figure 7.58

(6) Position heaters or heating coils towards the centre of the vessel if possible (*Figure 7.59*).

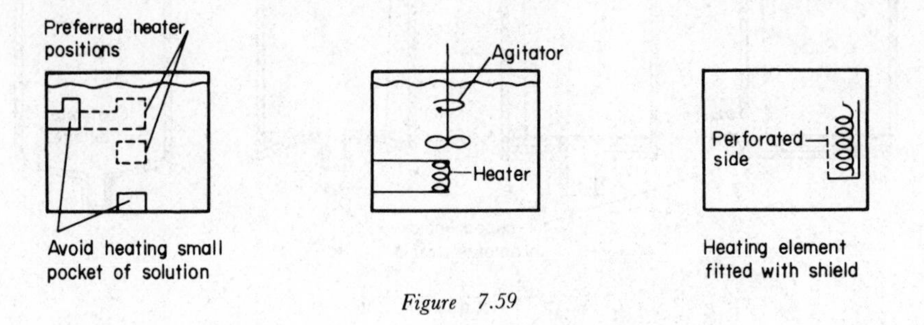

Figure 7.59

(7) Avoid in-tank protrusions of side inlets and outlets (*Figure 7.60*).

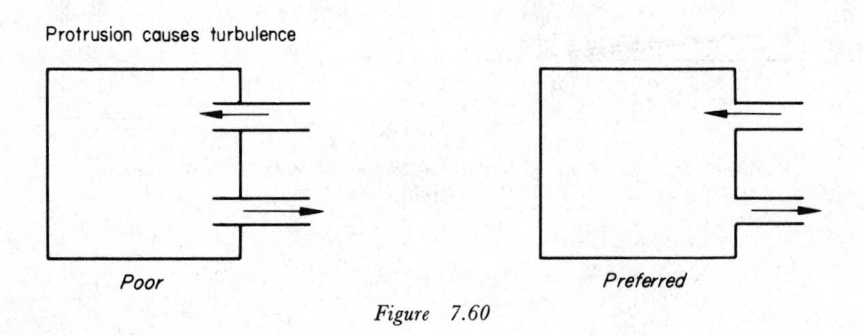

Figure 7.60

(8) Prevent crevice corrosion between the seating and tank (*Figure 7.61*).

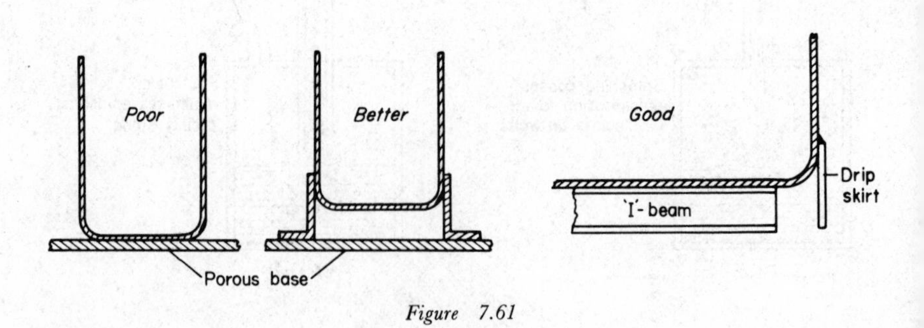

Figure 7.61

(9) Prevent adverse influence of haphazard insulation and avoid adverse effect of moisture entrapped in insulation (*Figure 7.62*).

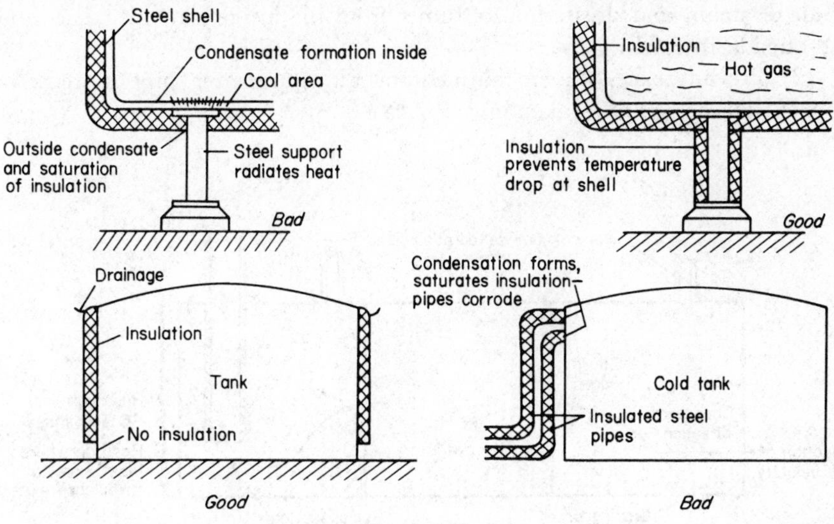

Figure 7.62

(10) Adjust geometry to the selected metals (*Figure 7.63*).

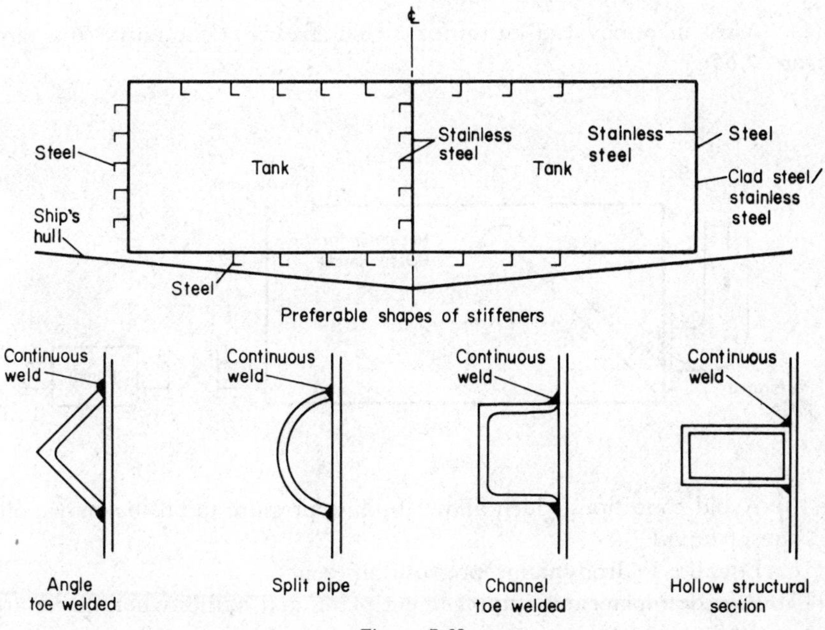

Figure 7.63

(11) Seal tanks holding hygroscopic corrodants well to prevent their breathing damp air.

(12) Seal tanks completely against uncontrolled leakage of liquids, blow of air or steam and dissipation of fumes from the inside outwards and from the outside inwards.

(13) Prevent excessive entrainment of air into water pipe systems by appropriate design of intakes (*Figure 7.64*).

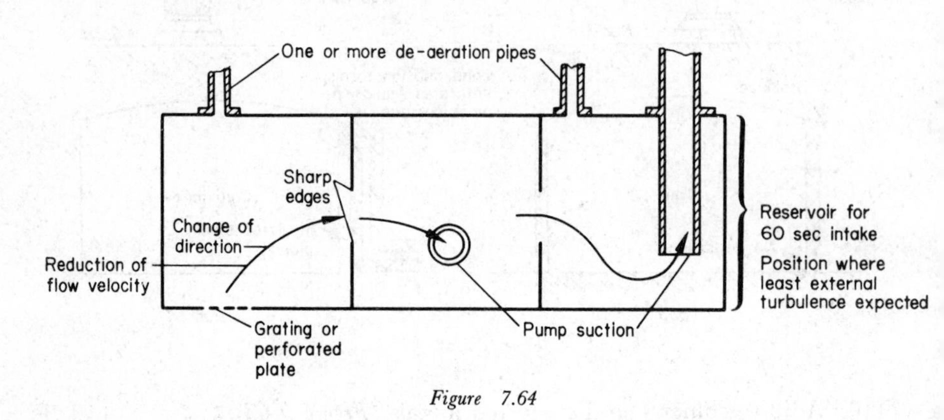

Figure 7.64

(14) Assist in removal of entrained air on stream at the entry to a tank (*Figure 7.65*).

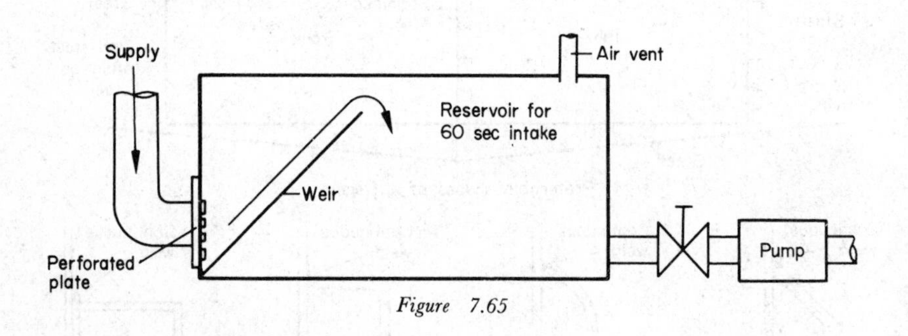

Figure 7.65

(15) Avoid conditions which allow absolute pressure to fall below vapour pressure of liquid.

(16) Equalise hydrodynamic pressure differences.

(17) Provide replaceable impingement plates and baffles where necessary (*Figure 7.66*).

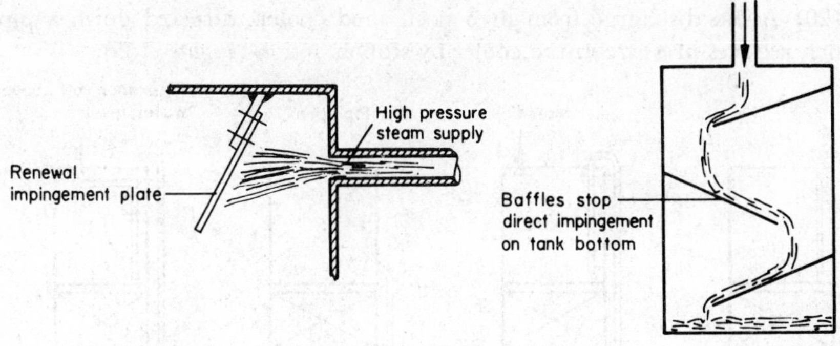

Figure 7.66

(18) Provide access to interior for inspection and removal of fabrication wastes and tools (*Figure 7.67*).

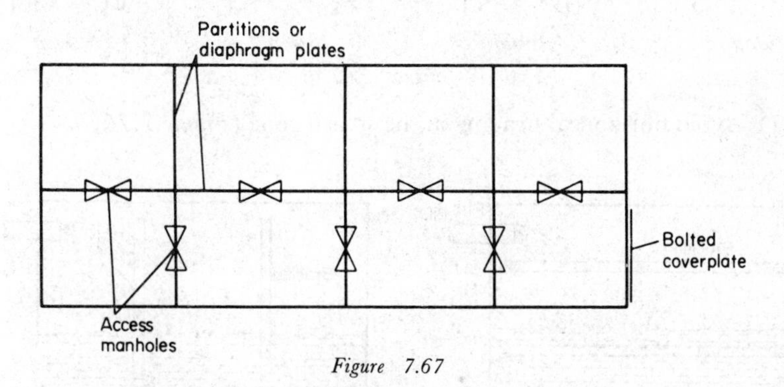

Figure 7.67

(19) Avoid machining which would aggravate corrosion (*Figure 7.68*).

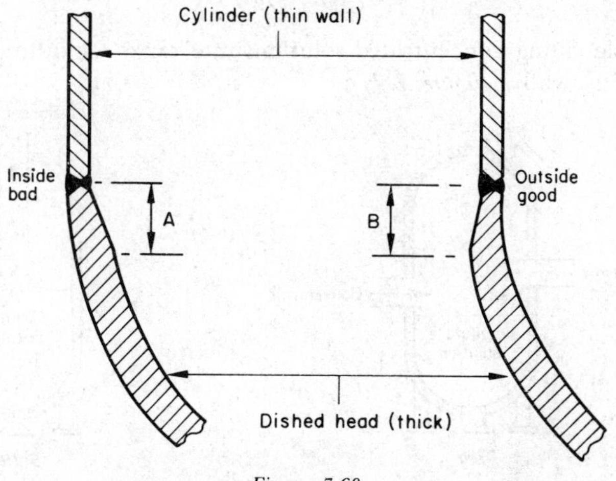

Figure 7.68

(20) Avoid discharge from high positioned coolers directed down a pipe which reduces pressure in the cooler by siphon action (*Figure 7.69*).

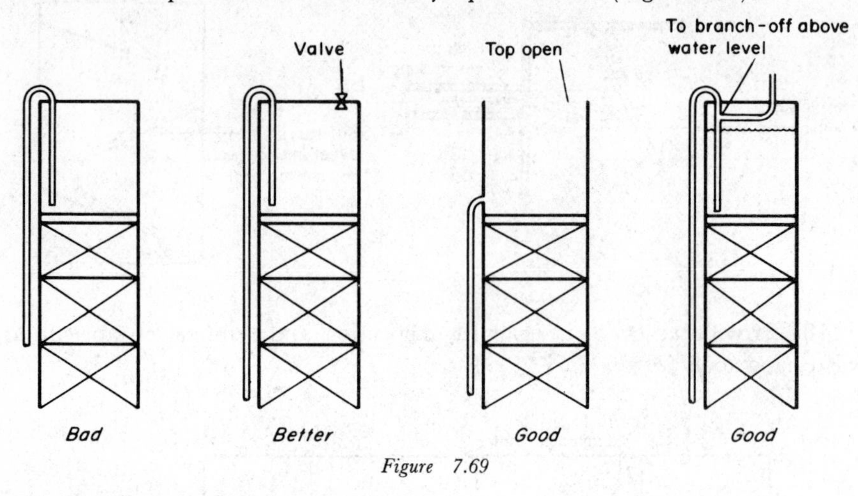

Figure 7.69

(21) Avoid horizontal bracing in the splash zone (*Figure 7.70*).

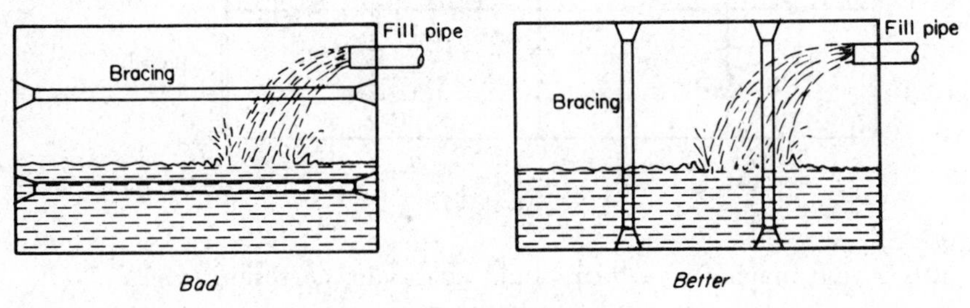

Figure 7.70

(22) Avoid filling concentrated solutions into tanks for dilution purposes along the side walls (*Figure 7.71*).

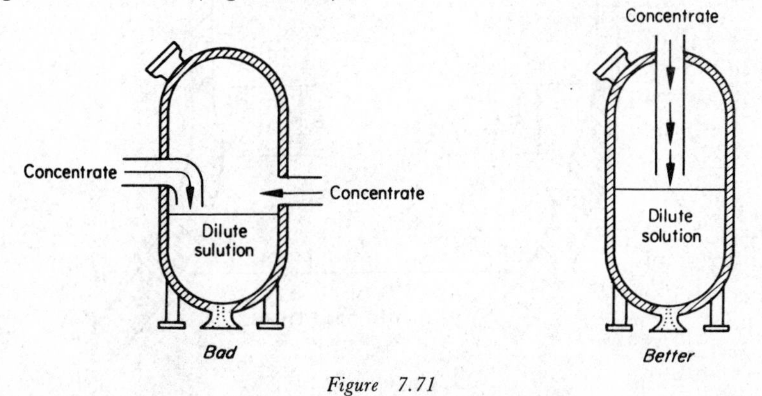

Figure 7.71

(23) Partially filled reaction vessels and storage tanks containing vapours of corrosive solutions should either by vented to the atmosphere (if possible) or provided with vacuum removal of vapour or with a condenser return to the vessel.

7.7 Electrical and Electronic Equipment

(1) Avoid or reduce corrosive effect of oxygen, moisture and airborne corrodants in the atmosphere within the equipment, especially in its critical spaces, i.e. where corrosion can impair its safe function:

(*a*) design equipment enclosures of appropriate air-, gas- or waterproof integrity allowable by their functional requirements;
(*b*) group suitable parts of equipment together for an easy and foolproof encapsulation, sealing or enveloping.

(2) Avoid inside and outside pockets, wells, traps and sump areas where water and condensed moisture could collect.
(3) Provide drainage paths for extraction of contaminants and condensed moisture away from the critical areas.
(4) Design the interior layout for an easy and efficient ventilation drying and extraction of humidity and contaminants. *Note*: the effect of interior heat generators is to be considered.
(5) Site the equipment away from air currents (i.e. ventilation blowers), collecting and carrying contaminants and excessive humidity; also away from excessive heat generators (i.e. over 50°C) unless especially provided for.
(6) Provide for continuous and impervious joints. Seal all crevices including washers, threaded fasteners and resistance welded joints.
(7) Keep spacing between conductors of different voltage potentials as wide as possible to avoid electrolysis and migration of silver.
(8) Threaded connections are not to be used in soft metals or plastics.
(9) Grounding of equipment is to be expertly engineered to avoid adverse galvanic and stray current effect on the structure and pipe systems.
(10) Plan the layout of cables to avoid corrosive areas or encase these in corrosion-proof conduits.
(11) The effect of welding, brazing and soldering temperatures on adjoining materials in the equipment will dictate selection of the jointing design.

8 Mechanics

8.1 Introduction

Any one of the known types of corrosion can lead to damage or breakdown of the mechanical integrity of the designed product; however, stress corrosion cracking, hydrogen damage, corrosion fatigue and fretting corrosion can result either in a critically sudden, vital and catastrophic breakdown of function or otherwise dangerously reduce the calculated strength of the design materials.

Their propagation is closely associated with the product's mechanical strength properties and so a solution to this threat gains in urgency and should be of considerable interest to the designers and to the whole corrosion-control team. Furthermore, the problems caused by the named types of corrosion are aggravated by the impossibility of timely detection and remedy—an insidious attack occurs largely inside metal or on hidden interfaces—and thus there is more or less only one effective remedy, that is to take appropriate steps at the design stage for preventive control.

Considering the corrosion/mechanical affinity of the project design, in particular its relation between the strength of materials and their stress loading under the given corrosive conditions, this appreciation should relate mostly to the tensile stresses (residual or externally applied) arising from the geometry of the component, stresses attributable to fabrication and assembly (including heat treatment and welding) and stresses caused by the operation. The mentioned stress loading can be either static or cyclic.

Other forces which can have an adverse effect on the corrosion of materials are those arising from vibration and fluttering, and last but not least the effect of shock should be considered.

Neither of the mentioned corrosion attacks has been ultimately defined by research and, where a critical design or materials are being considered, their suitability testing in a laboratory or as a pilot project is recommended.

Generally it may be said that, given the right environment, none of the metals or alloys used is completely free of the danger of stress corrosion except, perhaps, those in pure form. Some of the most susceptible alloys are those normally selected for highly loaded and critical applications, and it is known that present-day demands on the available strength of materials are supporting this trend. Many failures attributed to fatigue of metals, overloading or other physical causes are, in fact, caused by stress corrosion.

Non-metallic materials also suffer from phenomena similar to stress corrosion, e.g. the presence of moisture lowers the strength of glass, stressed plastics crack when exposed to specific organic solvents, etc.

The analysis of corrosion associated with mechanical strength will be naturally in very close relation to the appreciation of the prime engineering function and optimalisation of the designed product. Furthermore, one can qualify it as a functional analysis with a slant towards corrosion-control appreciation. Mechanical fault can initiate or aggravate corrosion incidence and corrosion *per se* can initiate or cause catastrophic failure.

Whilst an engineering product can fail due to stress, fatigue or friction in a benign corrosion environment, unless absolute perfection has been reached in design and fabrication through strict attention to the good practices of secure mechanical design, these optimal conditions are only very rarely obtained in general practice. It is truly advisable to pursue the sound policy of parallel appreciation of functional engineering and corrosion-control parameters by mutual consultancy to secure a safe product.

8.2 Scope

This part of analysis attempts to provide common lines of communication between the twin expert bodies of the design team—designers and corrosion specialists—in the field which up to now has been mostly claimed as the sole domain of the functional engineers. Whilst this subject is still very much in the melting pot, and due to the vast extent of knowledge involved in the mutual consultations, the following paragraphs claim only a reasonable chance to advise, indicate and initiate some of the possible ways and means to reach a common denominator between the designers and corrosion specialists in their endeavour to secure a safe design, and to assist either of the concerned parties in their recollection of the selective factors involved.

8.3 General

(1) Materials, stress level, environment, service temperature and design life are important parameters and should be considered in every design.

(2) Stress corrosion cracking is affected only by tensile stresses (residual or externally applied)—purely compressive stresses do not cause stress corrosion cracking. No stress corrosion cracking should occur at elastic stresses.

(3) Even where repetitive (cyclic) stresses can be met in relatively small doses, corrosion fatigue should be appreciated in relevant corrosive conditions and necessary design modifications made. Combination of fluctuating loads and stress raisers leads to corrosion fatigue.

(4) Limited friction, repeated relative motion or displacement between interfaces under load or subject to vibration can cause fretting corrosion. This corrosion does not occur on surfaces in continuous relative motion.

(5) Select correct materials (see Chapter 5). If possible, metals and alloys susceptible to stress corrosion or corrosion fatigue should not be specified

for highly loaded and critical structures and equipment in malignant corrosive environment.

(6) Preference should be given to materials which are resistant both to intergranular and stress corrosion, especially for applications involving residual and induced stresses. Alloys which are normally most resistant to intergranular corrosion are also more resistant to stress corrosion.

(7) Avoid metals subject to hydrogen embrittlement in critical structures and equipment.

(8) Avoid, where possible, high strength materials where corrosion fatigue may occur.

(9) Any selection of dissimilar metal couples, if absolutely necessary, should be confined to compatible couples in environments conducive to stress corrosion cracking, corrosion fatigue and fretting corrosion.

(10) Specify for improvement of ductility and impact strength.

(11) Within the requirements of the economic life of the product, specify for adequate control of heat treating and metal working processes to develop microstructure optimally resistant to specific environment.

(12) All bending, forming and shaping should preferably be performed on metal in an annealed condition and every effort made to use the lowest practicable stress level.

(13) Hydrogen embrittlement should be appreciated separately from stress corrosion when evaluating corrosion problems that can arise at cleaning, welding, treatment, cathodic protection and operating stages.

(14) Specify metal working, heat treating, flame and induction hardening, case hardening, carburising and nitriding (grain size refinement, metallurgical phase transformation, strain and dispersion hardening), whichever is required for increase of local strength or for improvement of fatigue strength or for introduction of compressive residual stresses into one or both of the rubbing surfaces.

(15) Carbide solution treatment of corrosion-resistant steels should be specified to minimise their sensitivity to intergranular corrosion.

(16) Specify suitable stress relieving (heat treatment, surface treatment, ultrasonic oscillators).

(17) None of the stress-relieving treatments can remove all internal stresses—it is the purpose of stress relieving to lower the stress to a level at which stress corrosion is less likely to occur.

(18) Select welding techniques that can produce sound welds. Defects (selective precipitation of phases, gas pockets, laps, undercutting, non-metallic inclusions, metallic alloying with prefabrication primers and other surface coatings, fissures and cracks) can act as sites of high residual tensile stress and thus lower the corrosion resistance.

(19) The chemical and metallurgical composition of welding rods should be compatible with the base metals, especially in the case of high strength metals.

(20) Select and specify appropriate welding rods and welding techniques that will not cause hydrogen embrittlement of high strength metals.

(21) If such alloys that do not show a detrimental change during welding cycles cannot be used in stressed structures or equipment, a post-weld heat treatment, stress relieving or baking is indicated.

(22) Careful and optimal preparation and finishing of welds for stressed structures and equipment is imperative.

(23) For prevention of stress corrosion cracking observe the precautions stipulated in Chapter 3, Section 3.16.

(24) Stress corrosion cracks grow in a plane perpendicular to the operating tensile stress, irrespective of its nature (applied or residual). Take appropriate precautions in design.

(25) Control the stress level by design. Time to failure depends on stress level, i.e. it tends to decrease rapidly as stress increases into a range of 50–90% of yield strength. Laboratory data, however, are not always reliable in practical conditions. Uncontrolled stress corrosion cracking could occur at stress levels considerably below the yield strength but active stresses would have to be great enough to cause some plastic strain (creep strain might be sufficient). *Note*: corrosion within stress cracks can develop pressure up to 1 ton/in^2 (1.6 kgf/mm^2).

(26) For prevention of hydrogen embrittlement observe precautions stipulated in Chapter 3, Section 3.10.

(27) For prevention of corrosion fatigue observe precautions stipulated in Chapter 3, Section 3.6.

(28) Balance strength and stress throughout the component (*Figure 8.1*).

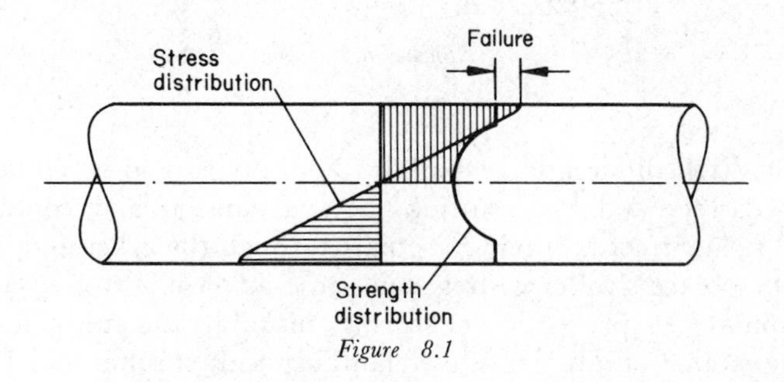

Figure 8.1

(29) Evaluate influence of stress distribution, for a given strength distribution, on the fatigue life of the product (*Figure 8.2*).

(30) Avoid weakening of load-bearing part at a point of maximum stress by notches, abrupt changes of section, sharp corners, grooves, keyways, oil-holes, screw threads, etc.

(31) Avoid deformation of materials round welds, rivets, bolt-holes, press fits or shrink fits.

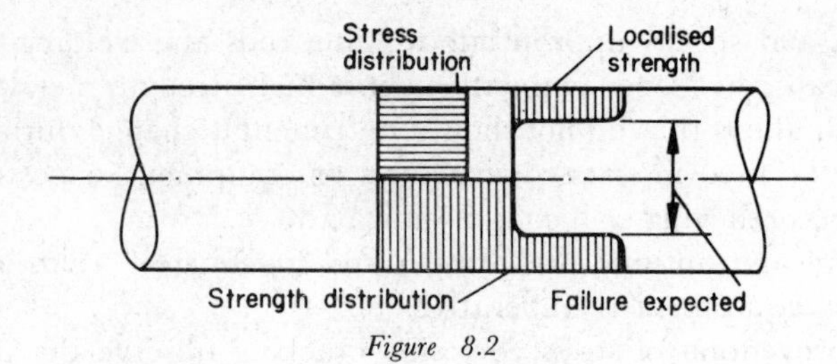

Figure 8.2

(32) Streamline fillets for various types of loading to obtain a decrease of stress concentration and to improve the stress flow (*Figure 8.3*).

(33) Where riveted joints and expanded-in tubes tend to be loosened by stress-relieving heat treatment, specify welding for attachment.

(34) For prevention of fretting corrosion observe precautions stipulated in Chapter 3, Section 3.7.

(35) Develop in design appropriate geometry, especially in critical sections (see Chapter 7).

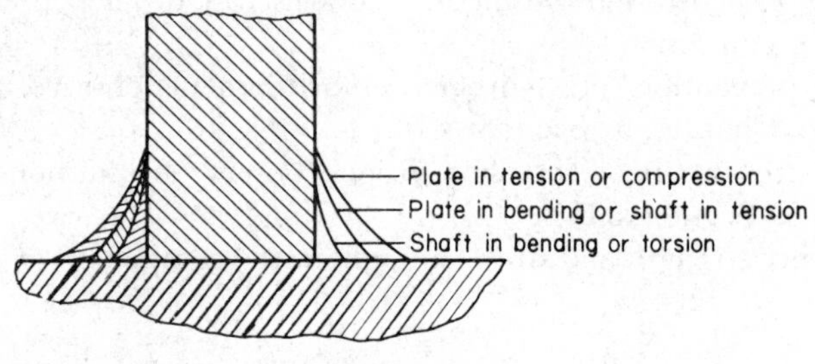

Figure 8.3

(36) Allow for differential expansion and pressure differentials. Select design for correct and exact fitting (note expansion and contraction of metals and strain creep). Forcing one part through the other and subjecting components to excessive local stress can cause adverse corrosive conditions.

(37) Stipulate for prevention of scoring, make arrangements for flushing of debris by motion of lubricants, use clamping pads of softer metal, etc.

(38) In the absence of a corrosive environment the magnitude of stress is considerably below the level causing damage. Identify corrodant (e.g. chlorides) and design for its removal, reduction or elimination even if its quantity is small, especially from the critical areas if possible. *Note*: stress corrosion cracking or corrosion fatigue may occur even in humid air or other mild corrosives.

(39) Prevent intermittent wetting and drying of critical surfaces if possible.

(40) Fatigue strength increases in vacuum or in inert atmospheres.

Oxygen and water vapour contribute to corrosion. Increase of intrinsic fatigue strength of material may not improve fatigue corrosion behaviour as much as an optimalised environment.

(41) Set design allowable stress that will minimise the rate of fatigue damage in service (*Figure 8.4*).

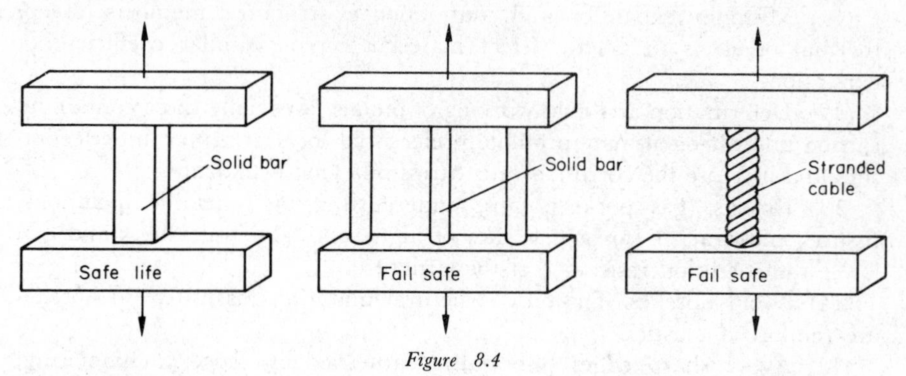

Figure 8.4

8.4 Structures

(1) Provide in design for sufficient flexibility of structures to prevent over-stressing by thermal expansion, vibration and working of the structures.

(2) Avoid riveted assemblies which can be subject to vibration.

(3) Stress analysis of complex structures by computer is recommended.

(4) Distribute stress with the metal's anisotropic characteristics in mind:

(*a*) either avoid exposing traverse planes or protect them;
(*b*) avoid or relieve residual stresses from quenching and fitting;
(*c*) select fabrication with special attention to the consequences of high localised stress pattern.

(5) Structural members in direct tension or compression are preferred to those subject to bending and torsion.

(6) Reduce the stress concentration factors in the structure as much as possible.

(7) Size and position the members within the structure to carry distributed loads. The smaller the member, the better it can distribute the stress.

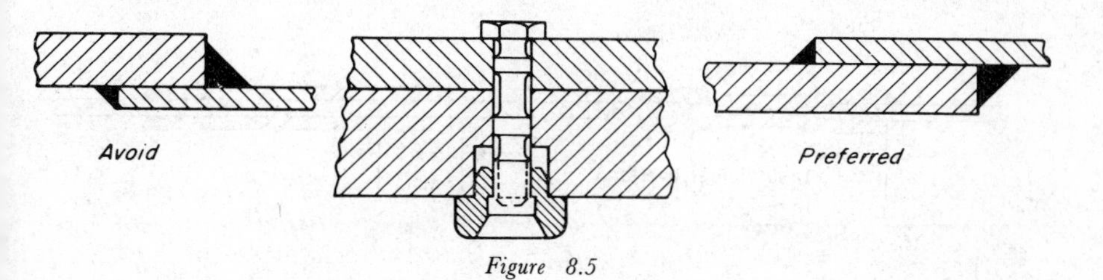

Figure 8.5

(8) Supplement strength by improved stress flow in changing the shape of critical sections.

(9) Provide generous fillets at internal and external corners.

(10) Balance the stiffness. Relative stiffness, where each member carries its share of load, improves the strength (*Figure 8.5*).

(11) Minimise expansion and contraction of structural members (creep, thermal or stress induced). Select materials having similar coefficient of expansion.

(12) Deformation and cold working of metals, especially those containing carbon and nitrogen, may promote preferential local attack at imperfection sites and increase the corrosion rate. Stress relieving is indicated.

(13) Defects (gas pockets, laps, undercutting, non-metallic inclusions, fissures and cracks) can act as sites of high residual tensile stress and can lower the corrosion resistance of the structure.

(14) Avoid notches. The only structural materials insensitive to notches are reinforced plastics.

(15) Avoid sharp edges (especially feather edges), specify chamfering, removal of burrs by grinding, milling or peening. Avoid sharp re-entrant corners.

(16) Select fabrication, machining and assembly operations imparting minimum residual stresses. Fillets should be streamlined if possible.

(17) Input stresses in structural reinforcements and adjacent parts of structure should be reduced as much as possible.

(18) A design allowing the exact assembly and fitting of individual members or units without undue stressing of one part by the other is preferred.

(19) Avoid misalignment of sections joined by riveting, bolting or welding.

(20) Parts penetrating or interfering with the main structure should withstand the same hydrostatic pressure and deformation loading as the main structure.

(21) The structures should be reinforced for loss of stiffness in the way of penetration.

(22) Lateral stiffeners should be as large as possible or practicable.

(23) Simple welded joints are preferred to those riveted or bolted for attachment subject to stress loading. Butt and fillet welding is preferred to lap or spot welding.

(24) Avoid insufficient overlap (*Figure 8.6*).

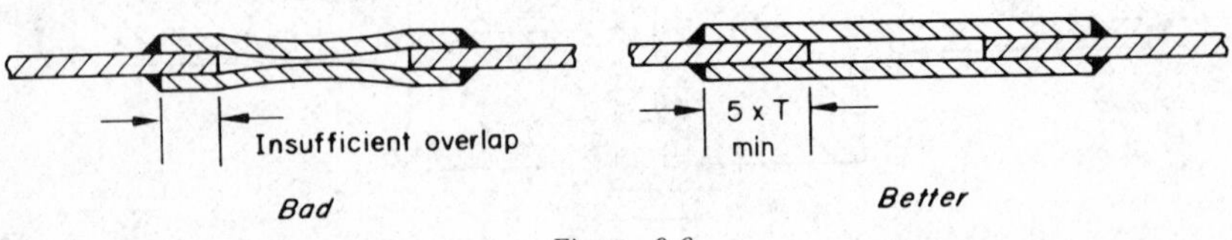

Figure 8.6

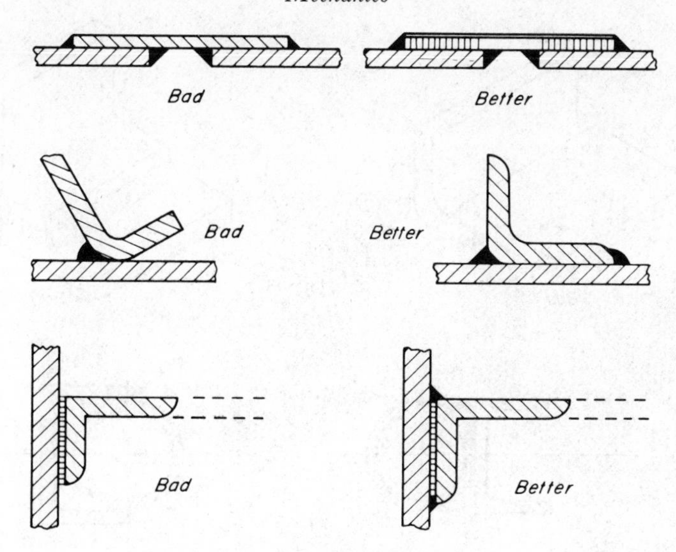

Figure 8.7

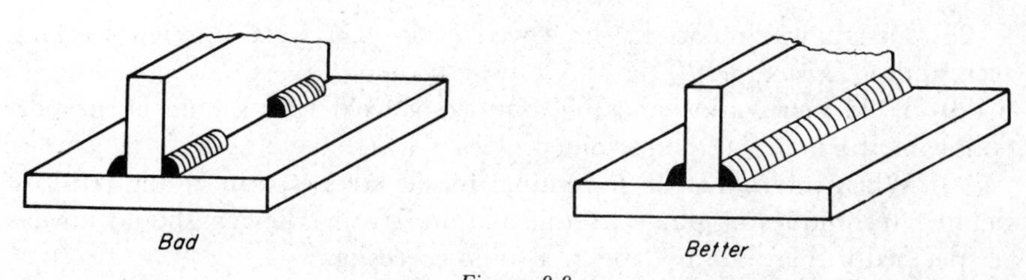

Figure 8.8

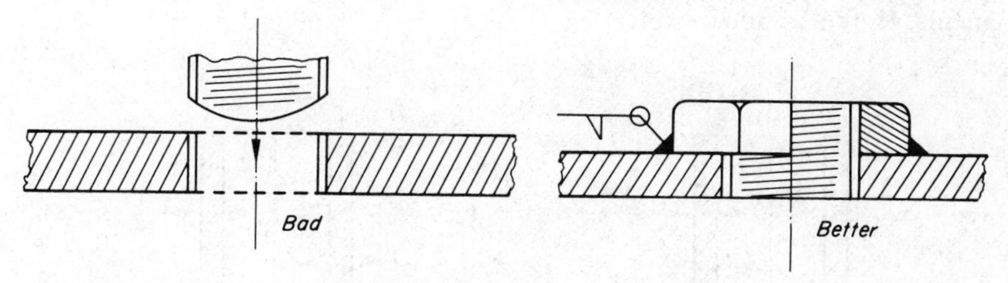

Figure 8.9

(25) Avoid incomplete welds (*Figure 8.7*).
(26) Avoid intermittent welds (*Figure 8.8*).
(27) Avoid tapped holes (*Figure 8.9*).

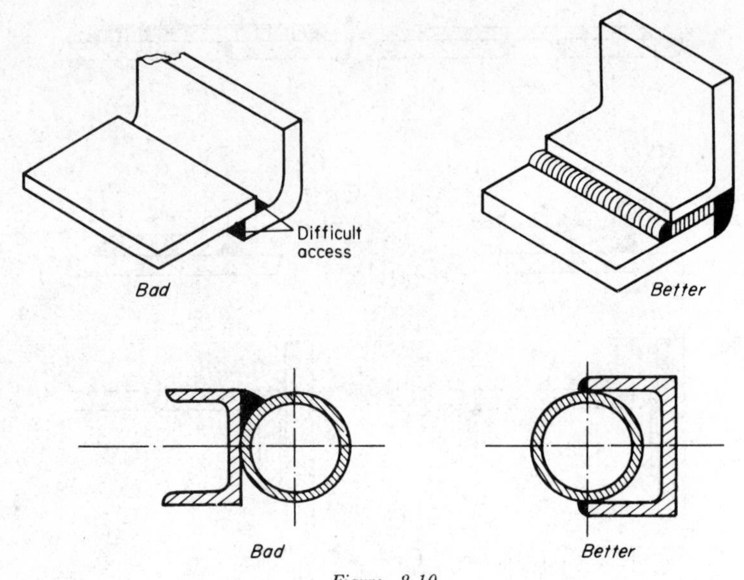

Figure 8.10

(28) Design weldments for improved quality of weld through working accessibility (*Figure 8.10*).

(29) Weld penetration and the number of weld runs should be proportionate to the materials being joined (*Figure 8.11*).

(30) When internal stresses (residual tensile stresses) can be the result of welding techniques or phase transformation, stress relieving should always be specified if there is any danger of stress corrosion.

(31) Evaluate preference of either safe-life or fail-safe design in a particular design.

(32) If preferable, select the optimum fail-safe geometry to fit the requirements of the applied stress:

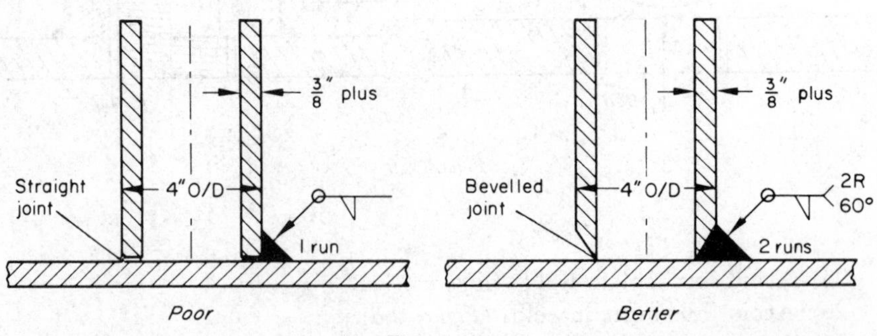

Figure 8.11

(*a*) Structures with multiple-load paths separated by spanwise discontinuities for unidirectional stress, with a load spectrum which includes some high peaks.

(*b*) Structures consisting probably of many elements (beware crevice corrosion), wide use of skin and flange doublers, high percentage of bonded joints and bolted or riveted joints designed to develop full compression in the skin for a biaxial stress (*Figure 8.12*).

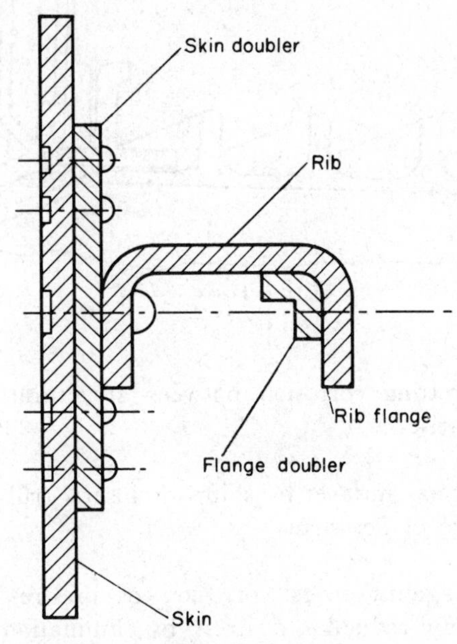

Figure 8.12

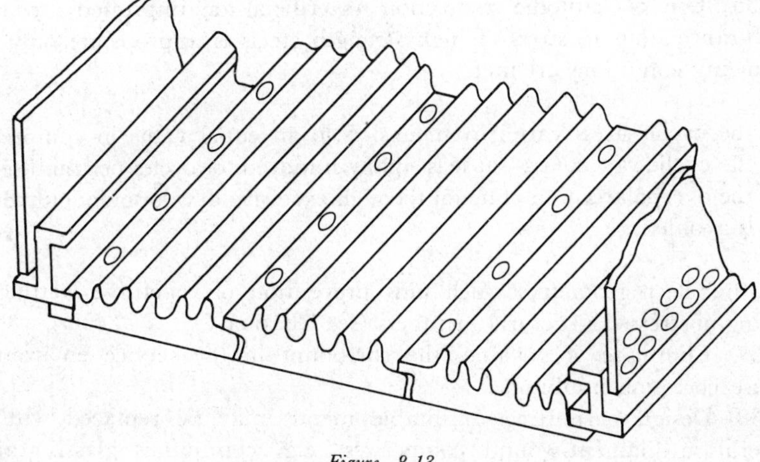

Figure 8.13

(*c*) Use of fail-safe structures for cases where multiple-load paths create a weight penalty and loss of inspectability is not recommended in corrosive conditions.
(*d*) Typical fail-safe construction for tension surfaces (*Figure 8.13*).
(*e*) Fail-safe shear web design (*Figure 8.14*).

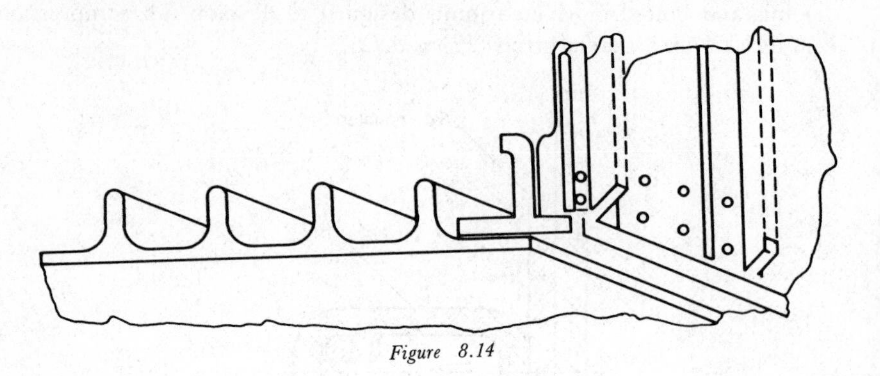

Figure 8.14

(33) Where fretting corrosion between structural members subject to vibration could arise:

(*a*) separate rubbing surfaces by shims or inserts (rubber, plastics);
(*b*) design for use of flex arms.

(34) Provide against stress corrosion of prestressed reinforcement in concrete, by careful reduction of stress, by elimination of corrosion through good concreting practice and by appropriate protection of embedded steel. *Note*: protect reinforcement cables awaiting full stressing and grouting.

(35) Use of cathodic protection (sacrificial or impressed) to restore endurance limit in stress of high strength steels is appropriate only if the following conditions are met:

(*a*) the cyclic stress varies from tension to an equal value in compression;
(*b*) the cyclic rate is fast—at least many hundreds of cycles per minute;
(*c*) the overpolarisation of metal through excessive development of hydrogen is avoided.

(36) Select geometry which aids prevention of repetitive wetting and drying of stressed structural members (see Chapter 7).

(37) Eliminate, if possible, the corrodant in the service environment, or use corrosion inhibitors.

(38) Design permitting unsuitable metals may be replaced with new generation filament-wound composites (e.g. continuous glass, graphite,

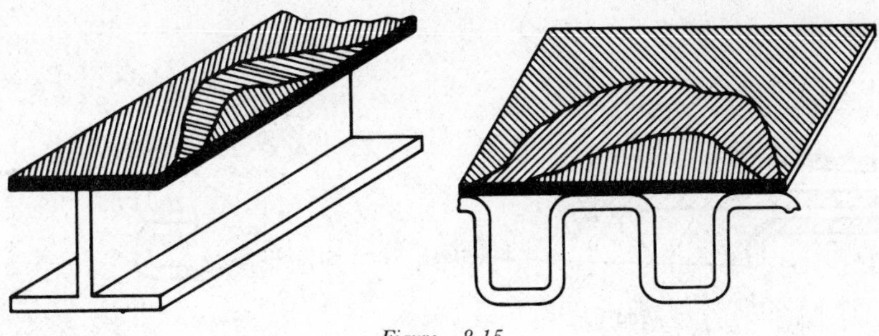

Figure 8.15

boron, beryllium, titanium alloy, steel, carbon, silicone filament or strip—unidirectional, bidirectional, multidirectional) (*Figure 8.15*).

(39) Avoid, in corrosive environment, initiation of stress raisers by excessive thermal shock (e.g. ambient to cryogenic temperatures) (*Figure 8.16*).

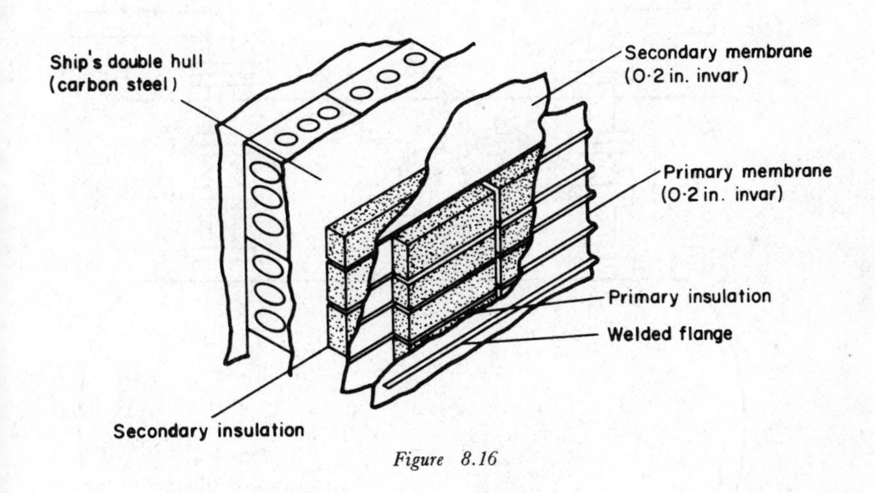

Figure 8.16

8.5 Equipment

(1) Machinery and equipment in a corrosion-prone environment should be mounted on seatings as stiff as functionally possible, with differing resonant frequency from the forcing frequencies initiated by the machine or equipment.

(2) Where the equipment is mounted on tubular seating, the seating should be in tension or compression.

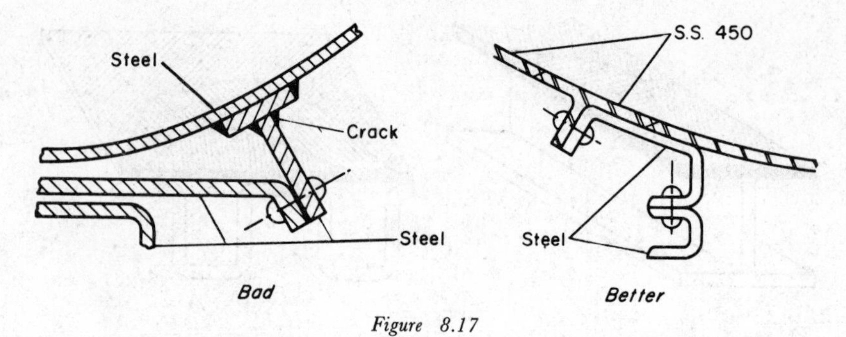

Figure 8.17

(3) Provide in design of equipment supports for sufficient flexibility and reduce stress concentration (*Figure 8.17*).

(4) Equipment subject to corrosive conditions should not be attached rigidly to both of two structures which can deflect relative to each other under shock loading (*Figure 8.18*).

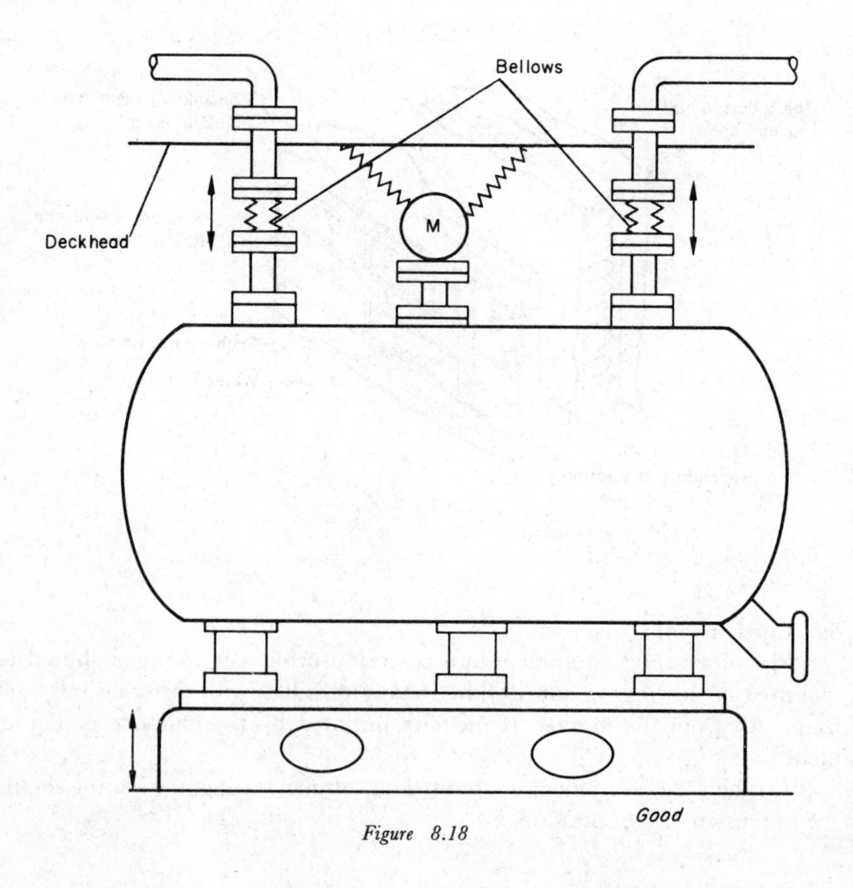

Figure 8.18

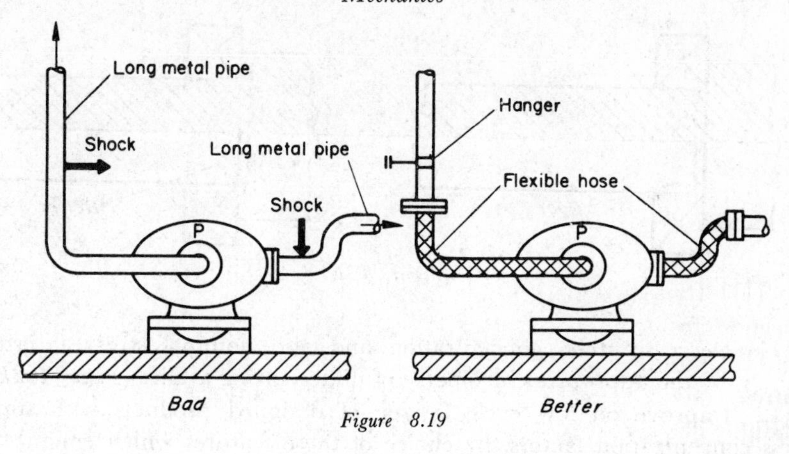

Figure 8.19

(5) Equipment subject to corrosive conditions should not be attached rigidly to both of two pipe systems, electrical conductors or ventilation ductings which can deflect relative to each other under shock loading (*Figure 8.19*).

(6) In high speed, high performance equipment subject to corrosion and resonance fatigue failure all component members, parts or groups should be considered together as one assembly, for prevention of bending stresses due to lateral vibration. The required lateral stiffeners should be as large as practicable.

(7) Maximum reliability of equipment is attained when all components have the same factor of safety, whatever their modes of fatigue.

(8) Distribute local stresses more uniformly within the critical equipment in the vicinity of stress raisers.

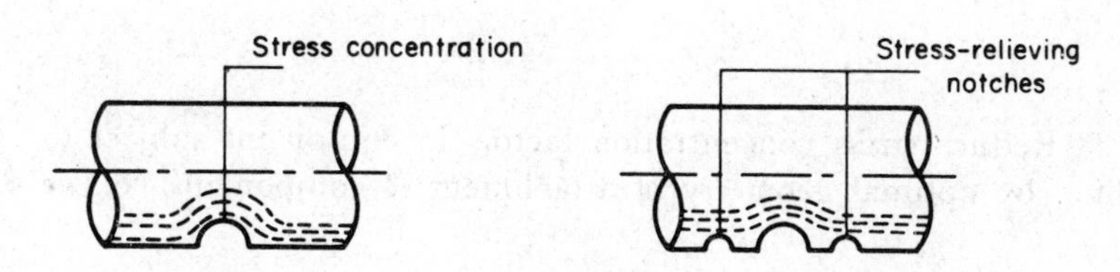

Figure 8.20

(9) Improve stress flow (*Figure 8.20*).

(10) Balance between the increase in size of critical sections and all other fatigue design parameters.

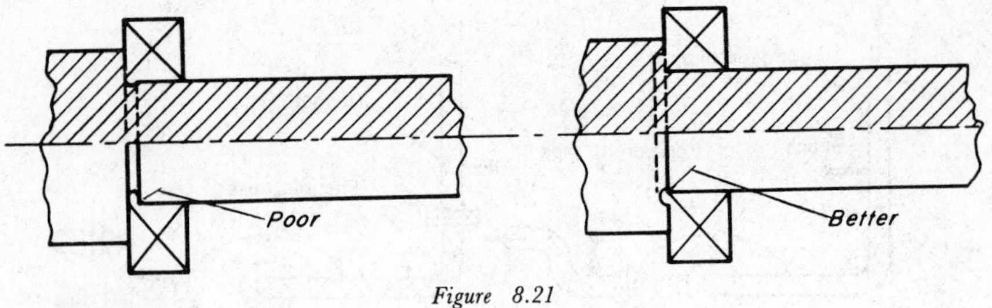

Figure 8.21

(11) Decrease stress concentration and also nominal stress at critical section, by the appropriate geometry of undercutting a fillet (*Figure 8.21*).

(12) Improve on low cycle fatigue of designed products with similar stress concentration factors, by choice of those features which enhance the fatigue life (*Figure 8.22*).

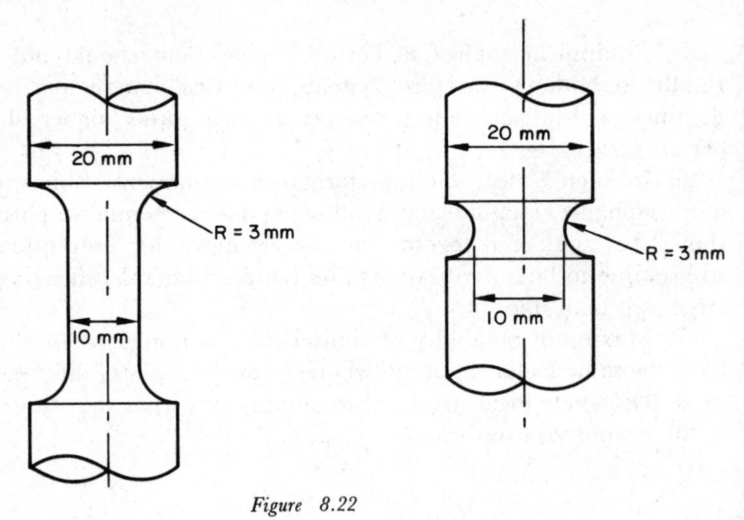

Figure 8.22

(13) Reduce stress concentration factors in equipment subject to cyclic loading, by optimal geometry of attachment of components (*Figure 8.23*).

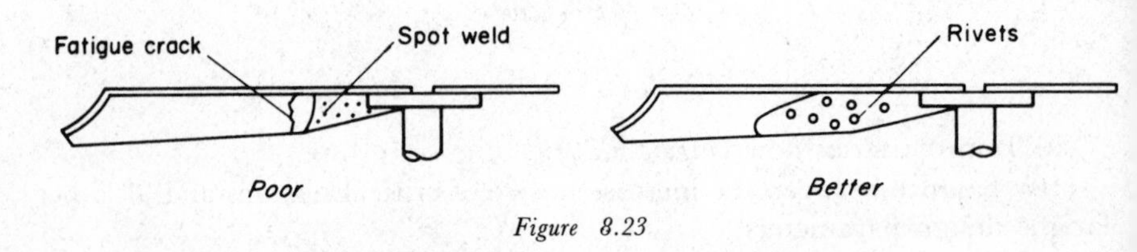

Figure 8.23

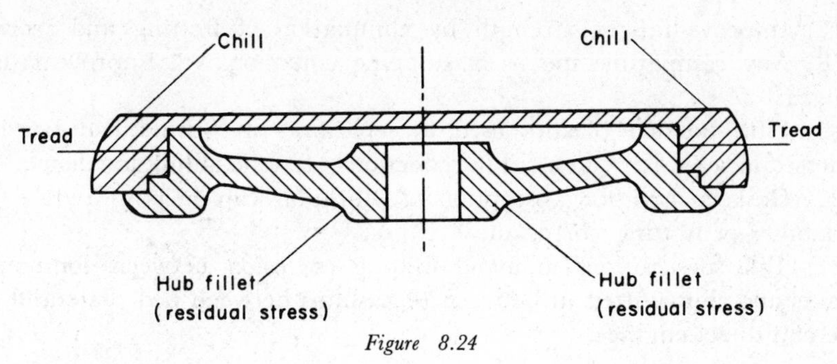

Figure 8.24

(14) Avoid residual stresses initiated by thermal effects causing non-uniform volume changes within a part in production (*Figure 8.24*).

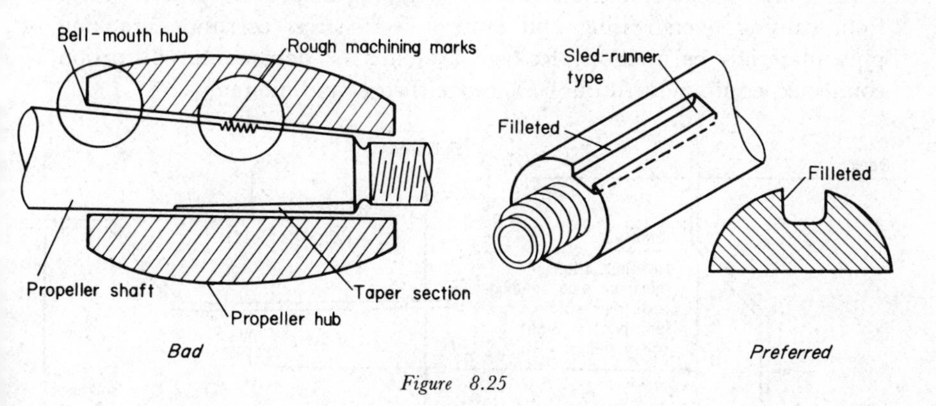

Figure 8.25

(15) Specify exact machining, breaking sharp corners, filleting interior corners and provision of sled-runner type ends to keyways to reduce stresses (*Figure 8.25*).

(16) Avoid cumulative effect of operational stresses additive to the existing residual stresses (*Figure 8.26*).

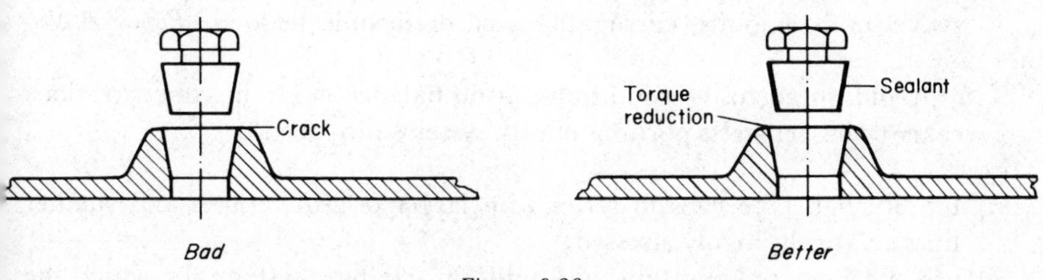

Figure 8.26

(17) Improve fatigue strength by elimination of fretting and scoring.

(18) Any compatible means of stopping corrosion will improve fatigue strength.

(19) Fibre-bonded plastics used as separators in bushes and bearings immersed in sea water can assist in reduction of fatigue failure incidence.

(20) Gaskets used for absorption of vibration can help to reduce the probability of fretting corrosion.

(21) Take precautions to avoid fretting corrosion between component surfaces and shims fitted in between (e.g. shims between bed plate and top plate of a diesel engine).

8.6 Piping Systems

(1) Piping systems can be adversely affected by thermal expansion, shock, vibration and working of the structures.

(2) Provide for sufficient flexibility of piping to prevent pipe movements from causing overstressing and failures from stress corrosion cracking of pipe materials or anchors, leakage at joints or detrimental distortion of connected equipment through excessive thrusts and moments:

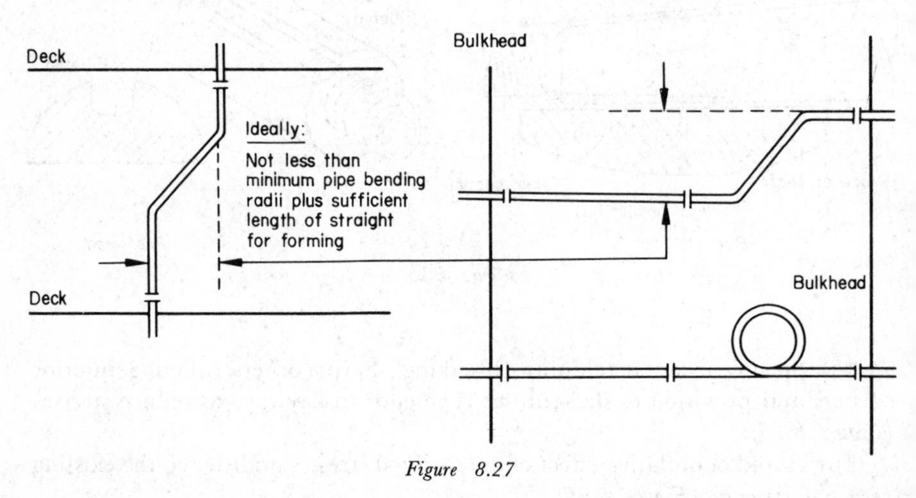

Figure 8.27

(*a*) change direction through use of bends, loops or offsets (*Figure 8.27*).

(*b*) provide for absorption of thermal movements by utilising expansion, swivel or ball joints, corrugated pipe or flexible bellows (*Figure 8.28*).

(3) Avoid, in corrosive conditions, an imbalance in strain concentrations of weaker or higher stress portions of pipe systems produced by:

(*a*) use of small pipe runs in series with larger or stiffer pipes and smaller lines relatively highly stressed;

(*b*) use of a line configuration, in a uniform size pipe system, for which the

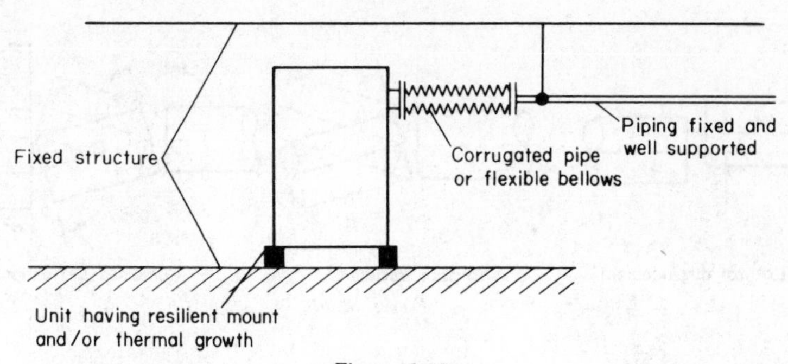

Figure 8.28

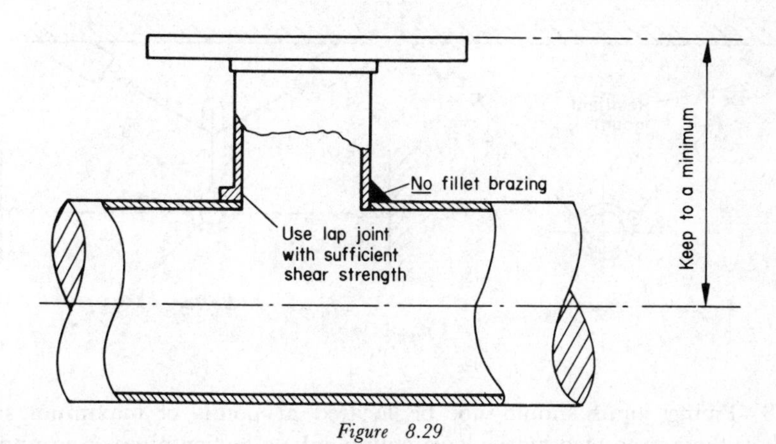

Figure 8.29

neutral axis or thrust line is situated close to the major portion of the line itself, with only a very small offset portion of the line absorbing most of the expansion strain;

(*c*) local reduction in size or cross-section or local use of weaker materials (*Figure 8.29*).

(4) Where expansion joints are subject to combination of longitudinal and transverse movements, both movements should be considered (*Figure 8.30*).

(5) Anchors, guides, pivots and restraints should be designed to permit the piping to expand and contract freely in directions away from the anchored or guided point.

(6) Hanger rods and straps should allow free movement of piping caused by thermal expansion and contraction and physical working of the supporting structure.

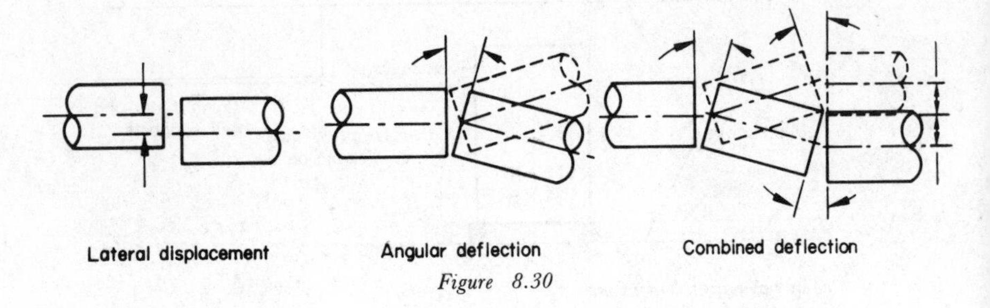

Figure 8.30

(7) Sway braces or vibration dampeners should be used to control the movement of piping due to vibration (*Figure 8.31*).

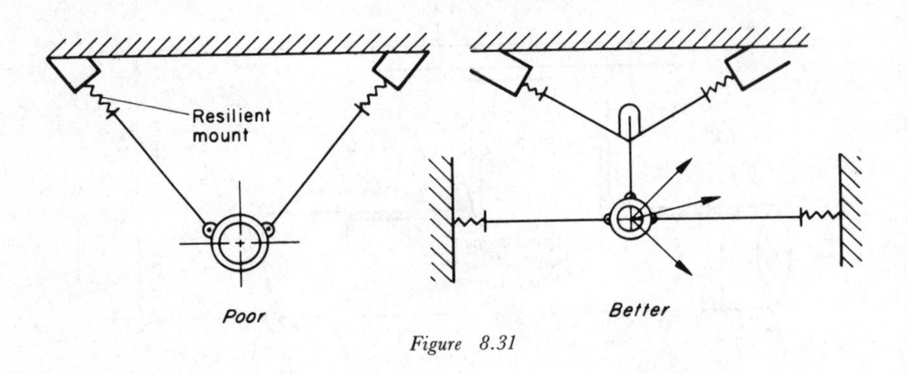

Figure 8.31

(8) Piping joints should not be located at points of maximum stress, such as those produced by the lever action of long flexing pipes or equipment.

(9) Where critical stresses are expected an appropriate geometry of pipe fittings should be selected. If such fittings are not available these areas should be adequately reinforced (*Figure 8.32*).

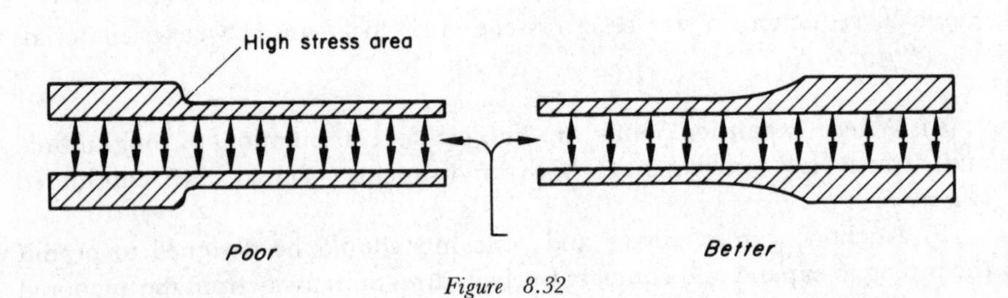

Figure 8.32

(10) Take-off connections should withstand all stresses of the piping system, including those induced by cyclic loading (*Figure 8.33*).

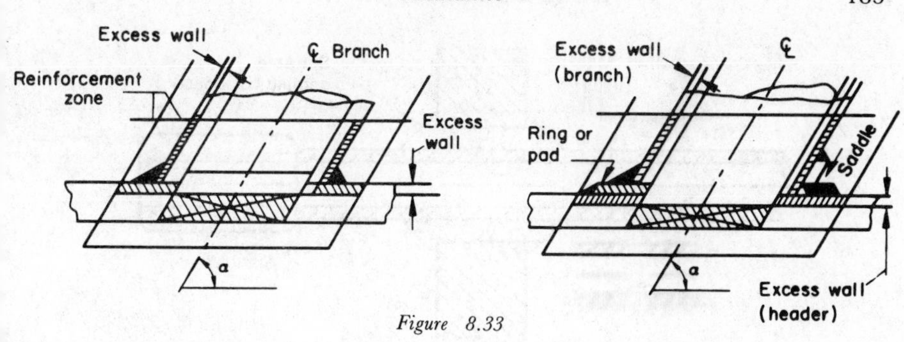

Figure 8.33

(11) Thermal shock to steam lines by contact with cold condensate return lines should be prevented by either lagging in take-off connections with steam main or a lengthwise metallic contact should be provided between the two mentioned parts (*Figure 8.34*).

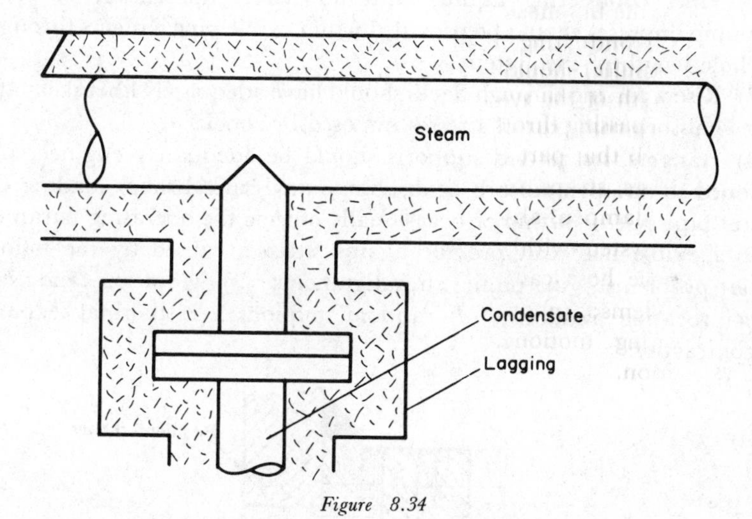

Figure 8.34

(12) Piping should be horizontally (by preference) offset between successive bulkhead penetrations by two bends of not less than 45 degrees. The offset distance of not less than four pipe diameters or 12 in (30.5 cm), whichever is greater, should be provided.

(13) Round or oval ducts are stronger and stiffer than the rectangular ones and therefore more effective in reducing vibration stresses.

(14) A pulsating pipe penetrating a non-watertight bulkhead should be

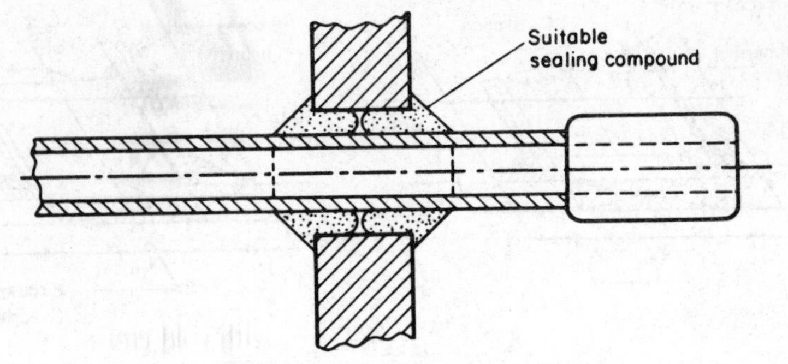

Figure 8.35

passed through an oversize cut hole—0.5 in (1.3 cm) oversize—and the clearance sealed with a sealing compound (*Figure 8.35*).

(15) A pulsating pipe penetrating a watertight bulkhead should be designed for bolting a resilient rubber, together with gasket, into a close-fitting hole in the bulkhead (*Figure 8.36*).

(16) Pipes conducting liquids with noticeable fluctuation of pressure (e.g. pump impulse) should be provided with flexible pipe hangers throughout the whole length of the system.

(17) Risers passing through decks should have adequately liberal expansion bends to absorb that part of stresses imposed by shock.

(18) Hangers, straps and supports should be adequately engineered and positioned to dampen, absorb or distribute any critical shock loading of the relevant pipe system within or occasionally outside the operating parameters.

(19) Flexible hose can provide against stresses caused by the following motion problems: (*a*) piping misalignment; (*b*) vibration and shock; (*c*) reciprocating motions; (*d*) random motions; (*e*) thermal expansion and contraction.

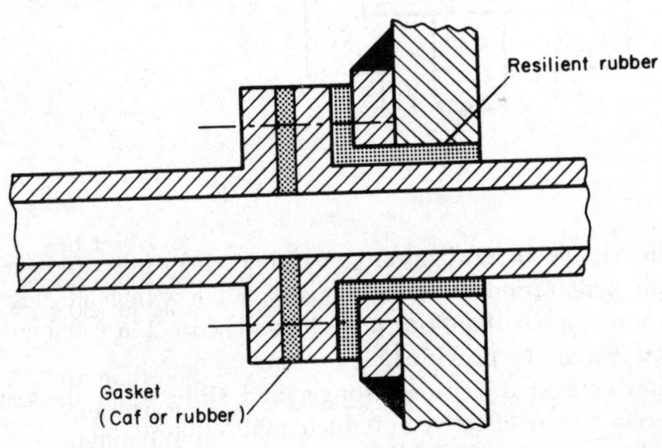

Figure 8.36

(20) Avoid sharp bends on flexible hose in corrosive conditions (*Figure 8.37*).

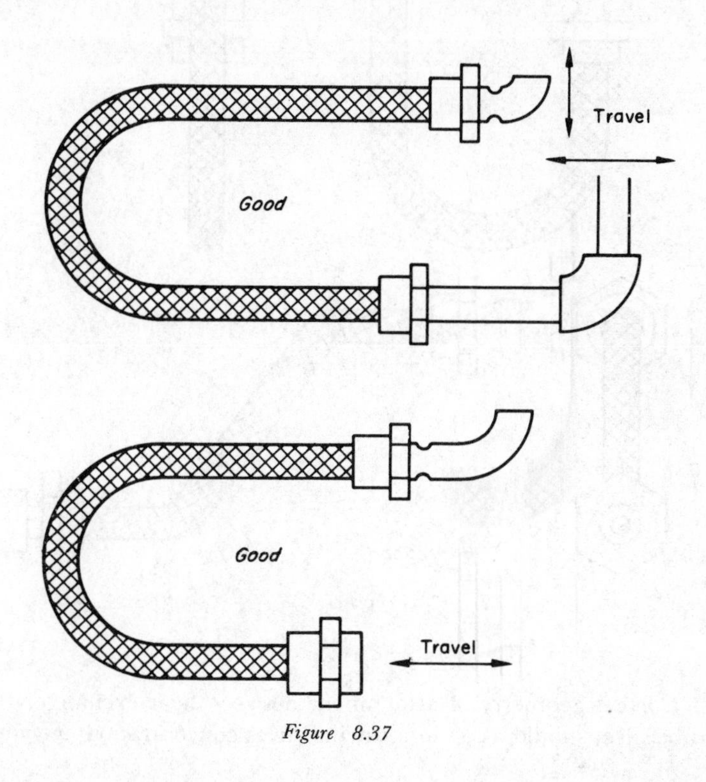

Figure 8.37

(21) Avoid subjecting flexible hose to torque by twisting on installation or on flexure (specify) in corrosive conditions (*Figure 8.38*).

(22) Snake underground plastic pipe in the trench to compensate for expansion and contraction.

(23) Where a valve installed in a continuous piping system is large and heavy compared to the piping itself, it is acceptable to support the valve by securing the piping adjacent to the valve.

(24) Regulating valves which project 1–2 ft (30–60 cm) from the pipe system in which they are installed should be supported to cater for athwartship shock stresses.

(25) Valves located at the end of a pipe should be supported by the valve flange vertically and athwartship to the nearest beam of the structure.

(26) Sea valves should be built into a corner between frame and longitudinal.

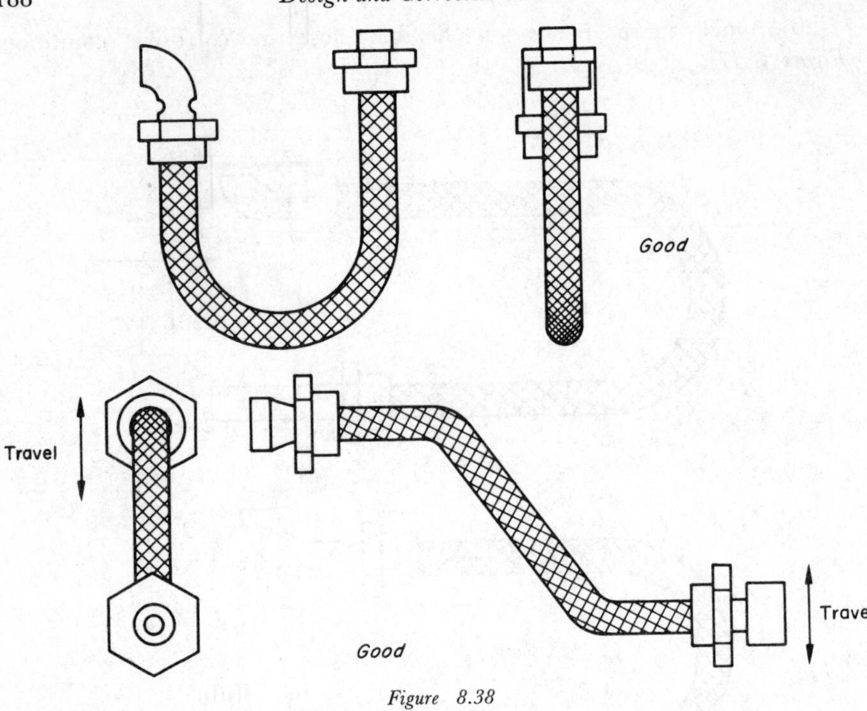

Figure 8.38

(27) Correct geometry of attachment between heat exchanger tubes and their tube sheet should assist in reducing stress concentration (*Figure 8.39*).

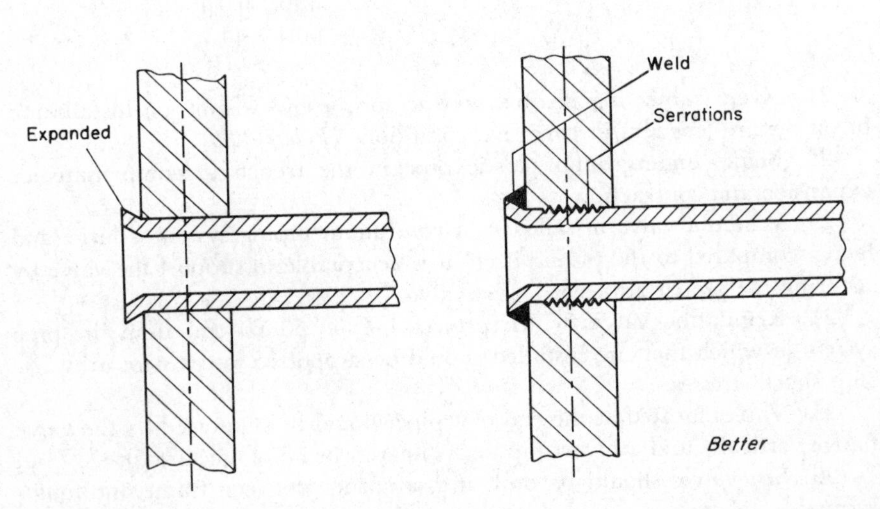

Figure 8.39

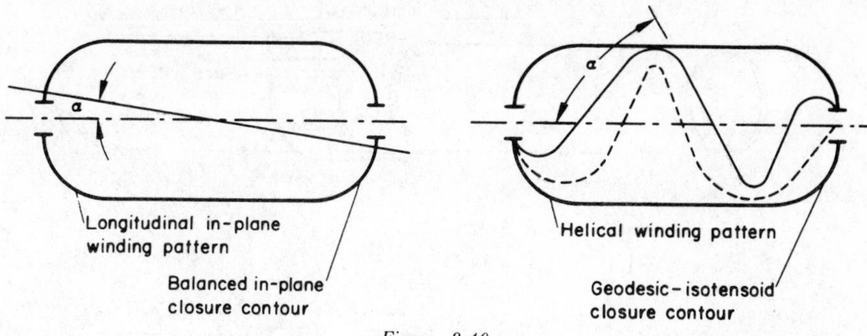

Figure 8.40

(28) Where inhibitors are used for reduction of stress corrosion cracking in closed piping circuits, adequate arrangement for replenishment of the inhibitor should be made.

(29) For design of extensive pipe systems in corrosive conditions their stress analysis by computer is recommended.

(30) Design permitting critical metallic pipes may be replaced by new generation suitable filament-wound composite pipes.

(31) Design allowing critical metallic pressure vessels may be replaced with suitable filament-wound composite vessels (*Figure 8.40*).

(32) Limit the low temperature or stress corrosion cracking of pipelines by crack arresters (*Figures 8.41–8.43*).

8.7 Vibration Transfer

(1) To minimise resonance corrosion fatigue, reduce vibration and fluttering on stressed structures or equipment in corrosive environment:

(*a*) By vibration isolation (*Figure 8.44*).
(*b*) By vibration absorbers (*Figure 8.45*).
(*c*) By vibration damping—design of mounting (e.g. sand-filled columns, etc.); inclusion of splinter silencers; lining with absorbent materials; application of damping coatings.
(*d*) By reduction of excitation magnitude—change of frequency (i.e. increase of natural frequency for reduction of resonance corrosion fatigue); regulating the stiffness of structures (e.g. increase of the amount of inertia of cross-section by using beads, ribs and flanges, using 'I', round or square hollow sections, etc.); modifying the mounting conditions by using angle braces to simulate built-in supports, rather than using the simple (pivotal) supports.
(*e*) By redistribution of mass.
(*f*) By reduction of effective length of a member by mounting struts parallel to the direction of vibratory motion.

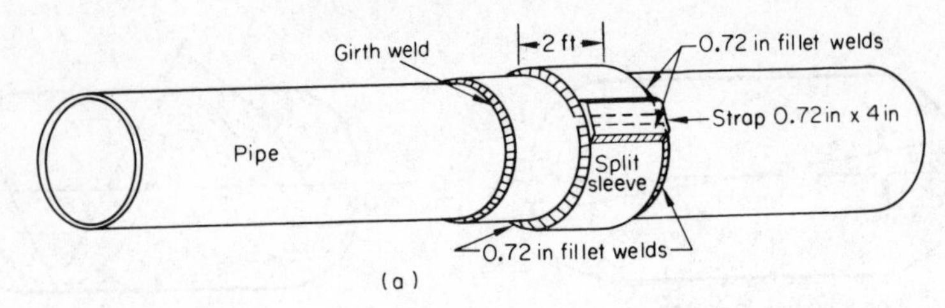

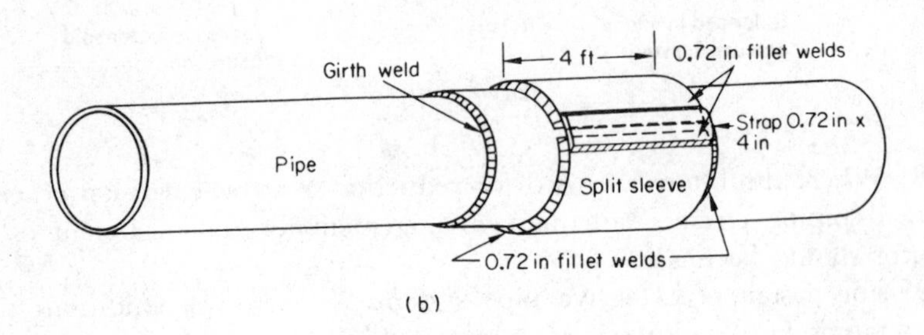

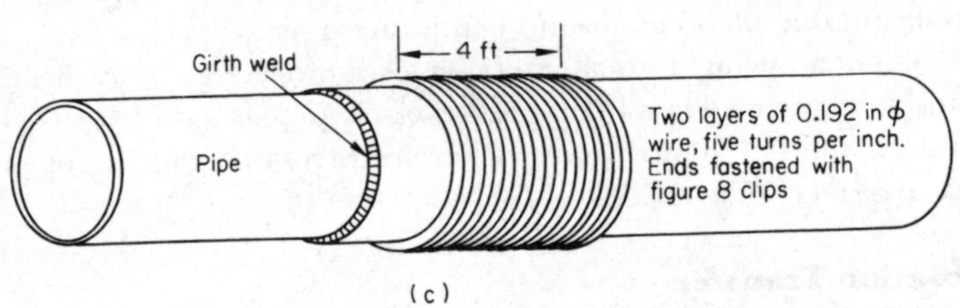

Figure 8.41. Crack arresters: (a) 2 ft split sleeve; (b) 4 ft split sleeve; (c) wire-wrapped pipe. See also Figures 8.42 and 8.43

(2) Vibration of equipment should be reduced or eliminated at its source.

(3) Ventilation trunking should be so routed that compartments with higher difference of resonance are not directly connected.

(4) Provide for installation of acoustic hoods where required.

(5) Specify pourable chocking compounds to reduce vibration transfer and improve on alignment of equipment seating and structural joints (*Figure 8.46*).

(6) Avoid cavitation fatigue in engine cooling systems:

(*a*) investigate and check for probable focal points of vibration in vicinity of vital components;
(*b*) investigate resonant frequency of the specified materials;
(*c*) select components made in material of higher fatigue resistance and with ability to work harden in cold-working action caused by cavitation;
(*d*) reduce dispersed air contents in fluid (bubbles 50 μm dia.);

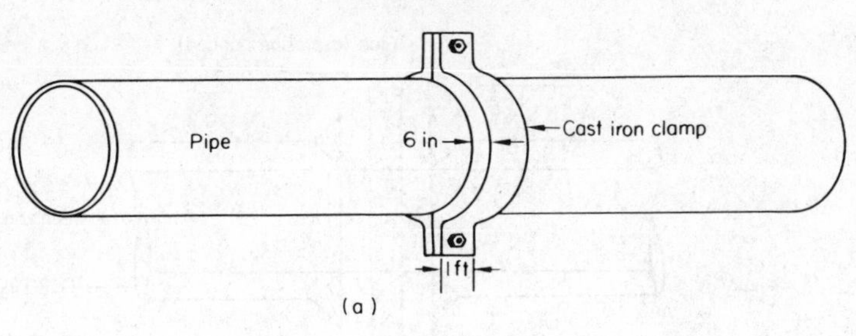

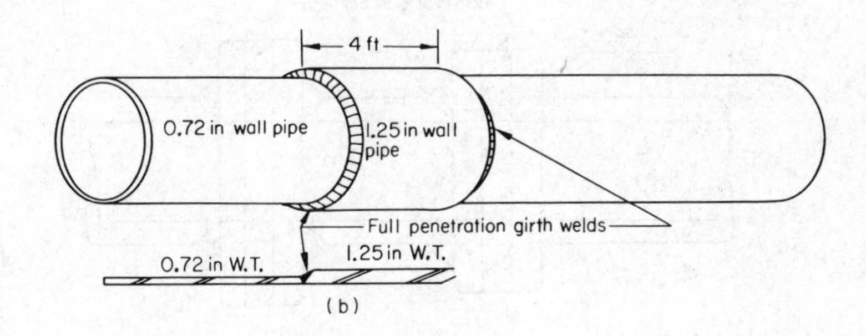

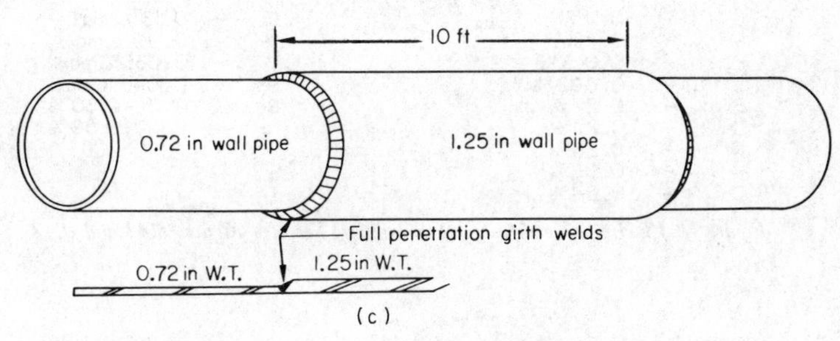

Figure 8.42. Crack arresters: (a) cast iron clamp; (b) 4 ft heavy wall pipe; (c) 10 ft heavy wall pipe. See also Figures 8.41 and 8.43

(*e*) inject or generate within the system larger size air or inert gas bubbles to buffer the mechanical cavitation process;

(*f*) prevent contamination of fluid by cathodic metals and corrosive agents (e.g. chlorides);

(*g*) inhibit the fluids and eventually use oxygen scavengers.

8.8 Surface Treatments

(1) Specify uniform and, in critical areas, top grade cleaning of surface (see Chapter 9).

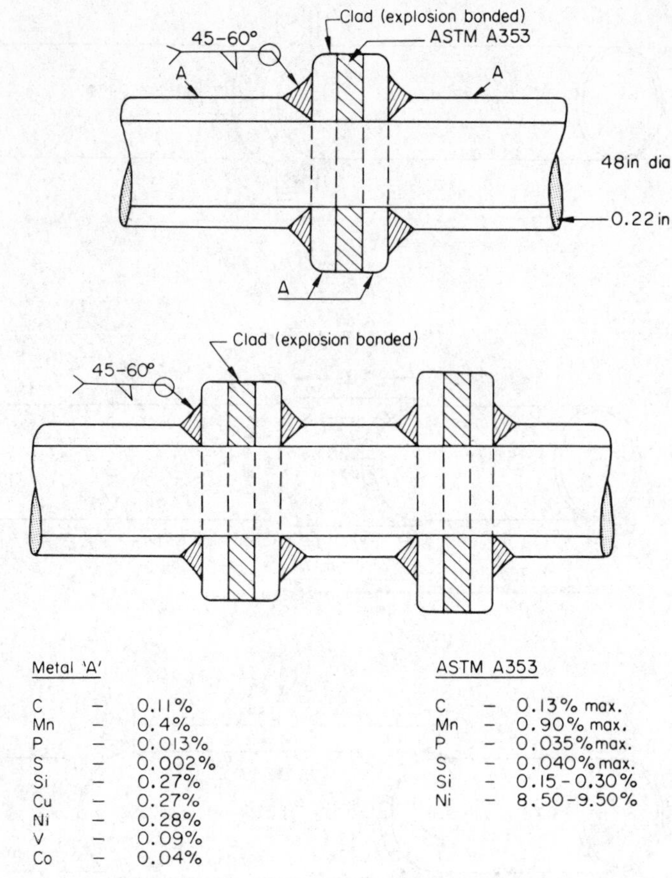

Figure 8.43. Preferred type of crack arrester. See also Figures 8.41 and 8.42

(2) Specify removal of oxidised, contaminated or decarburised surface layers.

(3) High strength steels should not be acid cleaned (except anodically) nor cathodically cleaned in alkaline bath. Select cleaning method which does not interfere with mechanical strength of a particular material in a given environment.

(4) Specify for avoidance of deep surface finish marks in production (or select appropriate fabrication technique) to avoid formation of stress raisers.

(5) To improve fatigue strength specify machine finishing with moderately light cut, gentle grinding, abrasive tumbling, etc.

(6) Reduce mean stresses by specifying input of compressive residual stresses at the surface of a component by work hardening (i.e. by shot peening

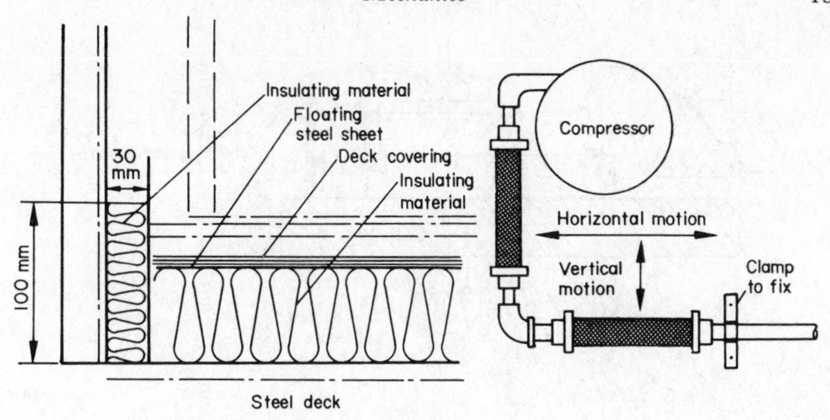

Figure 8.44

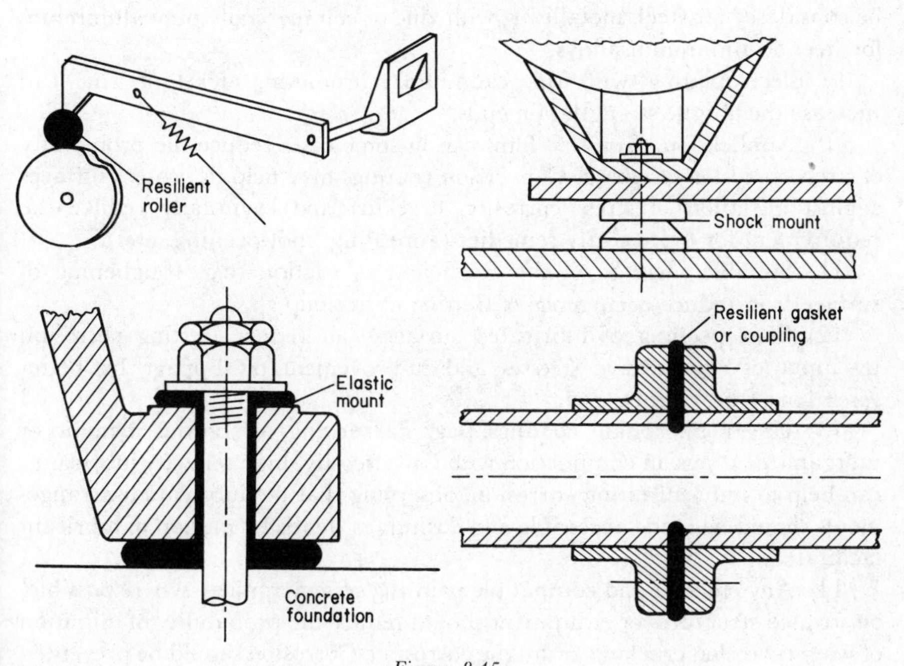

Figure 8.45

of stress concentrators and surfaces, by rolling of fillets, grooves and other surfaces, by vapour blasting, tumbling, burnishing and chemical peening).

(7) Specify for application of surface finishes and coatings, by techniques that do not produce tensile stresses nor cause hydrogen embrittlement.

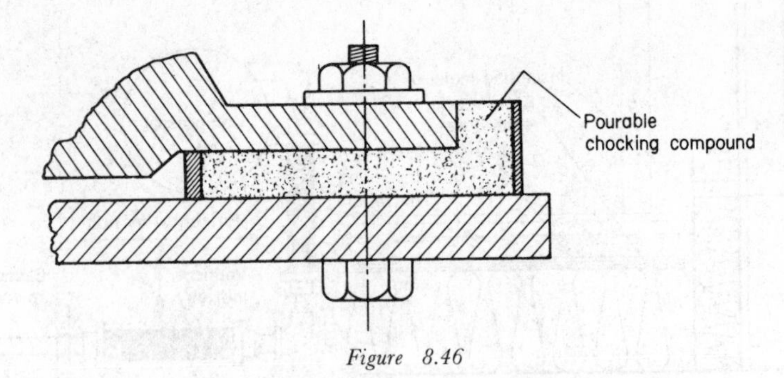

Figure 8.46

(8) Metal deposition (vacuum deposition, mechanical plating, metal spraying or electroplating in low hydrogen-producing plating baths) of stressed areas enhances mechanical strength of metals. Zinc deposition can be considered for steel, metallising with zinc or commercially pure aluminium for steel or aluminium alloys.

(9) Electroplating with tin, cadmium, chromium, nickel or zinc can increase the fatigue strength of metals.

(10) Application of passive films can in some cases reduce the probability of stress corrosion cracking. Conversion coatings may help to protect surfaces against initiation of stress corrosion cracking and eventually reduce the requirement for more costly remedies (annealing, shot peening, etc.).

(11) Suitable increase of the coefficient of friction (e.g. roughening of surface) can reduce occurrence of fretting corrosion.

(12) Blast peening of lubricated surfaces can reduce fretting corrosion by input of compressive stresses and improvement of slippage by better retention of lubricants.

(13) Use of phosphate coatings (e.g. Parkerising) or porous metallic or inorganic coatings, in conjunction with low viscosity, high tenacity lubricants, can help to reduce fretting corrosion, observing that the lubrication arrangements should be made accessible and flushing of debris by motion of lubricant facilitated.

(14) Any efficient and compatible painting system applied, where possible, on stressed structures or equipment should reduce the probability of initiation of stress corrosion cracking or fatigue corrosion. Corrosion should be prevented in all critically stressed components by all available means, including surface coatings.

(15) Coating the surfaces with organic coatings after case hardening, mechanical work hardening or metallising brings about improvement in resistance to stress corrosion cracking and in fatigue strength.

(16) Priming with a chromate primer containing not less than 20% zinc chromate should be specified for all fully heat treated alloys.

(17) The use of metallic, inorganic or organic coatings and linings in steel vessels where hydrogen embrittlement can occur is conditionally recommended, provided these vessels (or structures) are not fabricated of high strength steels, the structures are not under high stress loading and the coating does not contain reactive zinc or other metal which under specific environmental conditions could react electrochemically whilst development of gaseous hydrogen takes place.

(18) Steel, clad with austenitic stainless steel or nickel, can also be specified in an environment promoting hydrogen enbrittlement.

(19) Addition of selective inhibitors to the relevant surface environment can reduce the probability of stress corrosion, corrosion fatigue and fretting corrosion (see Chapter 10).

(20) The use of wide radii bends in corners of components for hot dip galvanising is recommended—this minimises local stress concentration.

(21) Whilst continuous sealed welds are preferred for hot dip galvanised components, whenever these are not practical staggered welding techniques should be specified to reduce thermal stresses.

(22) The assemblies which are to be galvanised should be preformed accurately to avoid using force to bring them into position.

(23) Welds should be stress relieved before galvanising.

8.9 Electrical and Electronic Equipment

(1) Select materials resistant to intergranular corrosion and stress corrosion cracking, where residual and induced stresses could affect the safe function of the equipment.

(2) Where metals are to be bent, formed or shaped, materials which are in an annealed condition should be used.

(3) Avoid, where necessary, metals subject to hydrogen embrittlement from acid cleaning or plating, or use low hydrogen-producing plating baths.

(4) Specify relieving of embrittlement immediately after plating for a minimum of three hours at 190° ±14°C.

(5) Specify mechanical stress relieve of parts prior to plating (shot peening).

(6) Specify appropriate preservation with organic coatings, vacuum deposition, mechanical plating, metal spraying or other processes not producing hydrogen; this in preference to electroplating or chemical plating where possible.

(7) Support lighting fixtures on resilient mounts where possible.

(8) Avoid rigid attachment of electrical equipment subject to corrosive conditions which can deflect relative to the conductors, whilst such equipment can vibrate or be exposed to shock loading.

9 Surface

9.1 Introduction

The optimal configuration, cleanliness, preparation, texture and pretreatment of internal and external surfaces and their electrical or electrochemical stability, in any of the expected environmental conditions, can considerably enhance the effectiveness of rationalised corrosion control in design. Furthermore, considering that corrosion usually originates at the surface, the establishment of appropriate and definitive surface parameters at the design stage should merit the high priority of concern it is prudent to adopt.

Although the close affinity of the product's geometric form to its individual surfaces is fully recognised, one may easily distinguish between the two-dimensional accent of the variegated surfaces and the three-dimensional involvement of the solid geometry, both applied to the common good, i.e. furthering of effective corrosion control contained in the complete design of the product.

9.2 Scope

This chapter indicates the relations between the chosen materials of the substrate and their geometry, on the one hand, and its logic input of optimum local surface conditions on the other. These surface conditions are being developed for the benefit both of functional fulfilment and corrosion control, and close co-operation between design engineers, corrosion specialists and technical experts in individual fields of surface-treatment technology is highly recommended.

9.3 General

(1) Simple, compact, smooth surfaces, optimally shaped, positioned and angled are preferred to haphazardly complex and rough textured configurations of planes, which are prone to accumulation and retention of dust, debris, contaminants and moisture, cause difficulties in rendering the requisite anti-corrosion precautions and which are severely affected by such adverse phenomena as impingement, turbulence, gas bubble formation and creation of concentration cells.

(2) Rounded contours and corners provide the best continuity of surface and are preferred to surfaces forming sharp angles (*Figure 9.1*).

(3) Hydrodynamically shaped surfaces are favoured in flowing sea water and other corrosive liquids and aerodynamically shaped surfaces

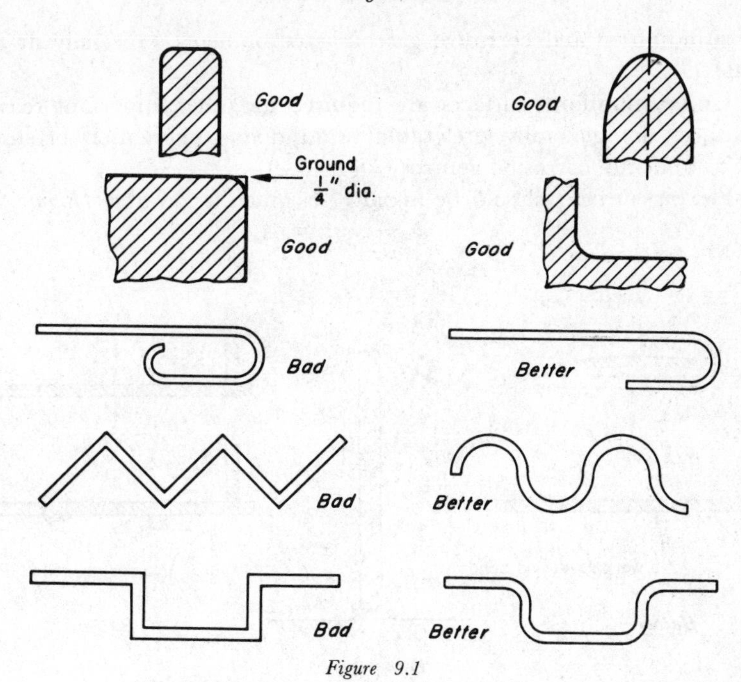

Figure 9.1

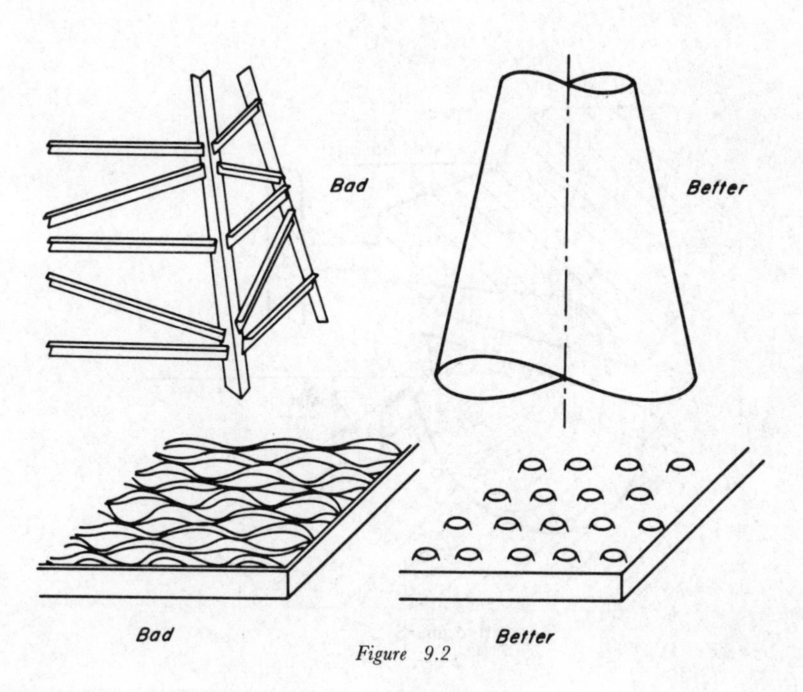

Figure 9.2

in the atmosphere and corrosive gaseous environments, especially at high velocities.

(4) Unless multiform surfaces are required for other important reasons, flat surfaces are generally preferable; a random combination of surface planes complicates corrosion control (*Figure 9.2*).

(5) Flexing surfaces should be avoided as much as possible (*Figure 9.3*).

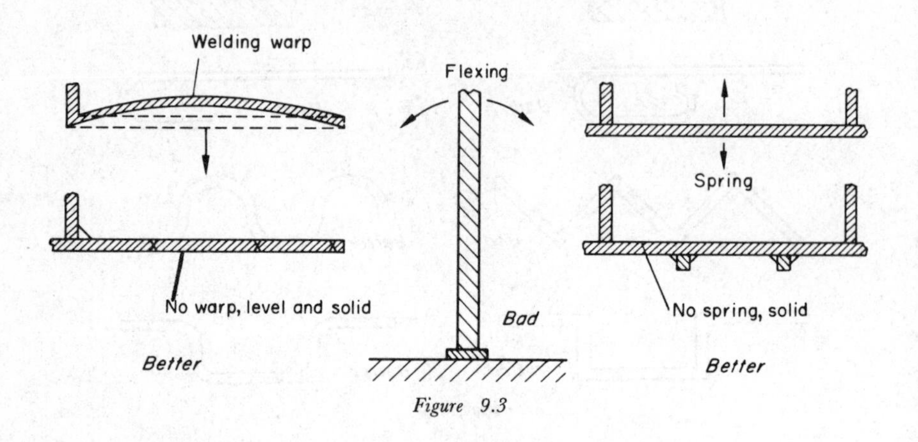

Figure 9.3

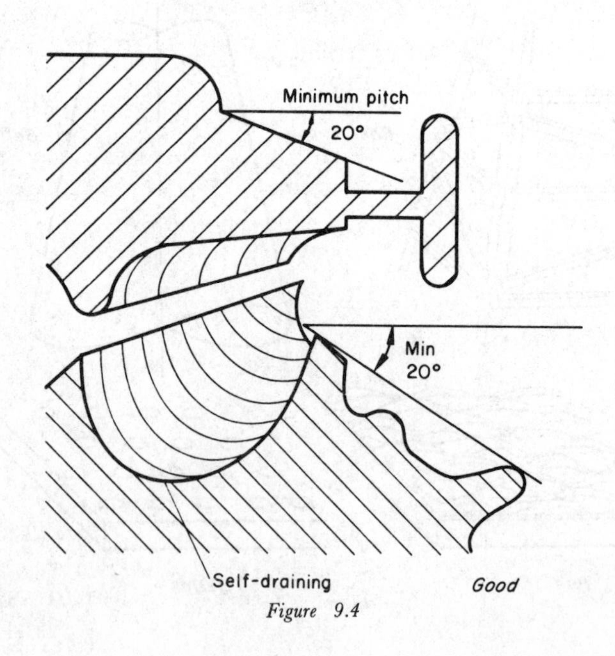

Figure 9.4

(6) Solid and hollow geometrical bodies are bound to have a number of surfaces and each of these could be exposed to environmental conditions with different corrosive potential. Separate evaluation may be required for each variant.

(7) Properly designed surfaces with optimum angle of incidence should assist drying and self-cleaning of materials forming the structure or equipment (*Figure 9.4*).

(8) Critical surfaces, such as welds or surfaces subject to high stress loading, should not if possible be contained in spaces of difficult accessibility or in areas where water can lodge (*Figure 9.5*).

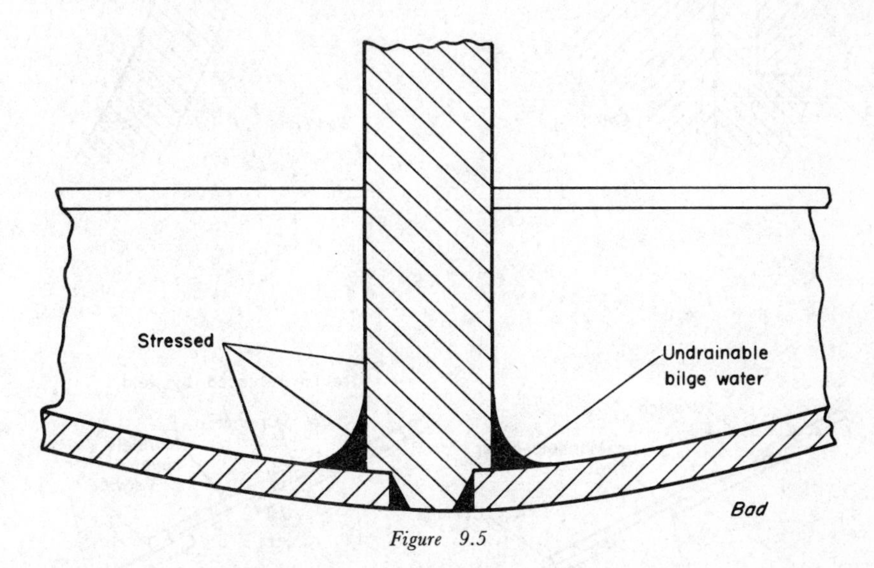

Figure 9.5

(9) The general configuration of surfaces within a utility can be considerably improved by judicious adjustment of the relative position of adjacent or mutually dependent planes and of their overall continuity of profile flow (*Figure 9.6*).

(10) The continuity of profile flow can be further secured with the help of the following design precautions:

(*a*) Reduction of crevices to a necessary minimum (*Figure 9.7*).
(*b*) Reduction of grooves to a necessary minimum.
(*c*) Reduction of faying surfaces to a necessary minimum (*Figure 9.8*).
(*d*) Judicious selection of open or closed joints (*Figure 9.9*).
(*e*) Arrangement of crevices and grooves for self-draining (*Figure 9.10*).
(*f*) Complete sealing—including all edges, to prevent moisture seeping around the edges—of crevices with suitable plastic materials or inhibited jointing compounds. Seal after the surfaces to be mated have been

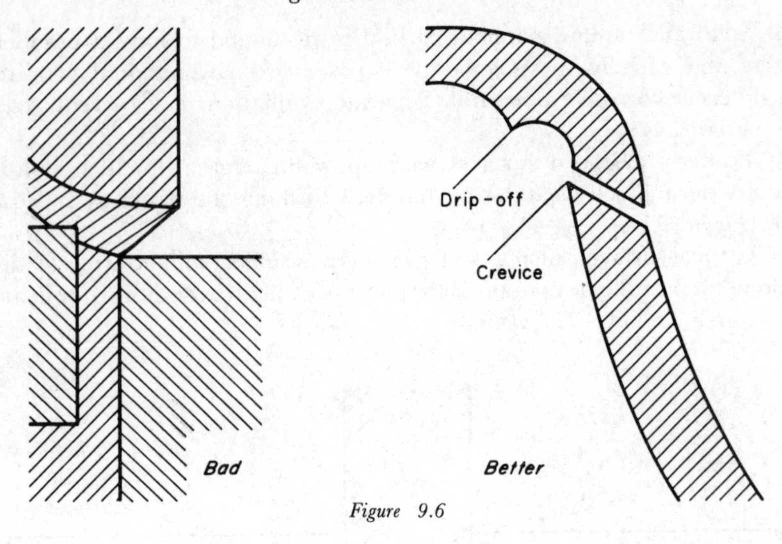

Figure 9.6

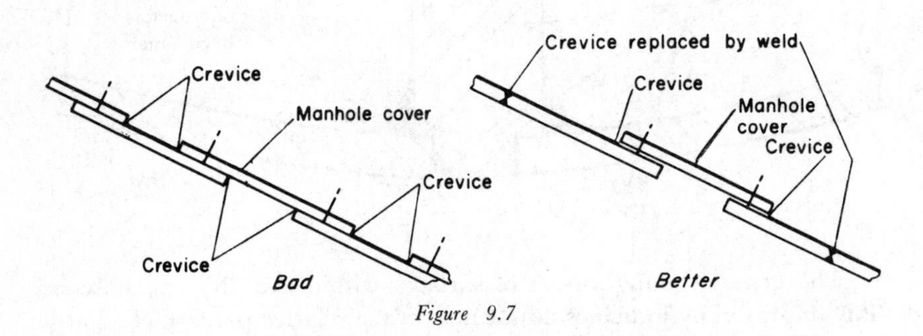

Figure 9.7

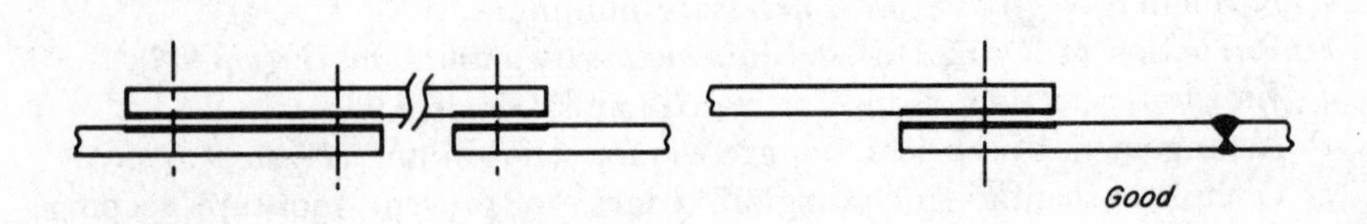

Figure 9.8

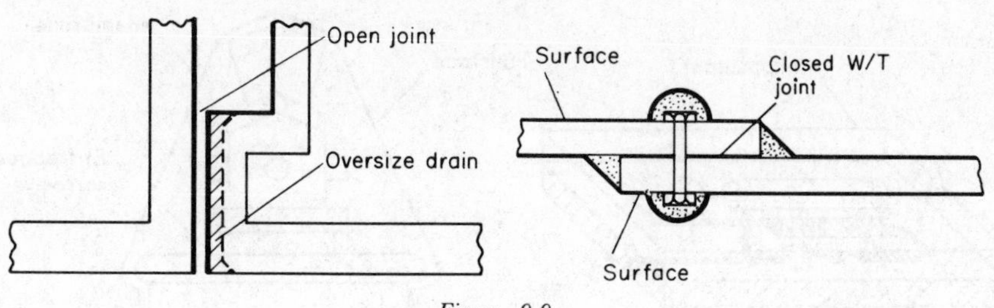

Figure 9.9

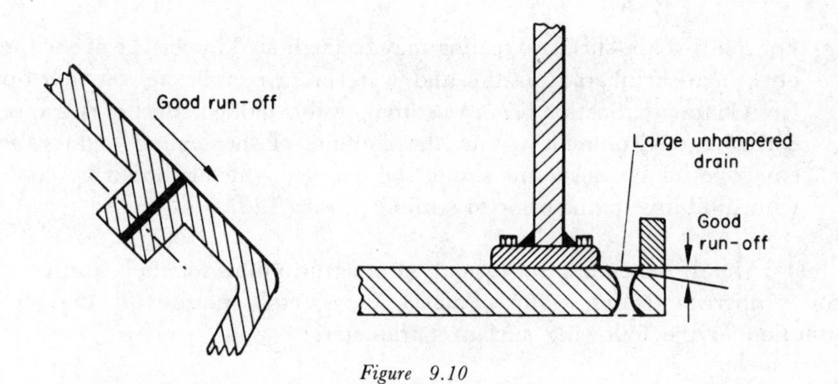

Figure 9.10

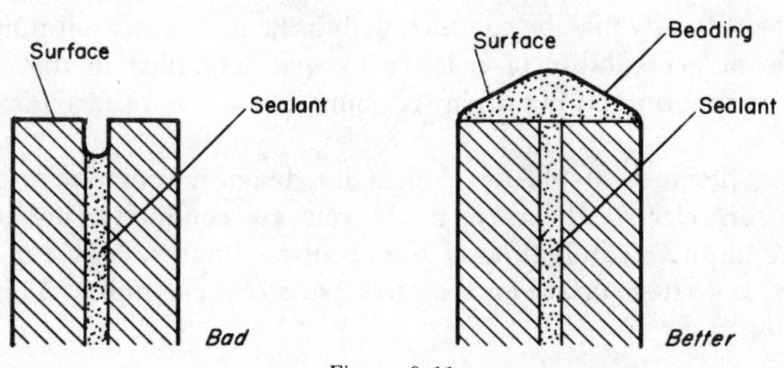

Figure 9.11

primed with inhibitive paint (e.g. zinc chromate primer). Crevices between components, one of which at least is stainless steel, may be sealed with petroleum jelly, approved anti-seize and separation compound (high temperature) or other compatible sealant (*Figure 9.11*).

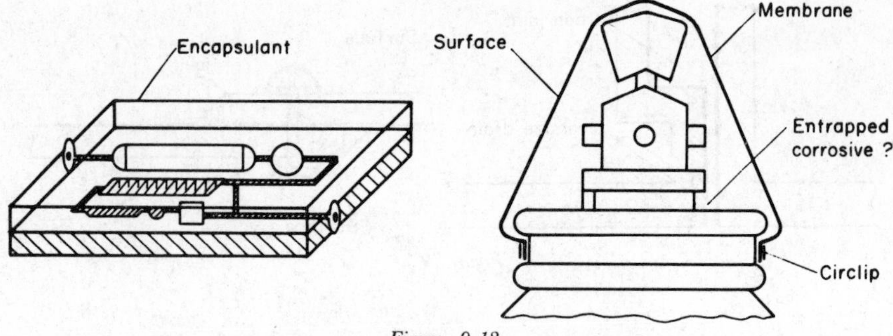

Figure 9.12

(*g*) For multi-shape surfaces, sealing may be facilitated by the use of encapulants, heat-shrinkable plastics and waterproof membranes or envelopes (see Chapter 6, Section 6.7). *Note*: the possible hidden effect of entrapped corrosive environment within the confines of the sealing membranes, envelopes or encapsulants should be avoided—metal should be coated with inhibitive paint prior to sealing (*Figure 9.12*).

(11) Metals depending on formation of surface films for their anti-corrosion properties (stainless steels, nickel alloys, etc.) require the designer's attention to the following surface parameters:

(*a*) beneficial conformation of surfaces;
(*b*) continuity of profile flow;
(*c*) total cleaning of surfaces as a preparation for formation of uninterrupted oxide film;
(*d*) uniform pretreatment of surfaces, if required, including those surfaces which eventually may be confined within the surface discontinuities;
(*e*) significant accessibility of reactive oxygen contained in the operating medium to form and maintain the sound protective surface film.

(12) The designer should develop in his design a conglomerate of such surfaces as are electrically stable in the relevant conductive medium. The ideal is the ultimate elimination of a concentrated adverse effect of one part of the bare or coated surface on the other parts of the complex. This may be achieved by:

(*a*) Selection of compatible materials (see Chapter 6).
(*b*) Selection of overall relative sizes of anodic and cathodic surfaces in the given environment (see Chapters 6 and 7).
(*c*) Avoidance of small anodic surfaces in conductive proximity to large cathodic surfaces within the critical part of the product's geometry (see Chapter 6).

(*d*) Specification of sound, continuous and efficient surface coverings and coatings to be applied on both anodic and cathodic surfaces. If only one surface can be coated, this *must* always be the cathodic one. Adequate inspection of the continuity of surface coatings (especially on anodic metals) should be specified on products to be used in a conductive environment. *Note*: sacrificial anodes are excepted.

(*e*) Provision for formation and re-formation of continuous protective films.

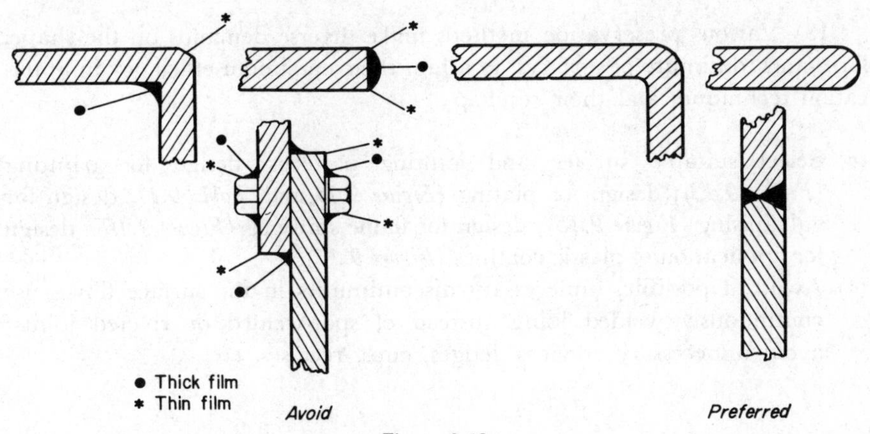

Figure 9.13

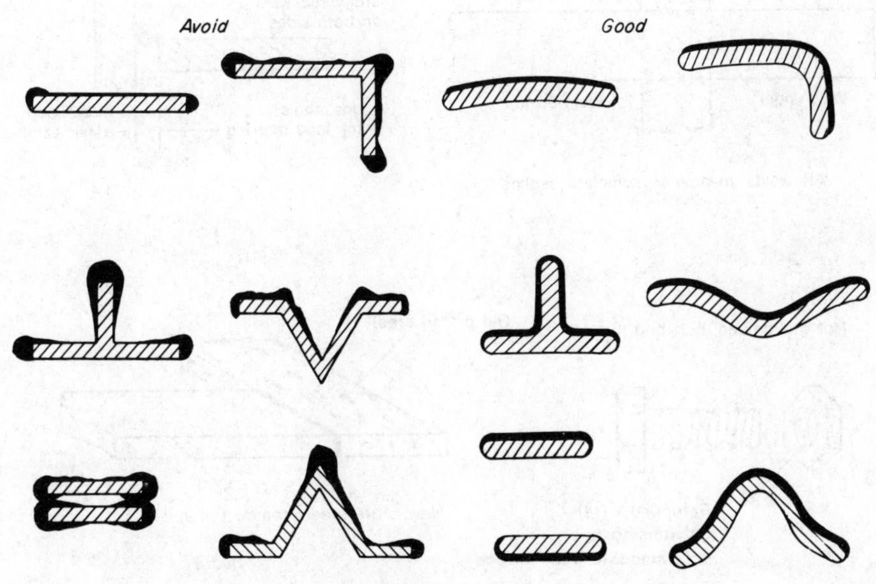

Figure 9.14

Table 9.1

Flat surfaces	Use 0.015 in/in crown to hide uneven buffing undulations
Sharply angled edges	Round the edges 1/32 in minimum radii
Flanges	Use generous radius on inside angles and taper the abutment
V-shaped grooves	Use shallow and rounded grooves
Ribs	Use wide ribs with rounded edges. Taper each rib from centre to both sides and round off edges. Increase spacing if possible
Spearlike juts	Crown the base and round off all corners

(13) Various preservation methods make diverse demands on the shape, form and continuity of surfaces, to attain their maximum efficiency in application techniques and their results:

(*a*) Select suitable surface and jointing patterns—design for painting (*Figure 9.13*); design for plating (*Figure 9.14* and *Table 9.1*); design for galvanising (*Figure 9.15*); design for flame spraying (*Figure 9.16*); design for application of plastic coatings (*Figure 9.17*).

(*b*) Avoid, if possible, unnecessary discontinuities in the surface flow—use continuously welded joints instead of spot-welded or riveted joints; avoid unnecessary crevices, ledges, cups, recesses, etc.

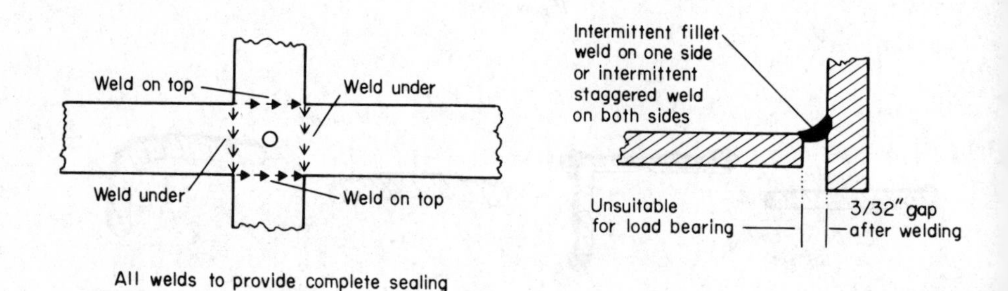

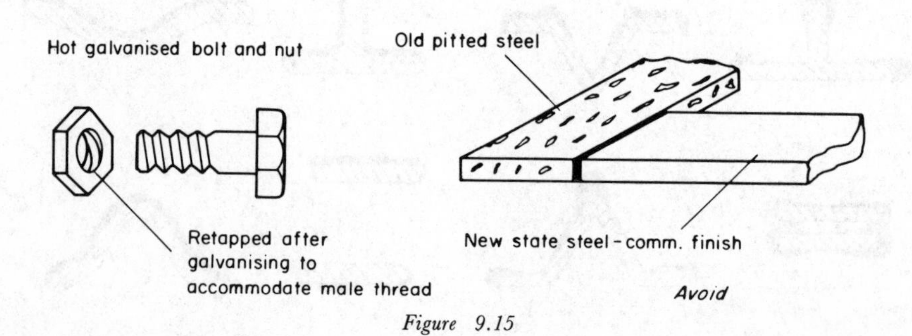

Figure 9.15

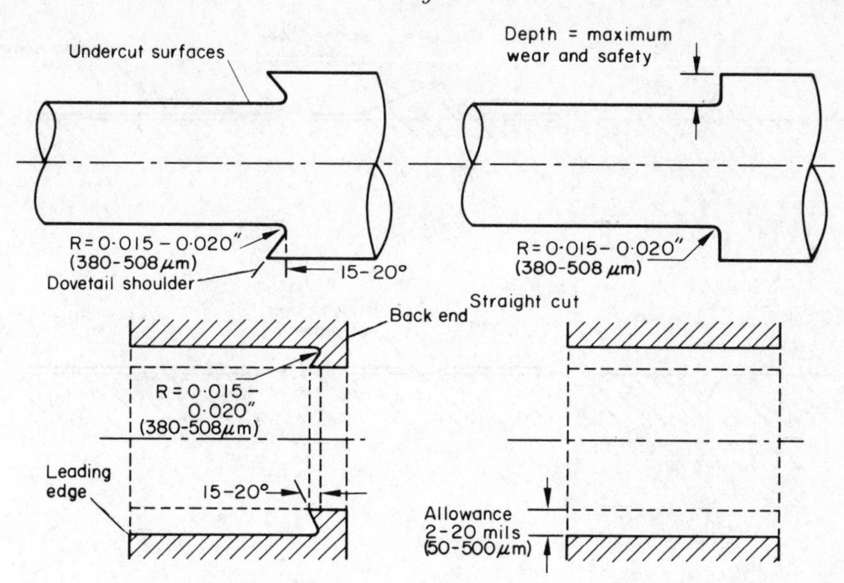

Figure 9.16

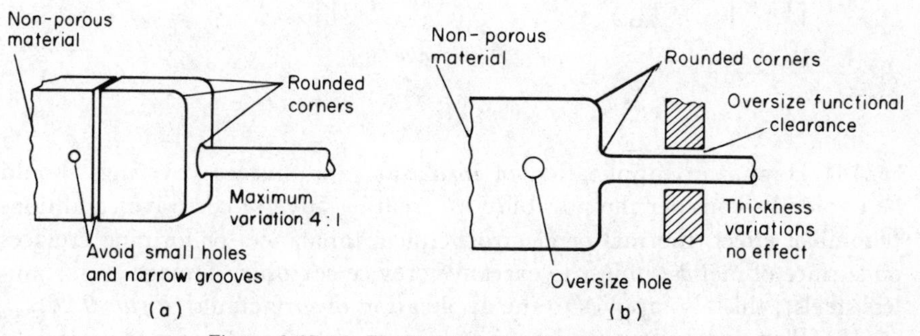

Figure 9.17. (a) Fluidised bed; (b) electrostatic spray

(*c*) Level out excessive roughness of surfaces—grind down any proud protrusions (*Figure 9.18*); fill in any hollows, creases and scratches with metal (e.g. lead, tin, etc.), plastic or plastic metal fillers (*Figure 9.19*).

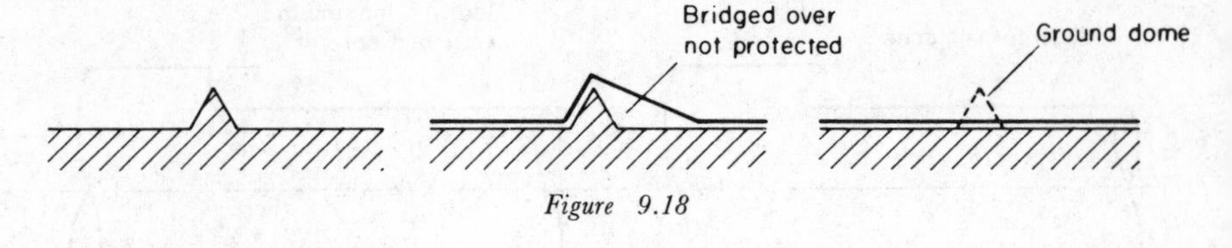

Figure 9.18

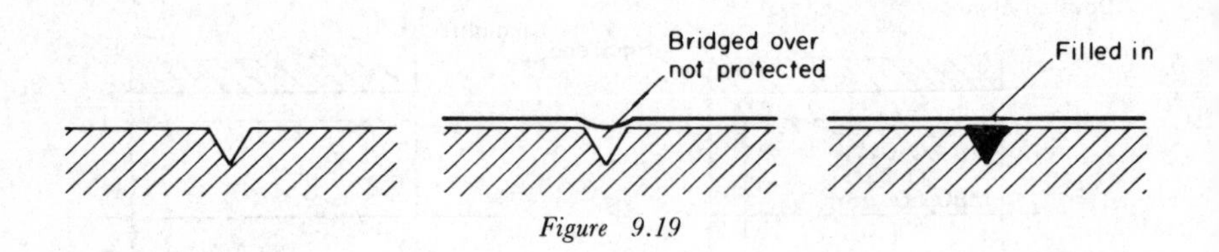

Figure 9.19

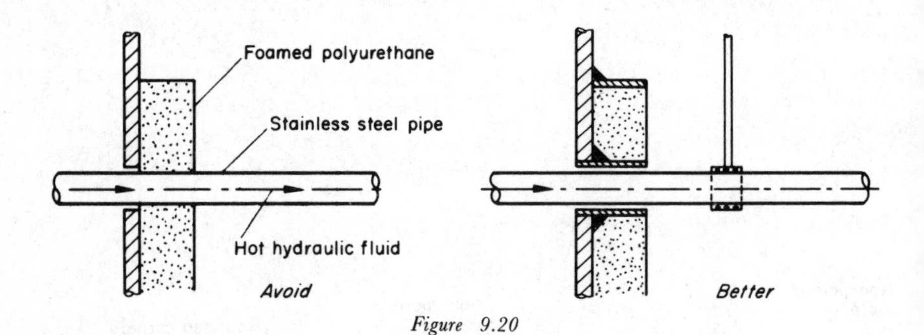

Figure 9.20

(14) Haphazard application of insulation and surface coverings should be avoided—consider the possibility of creating adverse corrosive conditions (chemical effect, thermal or electrochemical inbalance) or forming crevices on surface of metals subject to excessive crevice corrosion damage (e.g. stainless steels); this also applies to the application of surfactants (*Figure 9.20*).

(15) Plan precautions leading to reduction of surface damage to materials, products and components on storage, fabrication or erection (untreated, pretreated or fully treated). These precautions can either apply to the product itself or to the provision of ambient conditions from without the boundaries of the component.

(16) Where a surface damage by filiform corrosion on storage can be expected, provide for storing of coated metals in a low humidity environment; coat metals with brittle film; use low permeability permanent or temporary coatings.

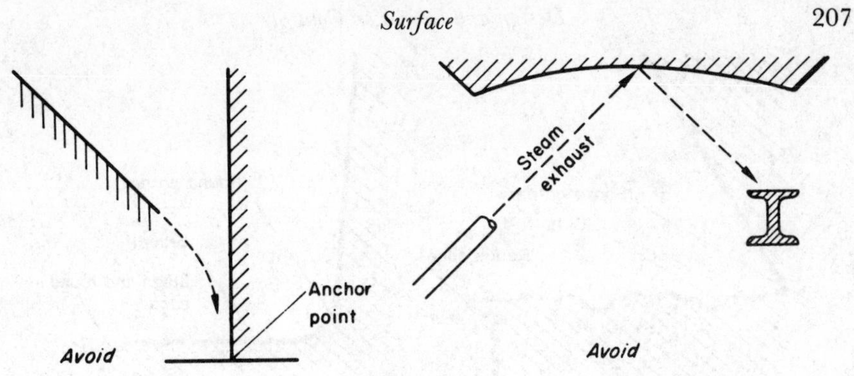

Figure 9.21

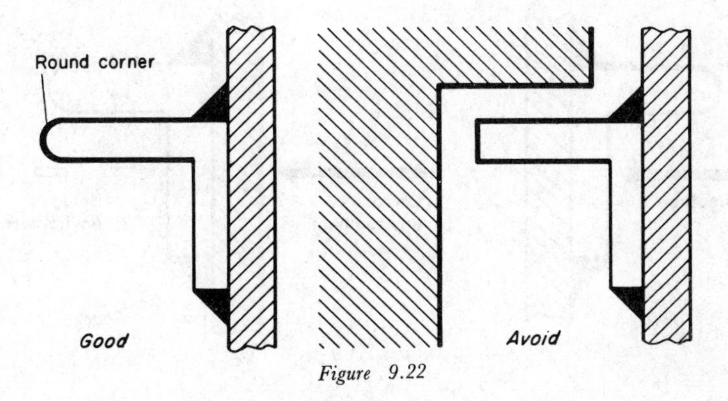

Figure 9.22

9.4 Structures

(1) Avoid adverse corrosive effect of the relative position and shape of adjoining surfaces on any of the individual strength members of the structure (*Figure 9.21*).

(2) Introduce rounded corners in design or specify round grinding where possible. The overall design should allow an easy access for grinding of corners (*Figure 9.22*).

(3) Avoid surfaces which support deposition and retention of dust—dust deposit causes metal to corrode (*Figure 9.23*).

(4) Where possible, change the location of strength members from surfaces exposed to a heavy corrosion loading to those which are subject to less corrosive conditions (*Figure 9.24*).

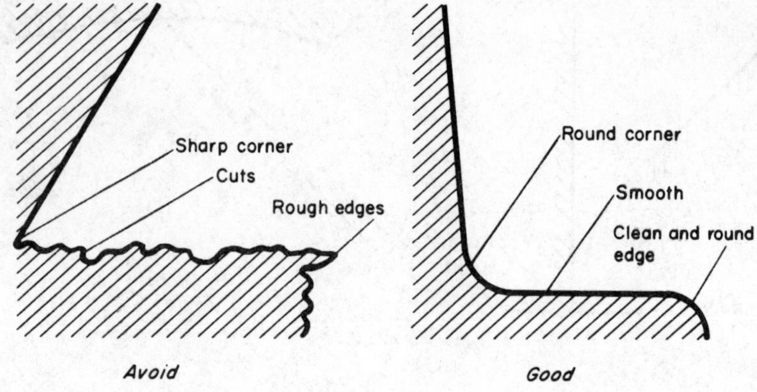

Figure 9.23

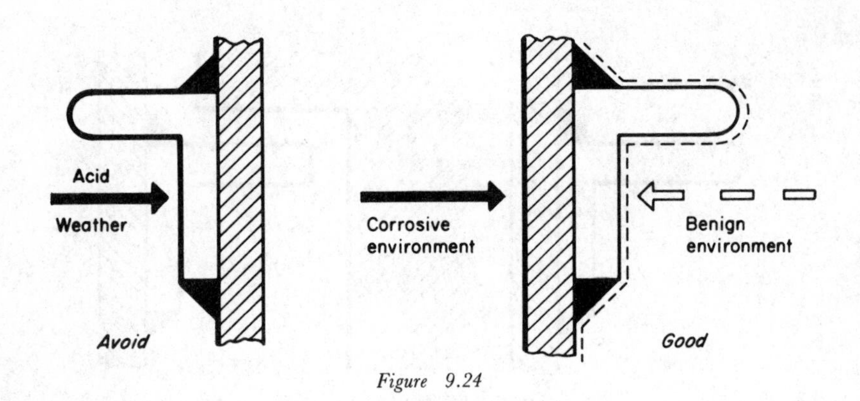

Figure 9.24

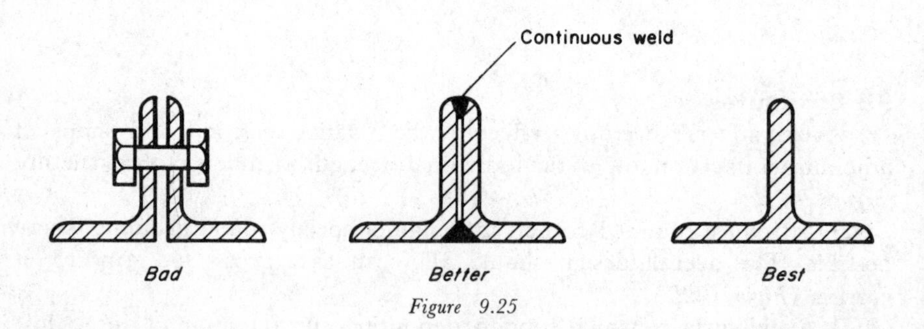

Figure 9.25

(5) Reduce the number of protruding fasteners (bolts, rivets) to a reasonable minimum. Preferred welded joints aid shaping of optimal surfaces. Monolithic components are best, if practicable (*Figure 9.25*).

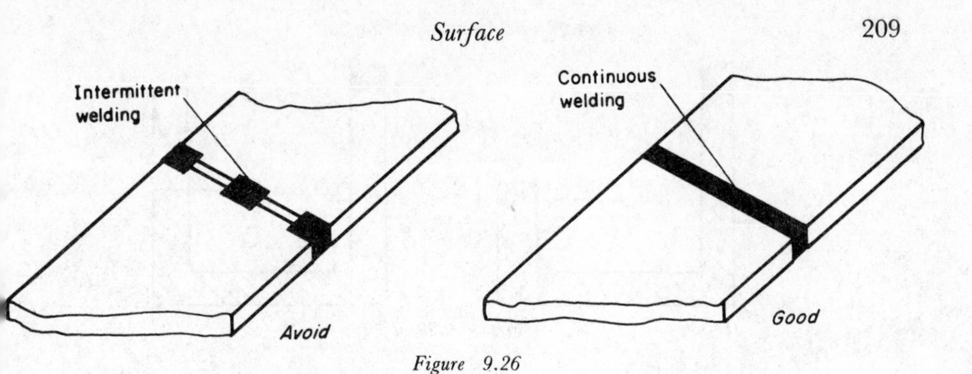

Figure 9.26

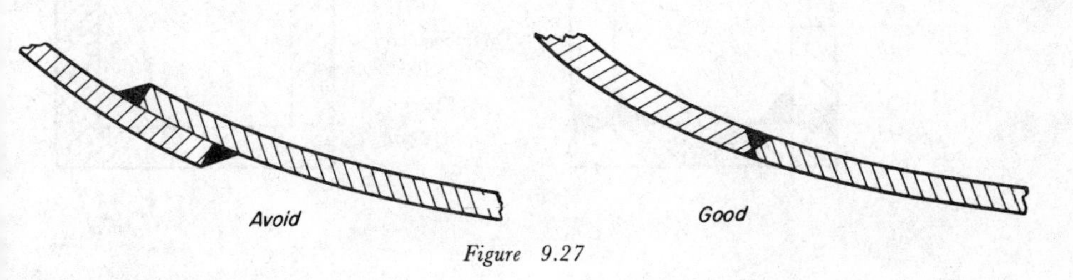

Figure 9.27

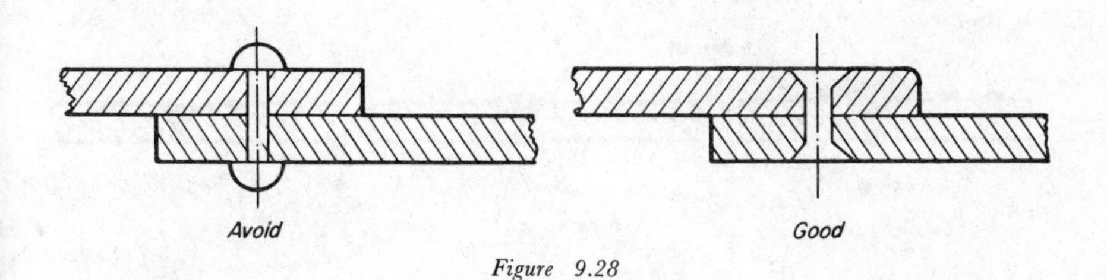

Figure 9.28

(6) Continuously welded joints facilitate optimalisation of surfaces—intermittent or spot welding should not be used in strength structures, unless necessary (*Figure 9.26*).

(7) Butt-welded joints provide better shape of surface than lap joints (*Figure 9.27*).

(8) Countersunk rivets or screws secure better surface profile than other types of corresponding fasteners (*Figure 9.28*).

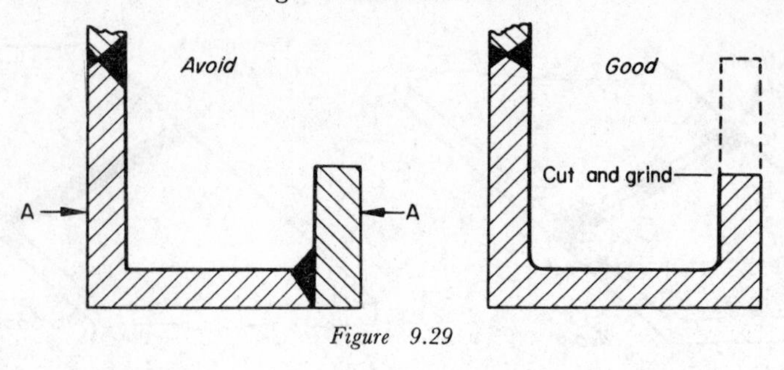

Figure 9.29

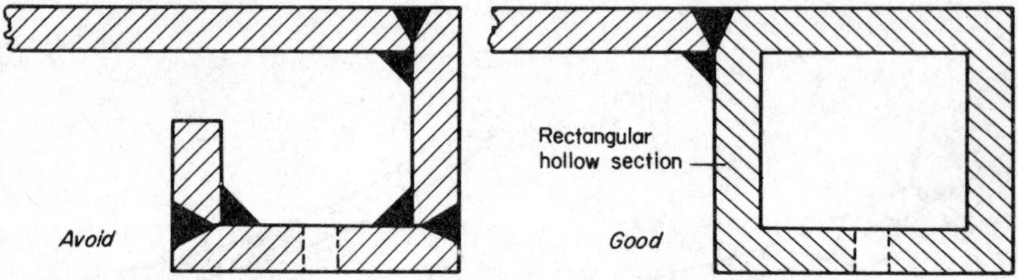

Figure 9.30

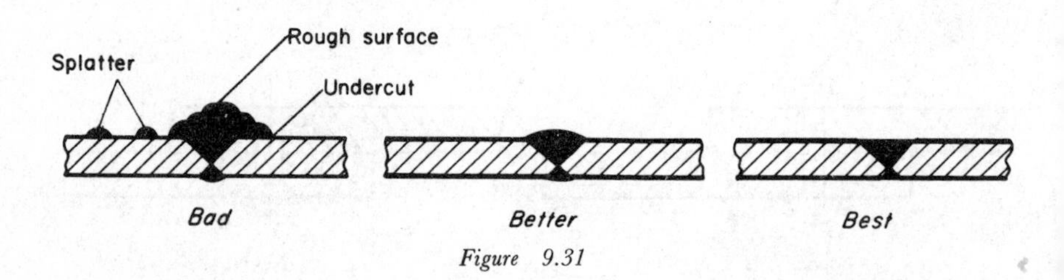

Figure 9.31

(9) Long horizontal runs of welding should not be used in structural channels and grooves where water can lodge (*Figure 9.29*).

(10) Avoid in design any welding in pockets which are not accessible for cleaning, grinding or blasting (*Figure 9.30*).

(11) Thorough finishing or smooth grinding of welds is of prime importance for securing a sound, clean surface. Specify the removal of flux, weld metal spatter, welding residue, burrs and other similar surface defects, whenever possible, prior to any type of overall surface cleaning (*Figure 9.31*).

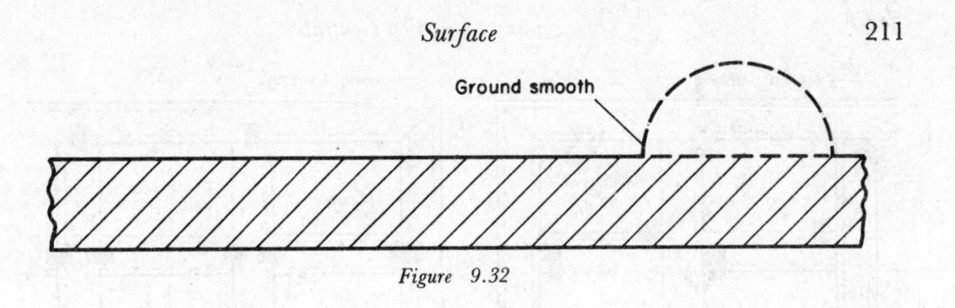

Figure 9.32

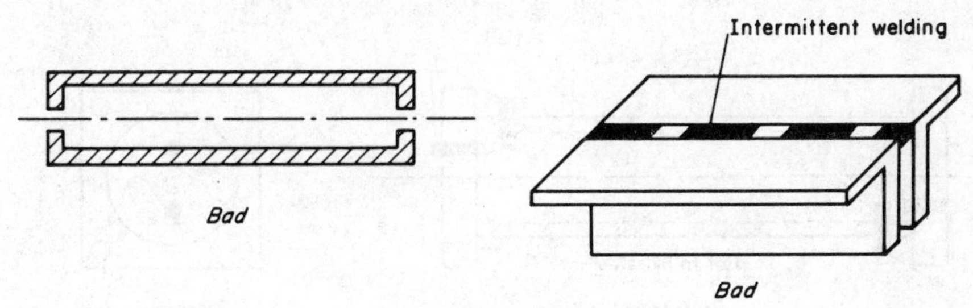

Figure 9.33

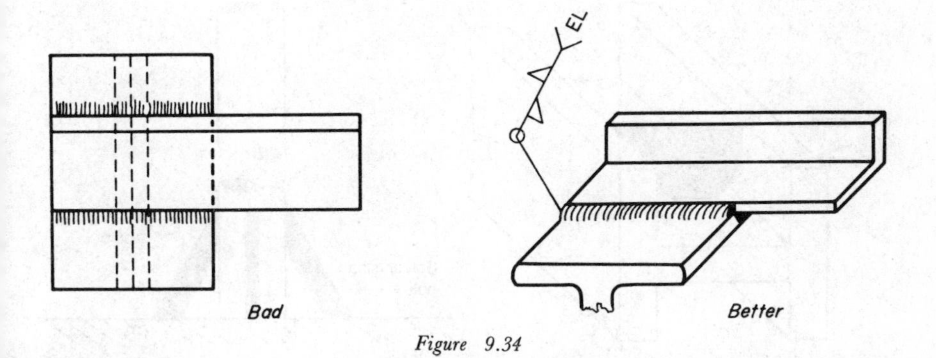

Figure 9.34

(12) Temporary lugs and brackets should be removed and their original positions ground smooth (*Figure 9.32*).

(13) For structural steel designated for pickling, the geometry and fabrication techniques should provide homogeneous continuity of surface without crevices, ledges, cups or recesses, where otherwise the pickling liquid can penetrate and be retained (*Figure 9.33*).

(14) Crevices appearing between joined structural members prepared for galvanising should be fully enclosed by sound, poreless and continuous welds (*Figure 9.34*).

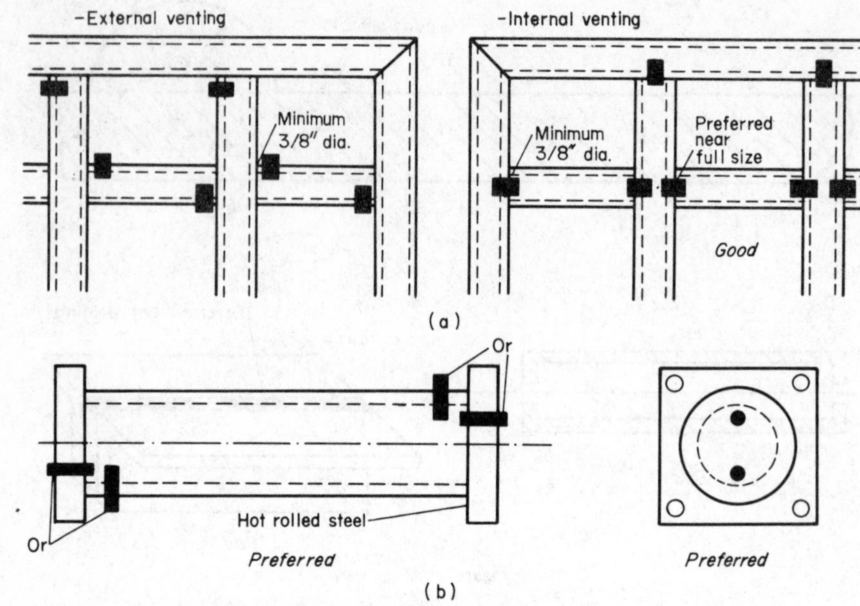

Figure 9.35. (a) Pipe structures and handrails; (b) pipe columns with top and bottom plates

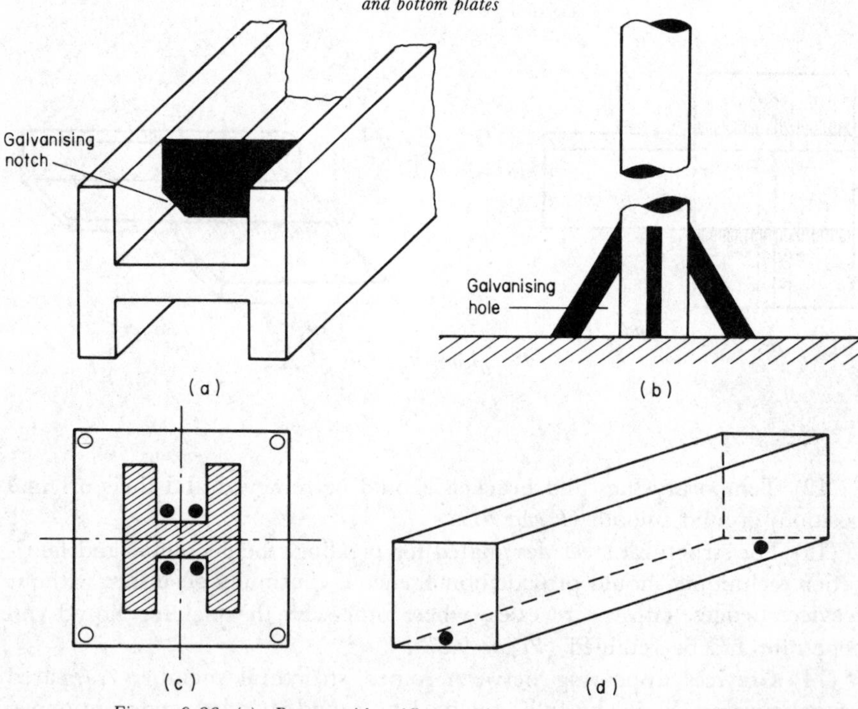

Figure 9.36. (a) Beams with stiffeners; (b) column with base supports; (c) column with base plate; (d) open top box

(15) Design welded pipe assemblies, which are to be galvanised, with full open mitre joints.

(16) No closed or blind sections of pipes (rectangular or round) should be immersed in the galvanising bath. Sufficient continuity of air/metal interface by openings of the requisite number and size should be provided, to allow escape of gases and moisture during pickling and galvanising (*Figure 9.35*).

(17) Provide notches and holes to permit free and uninterrupted flow of molten zinc on parts designed for galvanising (*Figure 9.36*).

(18) In planning for reliable, long-lasting sealed joints one should consider the stresses that may be imposed on the sealant by the movement in joint;

(*a*) Normally, the sealant in a wider joint will be strained less than in a narrow joint during expansion, if the sealant is filled to the same depth in either joint.

(*b*) If the joint movement amounts to 15–35% of the total joint width, a shallow sealant depth in a wide joint will minimise stress on the sealant and on its adhesive bond to substrate (this applies to expansion, butt, capping, and some floor, lap and corner joints).

(*c*) Generally, vertical joints will move more than horizontal ones, and will require shallower sealant application.

(*d*) Lap joint—*Figure 9.37(a)*.

(*e*) T-joint—*Figure 9.37(b)*.

(*f*) If a joint exceeds standard criteria it can be modified by the introduction of back-up material to build upon (polyethylene foam, closed cell urethane foam or clean jute). Back-up material, before insertion, should be from 25% to 50% wider than the joint. Surfaces of the substrate within the joint should be primed with an inhibited paint and the back-up material either contain inhibitor or be dipped in inhibited paint (e.g. zinc chromate primer) prior to assembly (*Figure 9.38*).

(*g*) Sealant performance is improved under stress if it adheres only to the sides of the joint and not to the bottom (*Figure 9.39*).

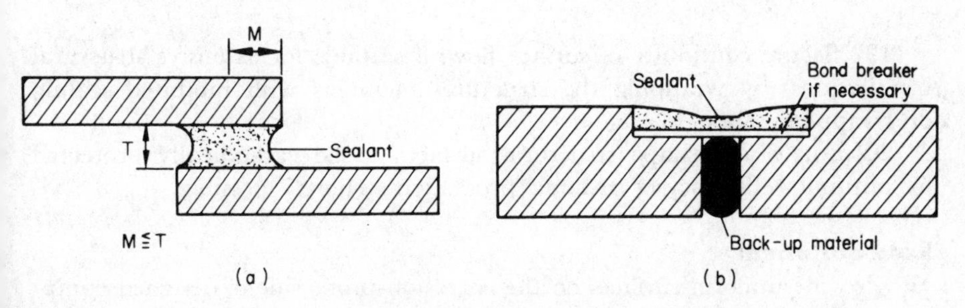

Figure 9.37

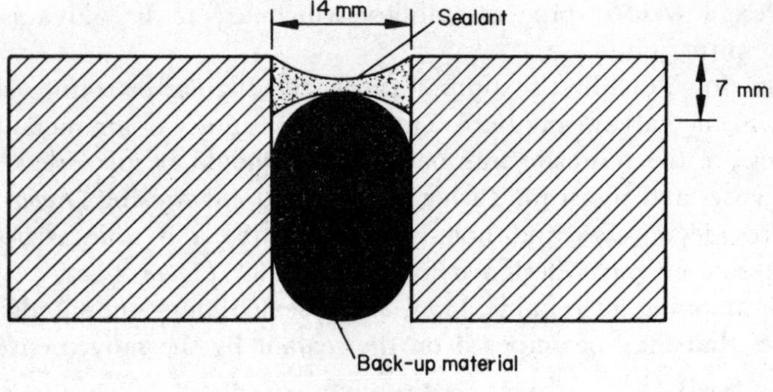

Figure 9.38

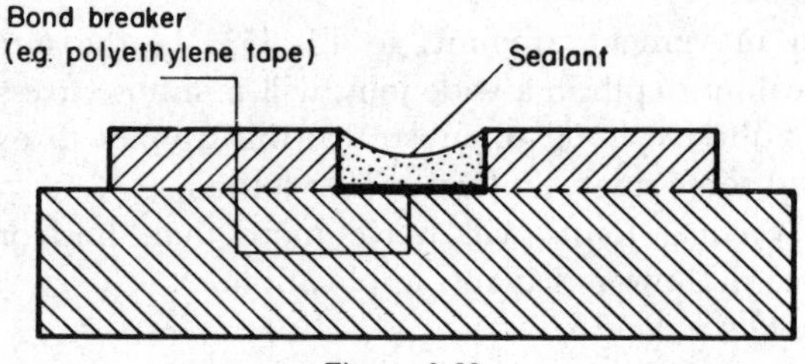

Figure 9.39

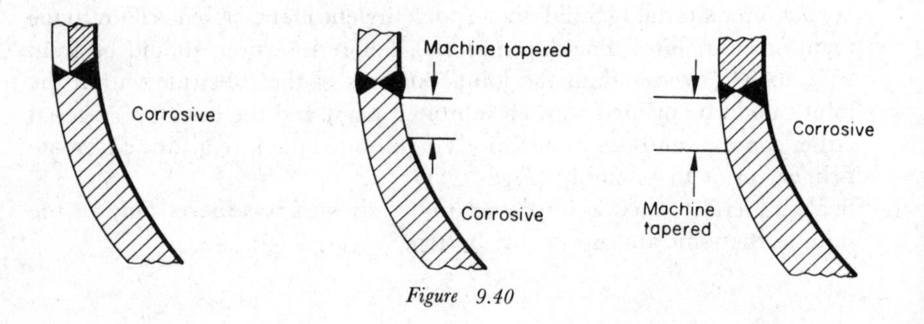

Figure 9.40

(19) Secure continuity of surface flow, if suitable for extensive structural installations, by wrapping the structural members with inhibited sealing or wrapping tapes.

(20) Angle and shape structural surfaces to be cathodically protected for optimum efficiency of cathodic protection system, if possible.

9.5 Equipment

(1) Provide uniform surfaces on the corrosion-prone side of designed equipment (*Figure 9.40*). *Note*: materials and fabrication processes.

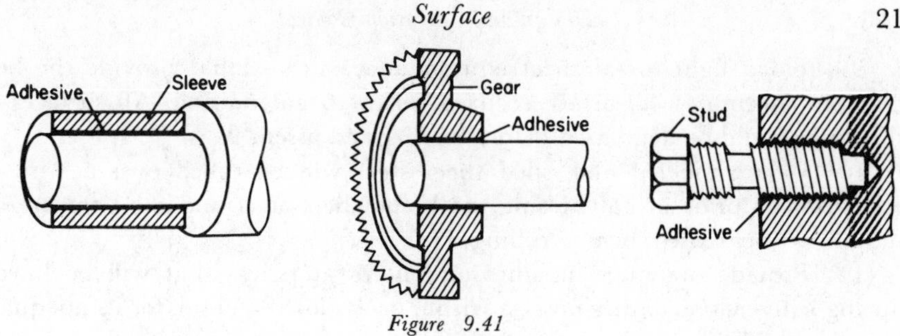

Figure 9.41

(2) Reduce the number of crevices, grooves and in-going pockets and sharp corners in the surface to a necessary minimum. If these are necessary, design for self-draining (see Chapter 7).

(3) In aggravated conditions, and design permitting, the complete equipment or its vital parts can be totally enclosed in watertight and airtight envelopes—possibly as self-contained units.

(4) Use of adhesives (e.g. structural, machinery, anaerobic adhesives, etc.) for joining individual components of an assembly can assist in the formation of smooth contours and the reduction of crevices—design permitting (*Figure 9.41*).

(5) To retain the lubricants and thus prevent corrosion, the surfaces of an equipment can be roughened by shot blasting (very fine), blast peening or application of various porous surfactants (electrodeposited porous metals, clad porous metals, anodising, phosphatising, ceramics deposition or lining).

(6) High polish rendered to surfaces can help to reduce the danger of corrosion fatigue (see Chapter 8).

(7) Access of selective organic solvents to critical plastic parts should be prevented to avoid crazing or other damage to their surfaces.

(8) Cut surfaces of reinforced plastics should be effectively sealed to prevent access of water and other adverse environments to the reinforcing fibres (*Figure 9.42*).

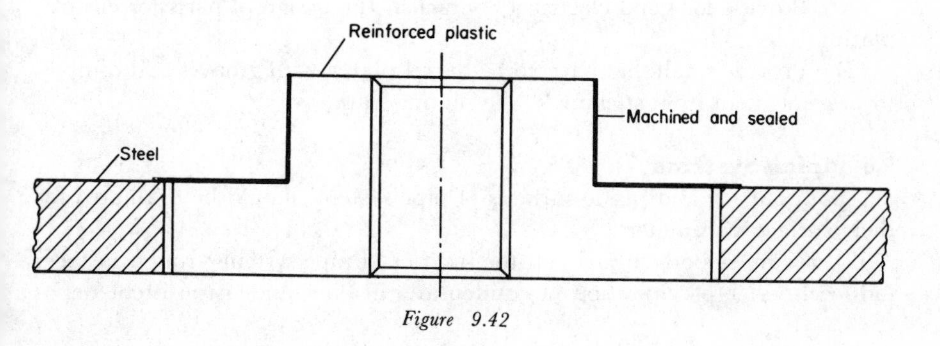

Figure 9.42

(9) Folded light metal sheet equipment casings should provide the best possible continuity of surface secured, prior to galvanising. All surfaces of sheeting should be degreased before folding and assembly.

(10) Where pickled and oiled sheet steel will be rolled over a wire or rod stiffener prior to galvanising, both the sheet steel and the wire or rod should be degreased before rolling.

(11) Provide openings, notches and holes at points that will be lowest during conversion coating process within each closed section, for its adequate draining, and so avoid inadequate rinsing between treatment stages, contamination of treatment baths by preceding stages and the incomplete coating of flooded sections.

(12) To prevent poorly applied conversion or production coatings provide a suitable method for hanging of parts on a finishing line, either by selecting a suitable shape of the part or by introducing in its design a permanent or temporary hanging device (flange, hook, ring, lug or hole).

(13) Avoid completely enclosed sections for components on which conversion coatings will be applied—cleaning and coating solutions cannot completely penetrate into these members even if small holes are spotted in several places.

(14) A further problem inherent in painting the interior of box sections is solvent reflux; even if paint film can be applied there the solvent entrapped within can wash off the wet paint film during the baking cycle.

(15) Self-cleaning surfaces and adequate drainage should be incorporated in components to be conversion coated.

(16) Closed joints should be conversion coated before assembly; open joints can sometimes be conversion coated after the assembly.

(17) Provide in design sufficient clearance to permit free movement between surfaces of movable parts after galvanising. Generally, a clearance of 1/32 in (0.8 mm) is sufficient.

(18) Design of parts to be electroplated or galvanised should be modified to provide adequate racking facilities.

(19) Small parts for barrel processing should be sturdy enough to withstand multiple impacts of barrel rotation.

(20) Provide for good electrical contact in the design of parts for electroplating.

(21) Provide small, flat parts to be barrel plated with grooves and dimples to prevent them from sticking in the plating bath.

9.6 Piping Systems

(1) Both outside and inside surfaces of pipe systems should be evaluated for their surface parameters.

(2) Secure smooth surfaces on the interior of pipe systems; rough surfaces induce heavier precipitation of condensate, heavier and inconsistent depo-

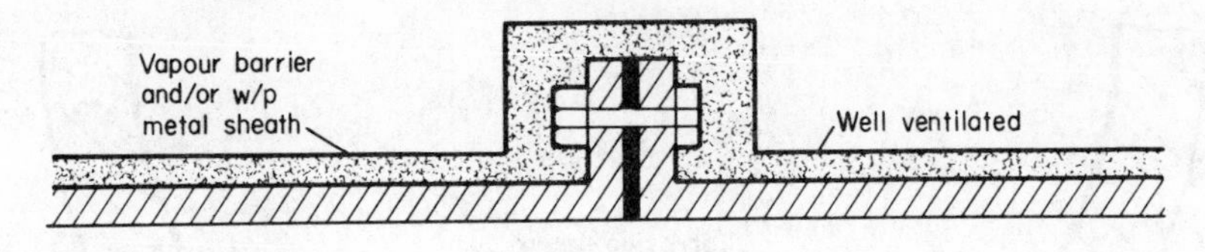

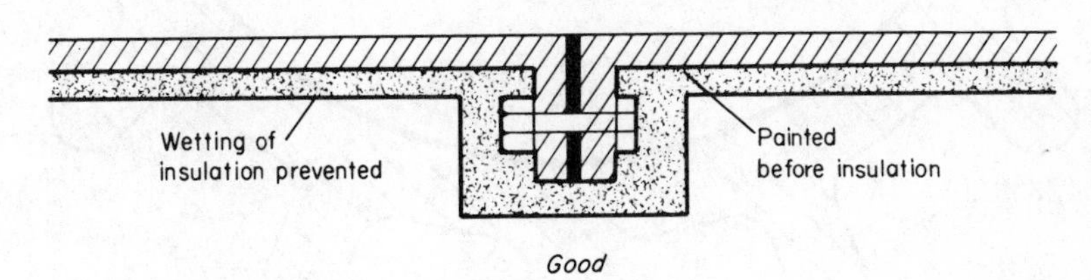

Figure 9.43

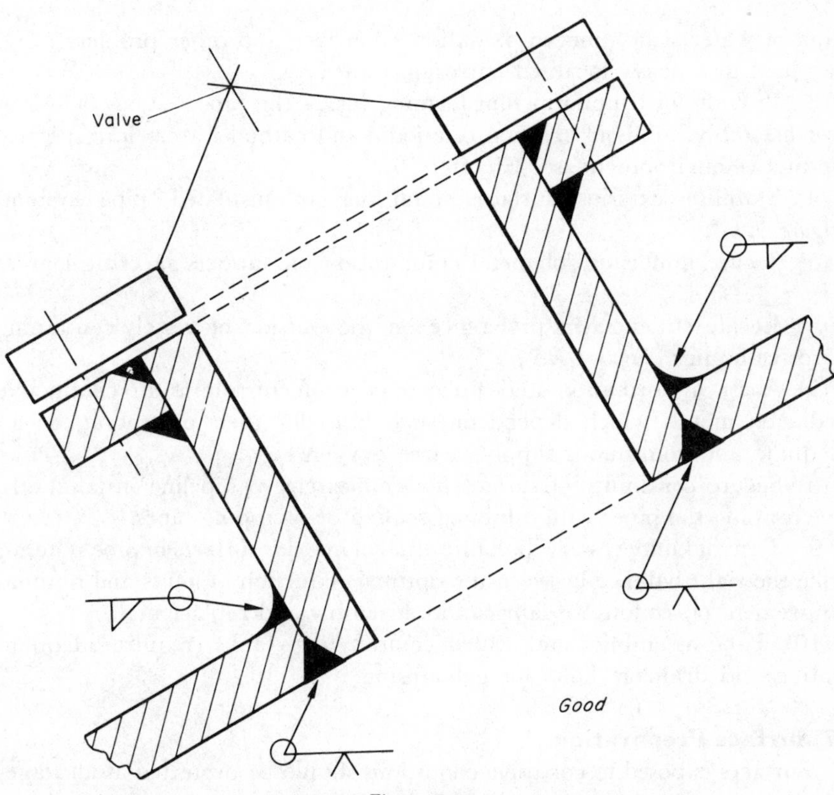

Figure 9.44

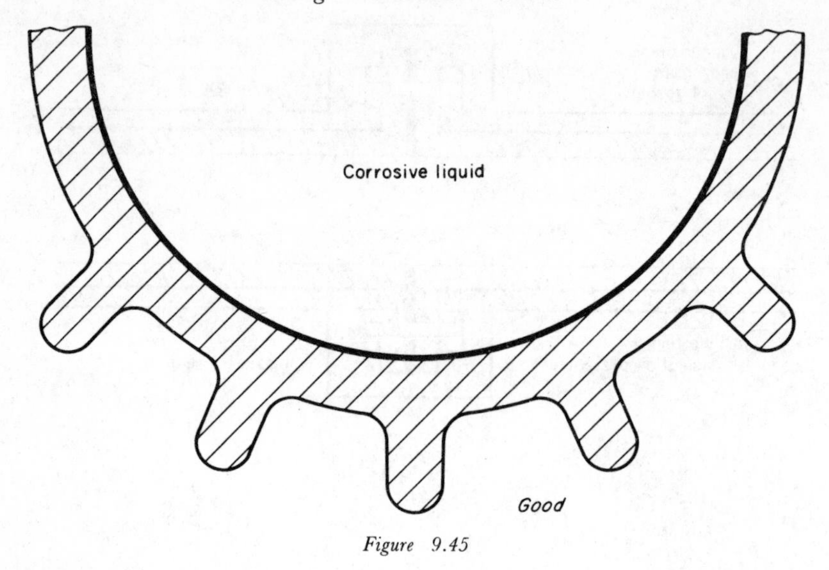

Figure 9.45

sition of water scale, uneven oxidation of surface and other problems, and may lead to a heavy localised corrosion attack.

(3) Provide for a uniform film forming inside the pipe systems before or after assembly, to avoid creation of anodic and cathodic areas in respective conductive environment (see Chapter 10).

(4) Stabilise exterior surface conditions of insulated pipe systems (*Figure 9.43*).

(5) Secure uniformity of metal composition on surfaces in critical areas (*Figure 9.44*).

(6) Locate stiffeners in preference on the outside of vessels containing corrosive liquids (*Figure 9.45*).

(7) Assist in formation and upkeep of protective films in conductive media on metals which depend on such films for their protection, by an adequate and continuous supply of free oxygen.

(8) Secure continuity of surface flow on extensive pipeline installations, by wrapping the pipes with inhibited sealing or wrapping tapes.

(9) To avoid unnecessary discontinuities of interior surfaces in pipe systems, strike the right balance between the optimal reduction of joints and optimal requirement of sections for fabrication, assembly and replacement.

(10) Tube assemblies and sealed cavities (e.g. tanks) require adequate venting and drainage holes for galvanising.

9.7 Surface Preparation

(1) Surfaces exposed to corrosive conditions should be protected at all stages of storage, fabrication, assembly and operation—temporary or permanent protective measures can be used.

(2) The texture of surfaces (surface finish) has considerable influence not only on the mechanical performance of the component, reduction of friction and control of wear but also on the extension of its economic life obtained through the efficiency of the relevant corrosion-control precautions. This applies whether the materials remain uncoated or any further finish be applied.

(3) The principal parameters in securing proper surface finish control are:

(*a*) machining or application cost control;
(*b*) friction reduction;
(*c*) wear control;
(*d*) lubrication control;
(*e*) durability;
(*f*) holding of tolerances;
(*g*) precise fittings;
(*h*) resistance to initiation of corrosion;
(*i*) economic permanency of corrosion control;
(*j*) application of protective coatings;
(*k*) final appearance;
(*l*) consistency of operation;
(*m*) reduction of vibration.

(4) Where a film of lubricant must be maintained between two moving parts (bearings, journals, cylinder bores, piston pins, bushings, pad bearings, helical and worm gears, seal surfaces, machine ways, etc.), the surface irregularities must be small enough to avoid penetrating the oil film under the most severe operating conditions but not too small to bring loss of lubricity in cases where boundary lubrication exists or where surfaces are not compatible (e.g. surfaces are too hard).

(5) Appropriate grain of surface finish should be specified for components subject to dry friction (machine tool bits, threading dies, stamping dies, rolls, clutch plates, brake drums, etc.).

(6) Smoothness and lack of waviness are essential on high precision pieces for accuracy and pressure-retaining ability (injectors, high pressure cylinders, micrometer anvils, gauges and gauge blocks).

(7) Smooth surfaces bring elimination of sharp irregularities which are the greatest potential source of fatigue cracks on highly stressed members subjected to load reversals (see Chapter 8).

(8) Smoothness of final appearance can also be controlled by production tools (rolls, extrusion dies, precision-casting dies).

(9) Surface finish control of such parts as gears may be necessary to secure quiet operation and to reduce vibration (see Chapter 3).

(10) The surface finish should be a compromise between sufficient rough-

ness for proper wear-in and sufficient smoothness for expected service life.

(11) Incorrect clearances between two surfaces in relative motion may result in local hot spots and high oil consumption (see Chapter 3).

(12) Excessively rough textured surfaces increase turbulence, retain more dust and lead to heavier precipitation, retention of condensate and deposition of water scale—all detrimental to proper corrosion control.

(13) Specify, where possible, for grinding of excessively rough surfaces to a smooth contour.

(14) Evaluate, in each individual case, which texture of surface gives the best anti-corrosion service and specify this degree of surface roughness in the design. Observe that it is not always sufficient to specify only the texture of the substrate but that the texture and consistency of preservation coatings or surfacing materials may also be required (*Figure 9.46*).

(15) Typical surface roughness is obtained by common production methods (*Table 9.2*).

(16) In the interests of corrosion control the designer should consider, at the design stage, whether the components should remain raw as supplied, untreated as machined, or whether they should be ground, honed, polished,

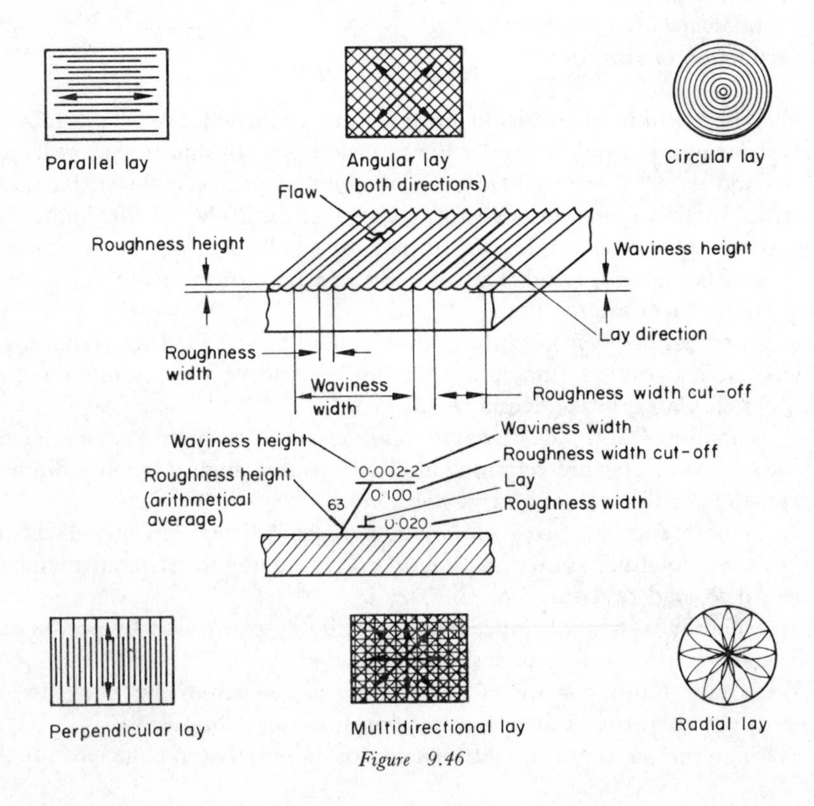

Figure 9.46

Table 9.2

Roughness height (μin)

Process	2000		1000		500		250		125		63		32		16		8		4		2		1		0.5
Flame cutting		†		*		†																			
Snagging		†		*		*		†																	
Sawing		†		*		*		*		†		†													
Planing, shaping				†		*		*		*		†		†											
Drilling, chemical milling						†		*		*		†													
EDM						†		†	*	*		†													
Milling				†		†		*		*		*		†		†									
Broaching, reaming								†		*		*		†											
Boring, turning				†		†		*		*		*		*		†		†		†					
Barrel finishing										†		†		*		*		†		†					
Electrolytic grinding													†	*		*		†							
Roller burnishing														†		*		†							
Grinding								†		†		*		*		*		*		†		†			
Honing												†		*		*		*		†		†			
Polishing														†		*		*		†		†		†	
Lapping														†		*		*		*		†		†	
Superfinishing														†		†		*		*		†		†	
Sand casting, hot rolling		†		*		†																			
Forging				†		*		*		†															
Permanent mould casting								†		*		†													
Investment casting								†		*		†		†											
Extruding						†		†		*		*		†											
Cold rolling, drawing								†		*		*		†		†									
Die casting										†		*		†											
Roughness height (μm)	51		25		13		6.5		3.2		1.6		0.8		0.4		0.2		0.1		0.06		0.03		0.01

*Average application.
†Less frequent application.

flash rusted, blast cleaned, blast peened, roughened, anodised, passivated, metallised, surfaced, sealed, prefabrication treated or painted.

(17) The maximum acceptable surface roughness compatible with the service and fabrication requirements should be specified preparatory to the application of protective coatings. Very smooth surfaces (e.g. new hot dip galvanising, polished components, etc.) on the other hand may require flash rusting, etching, phosphatising, anodising or abrasive blasting at various stages of fabrication or assembly to give optimum adhesion conditions.

(18) Surfaces roughened by very fine shot blasting or by application of porous coatings (electrodeposited porous metals, ceramics, anodising or phosphatising) can better retain lubricants and thus help to prevent corrosion.

(19) Surface conditions in design should be reconciled with the surface treatments to follow and their requisite application techniques—surfaces and their treatments are complementary to each other.

(20) All materials must be cleaned. Select and specify in design the mandatory method and standards in detail. Cleaning methods and techniques that render the best economic results within the whole life-cycle of the utility are preferred.

(21) Unless the specified cleaning operations on their own can automatically provide for the following, the removal of burrs, notches, flares, fluxes, weld metal spatter, etc., should *precede* the specified surface cleaning.

(22) Specify complete removal of mill scale on steel—partial removal is a waste of money.

(23) Select the economically advantageous removal of rust, considering the merit of long-term economy. The following methods can be considered: weathering and hand or mechanical cleaning, flame descaling, blast cleaning, vapour blasting, sodium hydride descaling, pickling or acid derusting.

(24) Specify removal of oil, grease, fingermarks, salt deposits and various organic and inorganic contaminants from the surfaces before or after the programmed physical or chemical cleaning to suit the purpose. The following methods can be considered: cold solvent cleaning, vapour and hot immersion cleaning, vapour degreasing, steam cleaning, liquor/vapour degreasing, emulsion cleaning, alkaline cleaning and ultrasonic cleaning.

(25) Cathodic cleaning of high strength steels in either acid or alkaline baths should be avoided—anodic cleaning is permissible.

(26) Flame cleaning should not be specified for removal of mill scale in a new unbroken state from steel.

(27) Blast cleaning is preferred to pickling for hot rolled parts with machined surfaces.

(28) All assemblies of cast iron, cast steel and malleable iron with rolled steel should be blast cleaned after assembly and prior to pickling (different pickling characteristics).

(29) Dissimilar materials (different analysis steels or different surface finishes of steel in an assembly) should be pickled and galvanised separately, and assembled after galvanising for uniformity of surface appearance.

(30) Corrosion control prefers, in general, the surface-cleaning methods given in *Table 9.3*.

Table 9.3

Material	*Preferred surface-cleaning method*
Steel	Abrasive blasting
Aluminium	Abrasive blasting—very fine grade abrasive
Copper	Mechanical cleaning, followed by wash with solution of 5% zinc chloride and 5% zinc muriatic acid at commercial concentration in water.
Nickel	Abrasive blasting—non-metal abrasive
Stainless steel	Abrasive blasting—non-metal abrasive
Zinc	Mechanical cleaning followed by wash with phosphoric acid solution, followed by removal of zinc salts

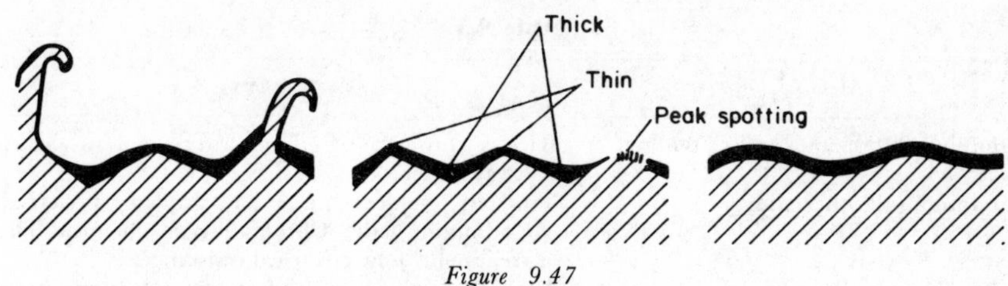

Figure 9.47

(31) Avoid specifying excessive roughness of surface for application of protective coatings (*Figure 9.47*).

(32) Specified blasting profile (amplitude and shape) should be adjusted to the thickness, consistency, external smoothness and adhesion of the coating which is to follow (*Table 9.4*).

(33) Surface hardening and hard surfacing of metals should be evaluated for a possible substantial aggravation of corrosion.

(34) Specify, if required, suitable surfacing materials (metals, ceramics, mastics, deck covering underlays, cements, fillers, noise damping and anti-condensation compounds, plastic and reinforced plastic linings and surfacers, potting compounds, rubber linings and metal-filled surfacers) for protection of relevant surfaces (e.g. against cavitation on propellers, cylinder liners, pumps, impellers, etc.) and for build-up of surfaces to uniform level, optimum surface profile (e.g. for drainability, improvement of contour and fairing and for improvement of appearance), or for fill-in of spaces which cannot be otherwise preserved. Degree of surface roughness of surfactants should also be indicated.

(35) Prefabrication treatment of steel should provide for adequate protection on storage, fabrication and assembly until such time as the final coatings can be applied (approximately 6–9 months).

(36) Pipe purchased for fabrication and galvanising should be ordered without mill scale or the mill scale should be removed by blast cleaning prior to pickling.

(37) Pipe fittings for galvanising should be of uncoated steel.

(38) Markings and lettering applied to surfaces to be galvanised or painted with zinc-rich paints should be made in water-soluble colours or otherwise be punched or embossed.

Table 9.4 RECOMMENDED MAXIMUM PROFILE AMPLITUDES (NORMAL CONDITIONS)

Application	*Amplitude* (mils)	μm
Prefabrication primer	2	50.8
Standard paints	3–4	76.2–101.6
High build paints	5	127.0
Sprayed metals	5–8	127.0–203.2
Electrodeposited metals	2	50.8
Removal of foreign matter (close tolerance surface)	nil	nil

Table 9.5

Metal	*Surface finish*
Aluminium 1100, 3003 and clad alloys	Bare or chromate-type film treatment. Low electrical resistance.
Aluminium (all other alloys)	Tin-lead (solder) plate or tinplate preferred. Cadmium plate. Chromate-type chemical film treatment. Low electrical resistance.
Copper, copper alloys	Bare. Tinplate or tin–lead (solder) plate preferred. Cadmium plate. Gold plate.
Cadmium	Bare or chromate treated
Iron and steel	Tinplate or tin–lead (solder) plate preferred. Cadmium plate
Magnesium	Bare. Clean immediately before and coat joint immediately after bonding
Nickel and corrosion-resistant steels	Bare. Difficult to bond because of adherent oxide film
Silver	Bare
Solder	Bare
Tin	Bare

9.8 Electrical and Electronic Equipment

(1) Specify and design for smooth surfaces without crevices, as far as practicable.

(2) Joints should be continuous and impervious, otherwise they should be sealed.

(3) Crevices, especially those in stainless steel (i.e. joints, under washers, at threaded fasteners, etc.), should be sealed with suitable sealants (e.g. polysulphide, polyurethane, epoxy or silicone rubber) or petroleum jelly.

(4) Resistance welded joints should be sealed.

(5) Non-hydroscopic insulation should be used.

(6) Marker tapes should be specified for use only on surfaces which have been treated previously with varnish.

(7) Proper, thorough and compatible cleaning methods should be specified before joining, coating, potting, impregnation and encapsulation of components.

(8) Flux residues should be removed after brazing and soldering.

(9) Welds should be cleaned after welding of scale, fluxes, spatter, oxidation and rough areas.

(10) Fingermarks should either be prevented or removed.

(11) Surface contaminants should be removed from conductor surfaces by an appropriate cleaning method. This should be followed by priming with de-ionised water and drying prior to application of an organic moisture barrier for protection.

(12) No aggressive cleaning methods should be used on printed circuit boards.

(13) The use of solid metals or plating with such metals as gold, rhodium

and platinum, which are inherently resistant to tarnishing, should be specified to ensure maintenance of maximum conductivity.

(14) Electromagnetic compatibility of electrically bonded metals should be secured by the selected surface finish (*Table 9.5*).

(15) Avoid using exposed soft solder at joints, prior to electroplating.

10 Protection

10.1 Introduction

The function of protection is, to a considerable degree, the upkeep of the optimum anti-corrosion factors built into the particular design itself. Protection on its own, therefore, cannot normally take the sole responsibility for preservation of a utility in a usable state. Both the intrinsic corrosion-control provisions and properties which are kept captive within the material boundaries of the designed structure or equipment, and the corrosion-protection activities which are applied from without, are complementary to each other. The demarcation of their respective boundaries will be largely governed by the rational trade-off of their comparative economic values.

High costing protection may favourably balance the appropriate replacement of more exotic materials or geometric forms with the cheaper ones; it may favourably compensate for reduction in strength, for less frequent maintenance, for better safety of operations, etc. Use of cheap protective measures may often prove false economy.

Protection should be tailored to the particular assembly complex and not to the individual composite parts, subassemblies or units. For optimum protection, consideration should be given to the geometry and location of the utility and its vital parts, ease of application and the effectiveness of the protective measures, these factors being reciprocally adjusted to suit each other. New or revolutionary protective measures and techniques should not be incorporated haphazardly in design—structures and equipment should be designed for their most effective use. The more inaccessible the surfaces, the better should be their protection. Active or passive ecological involvement of protective measures is of prime importance.

Only necessary, safe and economically feasible protection should be specified, preferably by methods and techniques applied under controlled or automated conditions, thus eliminating or reducing the adverse influence of human variant. The local obtainability of an efficient and expert labour force, as well as local climatic conditions at the initial production site and at the subsequent ports of call, will have a considerable influence on the selection of protective measures. Where these factors can have a critical effect on the efficiency of protection, preference should be given to those materials, methods and techniques that can give the best results being used at the consequential locality.

Basically, protection comprises those protective measures providing separation of surfaces from environment, those giving cathodic protection or anodic polarisation and those which cater for adjustment of environment. These methods can be used individually or in various combinations, the latter affording a greater degree of protection than the sum of individual effects.

10.2 Scope

To decide on the required, necessary and economically feasible protection the personnel engaged in this task have a vast variety of protective measures, systems, methods, techniques and especially competitive products to choose from. Although the products which are to provide protection may be defined in their basic generic groups,. it by no means signifies that all units of one generic group will provide, in the relevant case, the same application and performance characteristics and will therefore automatically qualify. Extensive engineering investigation, independent suitability testing and practical proof of effectiveness may be needed to precede the final choice, unless the word of an independent corrosion specialist or a reputable supplier can be taken on careful trust or with a calculated risk.

Since there is an abundance of volumes already published and a considerable number of expert branches of specialised science and technology are continuously expanding this most prolific field of information on the separation of surfaces from environment, cathodic protection or anodic polarisation and the adjustment of environment, there is no necessity to further enlarge on the technical details in this chapter. This ready information, however, can be reliable or much less reliable, depending on the highly competitive and intimate involvement of commercial interests. The designer therefore should seek enlightenment on the preselected specialities from dedicated and, if possible, independent specialists—in fact, co-operation is a necessity for a designer engaged in creative design. The designer, who is not a corrosion specialist, cannot acquire an encyclopaedic knowledge of all relevant disciplines.

For these reasons, and to allow the designer the intelligent insight necessary for the formulation of his design policy, only an outline of procedures is recorded in this chapter, the details being left to the correctly reasoned effort of all interested parties co-operating in the design team and the specialised information in hand.

10.3 General

(1) Separation of materials from the environment, provided by application of metallic coatings, by painting, coating with plastics or ceramics, by lining, sealing, enveloping, insulation and by application of temporary protectives (oils, greases, removable plastics, etc.) involves primarily a change in surface composition and this change is brought about normally by the addition of different materials (metallic or non-metallic) in the form of an outer

skin. Most of these processes involve a dimensional change (except perhaps diffusion coating) and a weight change.

(2) Ideal separation of surface requires total exclusion of air and moisture or other corrosive media from the protected surfaces. This is difficult to achieve due to the inherent porosity of various protective materials, due to the limited survival life of these materials and due to these materials being prone to application faults.

(3) To provide against the deficiency in effective separation of surfaces, recourse is normally made to multi-phase combinations of separation materials applied to surfaces in a form of protective systems, which combine several materials either of the same family or of several complementary categories.

(4) Make the basic decision as to the type of separation method to be used:

(*a*) which single or combined method can provide the optimum period of respite from repetitive maintenance and preserve the operational function and anti-corrosion integrity in the given environment;
(*b*) which methods are compatible with the materials to be preserved and, if a combination of separation methods is considered, whether the whole system will be compatible throughout;
(*c*) which methods will suit the considered geometric form initially and at repetitive maintenance; which method will suit requirements of frictional joints.
(*d*) what will be the effect of thermal shock, abrasion, impact, overheating and cryogenic temperatures on the selective system;
(*e*) will the system provide adequate heat transfer and electrical conductivity and will it satisfy dimensional and weight requirements within the established tolerances;
(*f*) which of the methods, together with their auxiliary requirements, are locally available and effectively applicable;
(*g*) what stage of fabrication and assembly may be critical for the optimum application of considered methods to fit well into the production programme;
(*h*) which of the preselected methods can secure the relevant requirements at the most economic cost?

(5) Metal-coating processes can be classified as anodic and cathodic. The anodic ones will protect substrate metal (even when porous or damaged) through their preferential corrosion, whereas the noble metal coatings, which are mainly used for their superior chemical resistance properties, will accelerate the corrosion of metallic substrate in such circumstances (see Chapter 6, Section 6.10 and *Table 10.1*).

(6) It is necessary to protect anodic coatings (particularly the porous

Table 10.1 TYPICAL LIST OF METAL COATINGS ON STEEL

Process	*Coating*	*Potential**	*DFT (mil and μm)*	*Use or limitation*
Aluminising	Aluminium	A	1–6; 25–152	Factory process
Brush plating	19 metals	V	0.01–6; 0.25–152	Waveguides, site work
Cathode sputtering	Metals Ceramics	V	4 μm; 4	Special applications
Chemical reduction	Cobalt	N	0.1–1; 2.5–25	Special applications
	Copper	N		Printed circuit boards
	Nickel	N		Special applications
	Palladium	N		
	M. alloys	V		
Detonation spray	Metals Ceramics	V	1–12; 25–305	Best quality, special applications, hard surfacing
Diffusion coating	Metals Silicates	V	0.01–30; 0.25–760	Special applications
Electrophoretic coating	Aluminium	A	1–10; 25.4–255	Small parts
	Molybdenum	N		
	Nickel	N		
Electroplating	Aluminium	A	0.25; 6.5	Wire, sheet, small parts
	Cadmium	A	0.15–0.5; 4–12.5	
	Chromium	N	0.005–20; 0.15–510	
	Copper	N	0.01–30; 0.25–760	
	Brass	N	0.07–0.1; 1.8–2.5	
	Gold	N	0.03–0.8; 0.75–20	
	Silver	N	0.1–1; 2.5–25	
	Iron	V	>125; >3175	
	Lead	N	0.25–50; 6.5–1250	
	Nickel	N	0.1–2.2; 2.5–55	
	Platinum	N	0.1–1; 2.5–25	
	Palladium	N	0.1–0.2; 2.5–5	
	Rhodium	N	0.01–0.02; 0.25–0.5	
	Tin	N	0.2–2; 5–50	
	Zinc	A	0.1–1; 2.5–25	
Explosion bonding	Metals	V	60–750; 1525–19050	Plates, tube sheets, strip
Flame spraying	Aluminium	A	4–8; 100–205	Porous, needs sealing
	Zinc	A	2–5; 50–125	Porous, needs sealing
	Tin	N	3–15; 75–380	
	Metals Ceramics	V	5–60; 125–1525	Low melting alloys
Fusion bonding	Metals	V	60–750; 1525–19050	Plates, tubes
Galvanising	Zinc	A	0.5–5; 12.5–125	Maximum length 80 ft (24 m)
Gas plating	Metals	V	0.01–70; 0.25–1780	Special applications
Immersion plating	Copper	N	0.05; 1.25	Special applications
	Nickel	N		
	Silver	N		
	Tin	N		
	Lead	N		
Ion plating	Metals	V	Thin film	Special applications
Lead coating	Lead	N	0.185; 4.7	Special applications
Metal cladding	Aluminium	A	10–300; 250–760	Sheets, plates, strips, tubes, transition joints, special applications
	Brass	N	60–750; 1525–19050	
	Copper	N	60–750; 1525–19050	
	Lead	N		
	Magnesium	A		

Table 10.1 (*Contd.*)

Process	*Coating*	*Potential**	*DFT* (*mil and μm*)	*Use or limitation*
	Nickel alloy	N	60–750; 1525–19050	
	Palladium	N		
	Platinum	N		
	Silver	N		
	Stainless steel	N	5–750; 127–19050	
	Tin	N		
	Titanium	N	31–400; 790–10160	
	Tantalum	N	20–125; 510–3175	
Oxyhydrogen spray	Ni-Cr-Al	N	0.5–1; 12.5–25	Special applications
Peen plating	Aluminium	A	2; 50	Special applications
	Cadmium	A		
	Lead	N		
	Tin	N		
	Zinc	A		
Plasma spraying	Metals Ceramics	V	0.01–100; 0.25–2540	Better quality, high temperature melting metals
Sherardizing	Zinc	A	1–3; 25–75	Small parts
Swab plating	Metals	V	0.01–6; 0.25–150	Special applications
Terne plate	Lead/tin	N	0.01–1; 0.25–25	Sheet steel
Tin dipping	Tin	N	0.006–1.2; 0.15–30	Special applications
Vacuum deposition	Metals Ceramics	V	0.01–3; 0.25–75	Special applications
Vapour deposition	Aluminium	A	0.5–1; 12.5–25	Special applications
	Chromium	N	0.1–1; 2.5–25	
	Iron	—	—	
	Nickel	N	1–100; 25–2540	
	Graphite	N		
Mechanical plating	Cadmium	A		Special applications
	Tin	N		
	Zinc	A		
Hard facing	Metals	V	30–400; 760–10160	Special applications

*A, anodic; N, noble/cathodic; V, various.

ones) with sealers or paints, especially when exposed to acids, marine environments or other corrosive conditions.

(7) Make the basic decision on the optimum coating metal and its method of application:

(*a*) which coating metal will withstand the expected environment;
(*b*) which of the coating metals can be applied to adequate thickness with good coverage and distribution by the available methods (*note*: the danger of microcracking of thickly applied chromium, rhodium or hard metals; corrosion rate of deposited metals versus economical and technical limitations; effect of differences in throwing power);
(*c*) which combination of metallic coating and substrate can provide optimum porosity and galvanic relationship;
(*d*) will the coating method change the physical properties of the substrate;
(*e*) will the coating metal allow the desired physical properties (appearance, colour, brightness, hardness, strength, wear resistance, temperature

resistance), electrical conductivity at the allowable cost and is the optimum technology readily available;

(*f*) what will be the finite purpose of the metallic coating;

(*g*) which one of the available methods of metal coating can secure most of the requirements?

(8) Typical detail appreciation of electrodeposition:

(*a*) which desirable physical, mechanical and chemical properties and what composition of deposited metal are required;

(*b*) what thickness of coating is required (*note*: the nature of substrate, nature of coating, environmental conditions and economics);

(*c*) what hardness of deposits is required;

(*d*) what precautions are necessary to reduce input of high tensile stresses in the deposits;

(*e*) what precoating is required to secure effectiveness of deposits;

(*f*) will the substrate be adversely affected by the process solutions (e.g. hydrogen embrittlement);

(*g*) which available method of application is suitable for the designed component (vat process, barrel process, brush plating, chemical reduction, etc.);

(*h*) what are the desired main and side effects of deposition (corrosion protection, decorative, specular and heat reflecting finishing, wear resistance, prevention of galling, stopping-off during carburising, electroforming, etc.), and which particular technique can provide the optimal results;

(*i*) which of the practical applications is most suitable for the composite of materials, geometry, surfaces and size of the component;

(*j*) what effect will the environmental conditions have on the deposited coatings and, if subject to abrasion, what will be the edge effect of deposited metal on the substrate;

(*k*) which will be the best way to secure non-porosity and uniformity of cathodic/noble coatings;

(*l*) which method of sealing will be compatible with the anodic metals deposits?

(9) Typical detail appreciation of hot-dip metal deposition:

(*a*) will the composite of materials, geometry, surfaces and size of the components suit the available practical application;

(*b*) will the basic metal be adversely affected by the pretreatment process solutions;

(*c*) what effect will the environmental conditions have on the deposited metal coating and, if subject to abrasion, what will be the edge effect of deposited metal on the substrate;

(*d*) what thickness or weight of coating is required to provide the optimum protection;

(*e*) will the reduction of coating thickness by rolling, wiping, centrifuging, etc., of molten metal be required to secure relevant thickness;
(*f*) will the improvement of properties or appearance of coating by chromatising, phosphatising, light rolling or roller levelling be required, and will the removal of palm oil or other post-metallising treatment be necessary on production;
(*g*) will any change of character of the coating by annealing and conversion, by anodising or dyeing, be required;
(*h*) will painting of the deposited metal be required;
(*i*) will any preparation or pretreatment of the deposited metal be required prior to further coating;
(*j*) will any joining be possible after metal deposition; which techniques can be used where hot-dip coatings are applied to raw materials prior to fabrication?

(10) Typical detail appreciation of hot metal spraying (corrosion prevention; sprayed lead for use in atmospheres containing sulphuric acid; tin for food vessels; stabilised stainless steel, nickel and monel for pump rods, impellers, etc., for build-up; hard facing; spray welding; etc):

(*a*) what will be the purpose and use of the metal-sprayed coating;
(*b*) which system of metal spraying will offer the optimum results (molten metal, metal powder, metal wire, electric arc, detonation spray, plasma spray or others);
(*c*) will the bond strength of the flame-spray applied coating exceed the design stress at the interface;
(*d*) will the surface roughness of the substrate be comparable with the particle size of the sprayed metal;
(*e*) will the composite of materials, geometry, surfaces and size of the components suit the available practical application;
(*f*) what effect will the environmental conditions have on the deposited metal coating and, if subject to abrasion, what will be the edge effect of the deposited metal on the substrate;
(*g*) what thickness of coating is optimal and can be applied to the substrate without obvious shear stress between dissimilar metals (shrinkage) which may arise, especially in environmental conditions of fluctuating temperatures, sustained vibration, etc.;
(*h*) will overall uniformity of thickness and minimum porosity be obtained;
(*i*) what hardness of the coating is required;
(*j*) what sealing will be necessary to counteract the porosity of the sprayed metals?

(11) For critical applications, and since thermal-sprayed coatings are not homogeneous materials, it is further advisable to consider:

(*a*) behaviour of melted particles on passage through the flame and the change in composition involved, pick-up of contaminants, embrittlement of layers and its influence on thermal expansion, thermal conductivity and strength of the coating;
(*b*) some metals show higher strength on the plane parallel to the surface than the one perpendicular to it;
(*c*) porosity influences the strength of ductile and brittle coatings, and therefore the pore size, shape and volume of porosity should be evaluated;
(*d*) porosity is influenced by variables such as torch to substrate distance, spray environment, substrate temperature and spray process;
(*e*) pore volume decreases the heat conductivity of the coating;
(*f*) the bond strength must exceed the design stress at the interface and this is relative to the true surface area, its roughness and the thickness of the coating.

(12) Typical detail appreciation of diffusion coating:

(*a*) will the composite of materials, geometry, surfaces and size of components suit the available practical application;
(*b*) will the basic metal be adversely affected by the pretreatment process solutions and by the heat of compression of the diffusion process;
(*c*) what effect will the environmental conditions have on the diffusion layer and what will be the edge effect when damage occurs;
(*d*) will the process secure the overall non-porosity of the coating;
(*e*) will the diffusion coating be adversely affected by normalising, air-hardening and other pretreatments, air or gas welding, brazing and silver soldering, etc.?

(13) The complete coating system is a complex multi-purpose finish, performing protective, sealing and decorative functions (it may also provide lubrication, conductivity, etc.). The system is the basic engineering unit of surface separation rendered wholly or partially by surface coatings or linings. The complete system comprises:

(*a*) preparation of surface to provide optimum interface;
(*b*) application of the required film thickness of the anti-corrosive medium (metallic or non-metallic), the thickness depending upon the service requirements of the coating system;
(*c*) application of the required thickness of sealing and/or decorative medium (sealer) to secure sufficient impermeability against the environment and thus to extend the functional readiness of the anti-corrosive medium;
(*d*) application of special-purpose coatings (anti-fouling, anti-condensation, noise damping, etc.).

(14) The most important parts of the coating system are the preparation of surface (see Chapter 9, Section 9.7) and the selection and application of anti-corrosive medium (various anodic metallic coatings, prefabrication primers, organic or inorganic corrosion-inhibiting primers, conversion coatings, anodising). Undercoats are only for improvement of appearance.

(15) Prefabrication primers are an important part of the whole preservation system. Their integrity should therefore be preserved throughout the process of manufacture, and every economically sound remedial action taken to repair any damage as soon as possible whilst fabrication proceeds and definitely prior to the application of the next coating; no area should remain untreated and open to corrosion for extended periods. All necessary activities should be included in the production planning.

(16) Prefabrication primers should satisfy following requirements:

(*a*) cover adequately the contours of the surface;
(*b*) allow an easy application by brush, roller, spray (all types including electrostatic spray) or by any other method available, required or suitable;
(*c*) secure a fast drying time—not more than 5 min for spray application and 20–30 min for brush or roller applications;
(*d*) have a reasonable pot life;
(*e*) supply a good temporary protection by a thin film both before and after fabrication, until such time as the full paint system can be applied;
(*f*) provide good, if possible permanent, base for widest range of subsequently applied coatings (*note*: solvent resistance);
(*g*) be free of toxic fumes on cutting and welding;
(*h*) will not interfere adversely with flame cutting and welding operations or the quality of the weld outside of established parameters—will also provide only for a minimum backburn without major damage;
(*i*) will withstand cold working of the metal without flaking;
(*j*) be electrically conductive where earthing in fabricated structure is required;
(*k*) possess a good resistance to abrasion and good adhesion to withstand fabrication, transportation and erection;
(*l*) be reliable when used under cathodic protection;
(*m*) be eventually tintable in various colours for marking different grades of basic construction steel, for marking distinctive sections of structures, etc.

(17) There is a large range of primers to choose from, differing in their purpose and quality. The differences, however, are not confined only to the variety of utility and quality within each generic group but also to the design of the coating regarding its method of application and the thickness of the applied film. Where the coating is to be applied to a relatively smooth surface,

with no sharp peaks and for a limited or temporal utility, then a thin film (e.g. prefabrication primer only, etc.) may suffice. Where the texture of the surface is rather more pronounced, where the corrosive conditions are more aggressive and where extended protection is needed, then a thicker film is required; in this case the original pretreatment should be extended by addition of one or several further coatings of primer to suit. Two-step application procedures should be used. Where the texture is even coarser, as on corroded steel, then a very thick film is required. In this case, high build primers can be used, the number of coats varying with the expected life and environmental conditions.

(18) A sealer primarily means any coating or lining which is applied on top of anti-corrosive compositions for the purpose of extending their utility in an efficient state for an economic period. The general requirements of a good sealer are as follows:

(*a*) good adhesion to the anti-corrosive composition (*note*: the use of barrier (tie) coats for improvement of adhesion);
(*b*) low permeability to water or other corrosive media;
(*c*) high film thickness;
(*d*) good chemical resistance;
(*e*) optimal resistance to abrasion;
(*f*) good weather resistance, including resistance to ultraviolet light.

(19) Where protection is required against atmospheric corrosion only (e.g. under rural conditions), it may not be necessary to use sealer provided an adequate film thickness of sacrificial metal contained, for example, in metallic coating or inorganic zinc coating, is applied. Otherwise an application of sealer is a necessity, observing that it is in the interests of the proprietor of a utility to avoid repetition of expensive overall preparation of surface. Sealer extends the effectiveness of anti-corrosive composition and the anti-corrosive composition prevents the onset of corrosion which had penetrated through damaged and porous sealer. Both are complementary to each other.

(20) Seventy-five per cent of the success of protection depends on adequate surface preparation and reliable application. The use of technically skilled industrial and approved applicators is recommended. Engineering planning, accurate specification and complete scheduling of the protection by protective coatings is a necessity.

(21) Protective coatings should only be used if it is more economical than use of corrosion-resistant metals and other materials.

(22) Care should be taken to ensure that all materials are stored, handled and maintained to avoid physical damage, contamination and deterioration of the protective coatings, and the requisite precautions should be planned.

(23) Protection by separation of surface from environment by protective

coatings should be evaluated together with relevant parts of Chapters 6, 8 and 9. Further, the problems and limitations of the applicator, climatic and working conditions, properties of materials in relation to procedures and schedules should be reviewed; application methods to suit the geometry chosen; systems which permit maximum application of money-saving practices, use of minimum number of different materials and least number of colours selected; maintenance practice anticipated.

(24) A typical list of protective coatings is given in *Table 10.2.*

(25) Typical basic appreciation of paint coatings/linings and their methods of application:

(*a*) which individual generic group of paints (or their combination) can withstand the environment;
(*b*) will the composite of materials, geometry, surfaces and maintainability suit the available practical application of a particular generic group of paint materials;
(*c*) what is the main purpose of paint application in the relevant case (general corrosion protection, alleviation of galvanic problems, improvement of mechanical strength, prevention of condensation, damping of noise, decorative purposes, insulation or any combination of the mentioned reasons);
(*d*) will the particular generic group of paint coatings and its application technique suit the planned fabrication processes, manufacturing programmes, location and the local weather conditions;
(*e*) will the coating of a particular generic group be applicable by a suitable method to adequate thickness with a good coverage and distribution;
(*f*) will the porosity factor and the galvanic or chemical relationships between the coating/lining and the substrate be satisfactory;
(*g*) what hardness and abrasion resistance is required of the coating and can the relevant generic group provide for the requirements;
(*h*) what requirements on labour can be expected from the contemplated generic groups of paint;
(*i*) what auxiliary precautions will be necessary for the application of relevant generic groups;
(*j*) what is the approximate comparative cost of relevant generic groups based on unit cost per year of the forecast effective life of the coating/lining?

(26) Typical detail appreciation of the coating system:

(*a*) what total thickness is required and which is the ideal number of coats to provide this thickness;
(*b*) what composition of individual films and their sequence will secure optimal protection;

Table 10.2*

	Alkyd	*Alkyd amine*	*Alkyd phenolic*	*Alkyd silicone*	*Alkyd urea*	*Styrenated alkyd*	*Acrylic*	*Bituminous*
Physical properties								
Sward rocker hard (8th day)	24	30	34	16–30	28	28	24	—
Flexibility	E	VG	G	VG	VG	G	E	E
Abrasion res. cyc. (Taber)	3.5K	>5K	> 5K	4K	>5K	>5K	2.5K	—
Max svc temperature (°C)	93	121	121	232	107	93	82	93
Toxicity	none	slight	none	none	slight	slight	none	—
Impact resistance	VG	E	G	G	E	G	E	E
Dielectric properties	G	G	VG	E	G	G	VG	—
Adhesion to:								
ferrous metals	E	E	E	VG	E	F	VG	E
non-ferrous metals	F	E	E	VG	VG	F	VG	E
old paints	VG	G	G	E	G	VG	P	—
Decorative properties								
Colour retention	G	VG	P	E	VG	G	E	—
Initial gloss	E	E	VG	E	E	E	E	P
Gloss retention	E	G	F	E	F	G	E	—
Chemical resistance								
Atmospheric, exterior	E	E	E	E	E	F	E	F
Salt spray	E	VG	E	E	G	G	E	E
Solvents, alcohols	F	G	G	G	G	G	P	P
Solvents, gasoline	G	E	E	E	E	F	G	P
Solvents, hydrocarbons	G	E	E	G	E	E	F	P
Ammonia	P	P	P	P	P	P	P	—
Alkalis	P–F	G–VG	P–F	G–VG	G	G–VG	F–G	E
Acids, oxidising	P	P–F	P–G	P	P–F	P–F	P–F	—
Water (salt, fresh)	F	G	G	G	F	G	E	E
Application								
Ease of application	E	bake	E	E	bake	E	VG	P–VG
Priming required	PR	none	none	PR	none	none	PR	none
Solvent for application	HyC	HyC	HyC	HyC	HyC	HyC	blend	—
Method	U	U	U	U	U	U	U	U
Cure	A or B	B	A or B	A or B	B	A or B	A	A
Baking temperature (°C)	135	160	177	177	160	149	—	—
Bake drying time	30 min	20 min	30 min	30 min	20 min	15 min	—	—
Air drying times:								
touch	2H	—	20 min	45 min	—	10 min	5 min	2H
handle	4H	—	60 min	2H	—	30 min	15 min	24H
re-coat	4H	—	6H	4–6H	—	4H	15 min	—

*A, air dried; B, baked; E, excellent; F, fair; G, good; H, high; L, limited or low; M, medium; MH, medium high; P, poor; S, slightly limited; U, unlimited; VG, very good.

Table 10.2 (*Contd.*)

	Alkyd	*Alkyd amine*	*Alkyd phenolic*	*Alkyd silicone*	*Alkyd urea*	*Styrenated alkyd*	*Acrylic*	*Bituminous*
hard	12H	—	6H	12H	—	4H	12H	24H
corrosion resistant	48H	—	48H	12H	—	48H	24H	—
Coverage (ft²/gal/mil; m²/l/μm)	450; 280	450; 280	450 280	500; 312	450; 280	400; 249	350; 221	— —
Average dry film thick. (mil; μm)	1.5; 38	1.5; 38	1.5; 38	0.6; 15	1.5; 38	1.5; 38	1.0; 25	3–250; (75–6350)
Cost		L					M	

	Epoxy amine	*Epoxy ester*	*Epoxy furane*	*Epoxy melamine*	*Epoxy phenolic*	*Epoxy urea*	*Furane*	*Phenolic*
Physical properties								
Sward rocker hard (8th day)	36	30	24	36	44	34	38	38
Flexibility	F	E	E	VG	VG–E	VG	F	G
Abrasion res. cyc. (Taber)	>5K	>5K	—	>5K	>5K	>5K	—	>5K
Max svc temperature (°C)	204	149	177	204	204	204	149	177
Toxicity	none	none	none	none	none	none	none	none
Impact resistance	G	E	G	VG	VG	G	F	G
Dielectric properties	VG	VG	G	VG	VG	VG	F	E
Adhesion to:								
ferrous metals	E	E	E	E	E	E	F	E
non-ferrous metals	E	E	E	E	E	E	F	E
old paints	G	VG	E	P	P	P	E	G
Decorative properties								
Colour retention	F	G	G	G	P	G	G	P
Initial gloss	E	E	G	VG	VG	VG	E	VG
Gloss retention	F	G	G	F	F	F	F	F
Chemical resistance								
Atmospheric, exterior	G	E	E	E	E	VG	G	E
Salt spray	VG	E	E	E	E	E	G	E
Solvents, alcohols	G	F	E	E	E	E	E	E
Solvents, gasoline	E	E	E	E	E	E	E	E
Solvents, hydrocarbons	E	VG	E	E	E	E	E	E
Ammonia	G	P	E	P	F	P	E	P
Alkalis	E	E	E	E	E	E	E	P
Acids, oxidising	P–G	P–F	F	P–G	P–E	P–F	P	P–F
Water (salt, fresh)	G	VG	E	G	E	G	E	E
Application								
Ease of application	cat	E	E	bake	bake	bake	E	E
Priming required	none	PR	none	none	none	none	PR	none

Table 10.2 (*Contd.*)

	Epoxy amine	*Epoxy ester*	*Epoxy furane*	*Epoxy melamine*	*Epoxy phenolic*	*Epoxy urea*	*Furane*	*Phenolic*
Solvent for application	blend	HyC	ket	blend	blend	blend	ket	alc
Method	L	U	U	L	L	L	U	U
Cure	A	A or B	B	B	B	B	A or B	A or B
Baking temperature (°C)	—	160	177	177	204	177	149	177
Bake drying time	—	30 min	30 min	30 min	30 min	30 min	30 min	30 min
Air drying times:								
touch	60 min	1H	—	—	—	—	1H	10 min
handle	2H	2H	—	—	—	—	4H	30 min
re-coat	6–8H	8H	—	—	—	—	6H	30 min
hard	12H	8H	—	—	—	—	24H	4H
corrosion resistant	7–10D	5D	—	—	—	—	48H	24H
Coverage (ft^2/gal/mil; m^2/l/μm)	500; 312	450; 280	450; 280	500; 312	450; 280	500; 312	400; 249	350; 221
Average dry film thick. (mil; μm)	1.8; 45	1.5; 38	0.5–1.0; 12–25	1.8; 45	1.8; 45	1.8; 45	3–5; 76–127	1.5; 38
Cost	H							

	Polyamide (nylon)	*Polyester*	*Silicone*	*Polyethylene*	*Chlorinated rubber*	*Neoprene rubber*	*Hypalone rubber*	*Viton*
Physical properties								
Sward rocker hard (8th day)	—	30	16	F	24	<10	<10	<10
Flexibility	G	G	F	E	VG	E	E	G
Abrasion res. cyc. (Taber)	—	3.5K	2.5K	—	>5K	5K	5K	1K
Max svc temperature (°C)	149	93	288	93	93	93	121	288
Toxicity	—	none	none	none	slight	none	—	slight
Impact resistance	VG	F	F	F	G	E	E	E
Dielectric properties	G	G	E	E	E	G	VG	G
Adhesion to:								
ferrous metals	VG	F	F	E	F	VG	VG	VG
non-ferrous metals	VG	P–F	E	E	VG	VG	VG	VG
old paints	—	P	E	—	—	—	—	—
Decorative properties								
Colour retention	—	G	E	VG	G	G	E	G
Initial gloss	G	G	E	VG	F	P	P	E
Gloss retention	—	F	E	VG	F	F	F	F
Chemical resistance								
Atmospheric, exterior	P	VG	E	P	E	E	E	E

Table 10.2 (*Contd.*)

	Polyamide (nylon)	*Polyester*	*Silicone*	*Polyethylene*	*Chlorinated rubber*	*Neoprene rubber*	*Hypalone rubber*	*Viton*
Salt spray	F	G	E	VG	E	E	E	E
Solvents, alcohols	G	G	F	E	E	E	—	E
Solvents, gasoline	G	E	F	P	G	G	G	E
Solvents, hydrocarbons	—	G	VG	VG	—	—	—	—
Ammonia	G	P	P	E	G	G	G	E
Alkalis	G	P	F–E	P–G	E	E	E	E
Acids, oxidising	—	P	P	VG	F–E	F–P	F–G	E
Water (salt, fresh)	F	G	E	VG	E	E	E	G
Application								
Ease of application	G	F	E	E	G	VG	VG	VG
Priming required	none	PR	PR	none	PR	none	PR	PR
Solvent for application	—	styr	HyC	—	HyC	HyC	HyC	blend
Method	L	L	U	SL	U	U	U	L
Cure	A	A or B	A or B	B	A or B	A or B	A or B	A or B
Baking temperature (°C)	—	149	232	232	149	149	149	—
Bake drying time	—	15 min	1H	15 min	15 min	15 min	15 min	—
Air drying times:								
Touch	—	1H	45 min	—	45 min	15 min	15 min	5 min
handle	—	1H	2H	—	2H	30 min	30 min	15 min
re-coat	—	1H	4–6H	—	4–6H	4H	4H	12H
hard	—	12H	12H	—	4–6H	4H	4H	12H
corrosion resistant	—	7–10D	12H	—	24H	7–10D	7–10D	12H
Coverage (ft²/gal/mil; m²/l/μm)	— / —	800; 498	350; 221	560; 348	450; 280	300; 186	250; 155	200; 125
Average dry film thick. (mil; μm)	2–30; 50–762	2.0; 50	1.0; 25	3–10; 76–254	1.5; 38	2–10; 50–254	2.0; 50	1.0; 25
Cost		M						

	Urethane	*Vinyl*	*Vinyl alkyd (1:1 approx)*	*Vinyl organosol*	*Vinyl plastisol*	*Nylon powder*	*Cellulosic*	*Chlorinated polyether*
Physical properties								
Sward rocker hard (8th day)	35–65	20	26					
Flexibility	E	E	E	E	E	G	G	F
Abrasion res. cyc. (Taber)	>5K	>5K	2.5K					
Max svc temperature (°C)	149	66	82		93	82	82	121
Toxicity	slight	none	none					
Impact resistance	E	E	E	E	E	VG	E	G
Dielectric properties	E	E	G		VG	G	VG	VG
Adhesion to:								
ferrous metals	E	G	VG	E	E			

Table 10.2 (*Contd.*)

	Urethane	*Vinyl*	*Vinyl alkyd (1:1 approx)*	*Vinyl organosol*	*Vinyl plastisol*	*Nylon powder*	*Cellulosic*	*Chlorinated polyether*
non-ferrous metals	E	VG	G					
old paints	—	—	—					
Decorative properties								
Colour retention	G	VG	E	G	VG	VG	E	G
Initial gloss	E	G	E	P	VG	G	E	G
Gloss retention	F	E	E		G	—	E	—
Chemical resistance								
Atmospheric, exterior	E	E	E	E	E	F	E	F
Salt spray	E	E	E	E	E	G	E	E
Solvents, alcohols	VG	F	G	F	E	G	F	E
Solvents, gasoline	F–G	E	E	F	E	E	G	VG
Solvents, hydrocarbons	—	—	—	—	G	E	G	E
Ammonia	P	E	P	P	E	G	P	E
Alkalis	F–VG	E	P–G	F	E	G	F	E
Acids, oxidising	P–G	G–E	P	P	E	P	P	E
Water (salt, fresh)	E	E	E	E	E	F	VG	E
Application								
Ease of application	E	F	VG	G	G			
Priming required	PR	PR	PR					
Solvent for application	blend	blend	blend	blend	none			
Method	U	L	U	L	L	L	L	L
Cure	A or B	A or B	A					
Baking temperature (C)	163	149	—					
Bake drying time	30 min	15 min						
Air drying times:								
touch	45 min	15 min	5 min					
handle	1–2H	30 min	15 min					
re-coat	4–6H	4–6H	15 min					
hard	18H	4–6H	12H					
corrosion resistant	5–7D	24H	24H					
Coverage (ft^2/gal/mil; m^2/l/μm)	—	250; 155	200; 125					
Average dry film thick. (mil; μm)	1.4; 36	1.0; 25	1.0; 25					
Cost		MH	M	L	L			

(*c*) what primers or anti-corrosive coatings are to be used and what will be their effect on the substrate;
(*d*) what fillers and undercoatings or intermediate coatings (barrier or tie coats) will be required to secure inter-coat adhesion and appearance of the finished product;
(*e*) what top coatings will be required and what will be their purpose (anti-corrosion, sealing, reflection, reduction of impact, insulation, anti-fouling, decorative, etc.);
(*f*) what colour differentiation between individual coats will be required for easy inspection;
(*g*) will the individual coats be mutually compatible;
(*h*) what will be the relation between the sequence of application of individual films, the manufacturing processes and their locality;
(*i*) what time lapse between individual films can be expected and what will be the effect on production planning and the final effectiveness of the coating system;
(*j*) what will be the effect of the weight of the coating system and its distribution on the overall design parameters;
(*k*) which part of the system can be applied at prefabrication stage, during fabrication and, finally, after assembly;
(*l*) what will be the approximate forecast life of the coating system *in situ* and, if this period is shorter than the life of the designed utility, can it be renewed in the probable location at the time of failure (*note*: include the requirement of surface preparation)?

(27) Typical detail appreciation of the product (e.g. proprietary paint):

(*a*) what wetting properties does the product possess and are these suitable;
(*b*) will the particular paint film be elastic enough to expand and contract with the substrate over a long period;
(*c*) will all the products in the system be compatible—has the primer an affinity with the substrate metal or non-metal;
(*d*) will the particular primer have a good adhesion to the various substrates within the assembly with the considered preparation of surfaces;
(*e*) what will be the expected detail environmental resistance and behaviour of the product under expected conditions (if possible based on tests);
(*f*) what will be the hardness and abrasion or impact resistance of the product under expected physical conditions—its friction coefficient;
(*g*) what will be the flexibility of the product in relation to the expected stresses and variable thermal loading;
(*h*) what will be the toxicity of the product at application and in the cured state and does this suit the application and operation arrangements;
(*i*) will the application of the product influence the electrical conductivity of the structure or equipment (when required or for welding);

(*j*) what are the decorative properties of the product and do these satisfy the demand (colours, colour retention, gloss, gloss retention, powdering, self-cleaning, etc.);
(*k*) what are the detail requirements for the surface preparation and application; what plant, accessories and experts will be required; are approved applicators available, are guarantees obtainable;
(*l*) what thinners and solvents will be required for thinning and cleaning;
(*m*) what drying methods will be required (air drying, stoving, etc.);
(*n*) what time and drying temperatures does the product require (touch dry, handling dry, overcoat dry and cured)—do these fit into the production plan;
(*o*) are there any limitations regarding temperatures and environmental conditions stated for the product—corrosivity of outgassing;
(*p*) what is the theoretical and practical coverage of the product;
(*q*) what is the optimal practical coat thickness of the product (wet or dry);
(*r*) will the above-mentioned detail parameters suit the designed utility?

(28) Typical detail appreciation of proprietary prefabrication primers (further to previous statements):

(*a*) will there be suitable and effective facilities for prefabrication priming available; can the prefabrication primed metal be supplied ex stock;
(*b*) will the substrate metal be suitable for prefabrication priming (type and thickness);
(*c*) will the handling, storing and fabrication facilities and programme be attuned to the proprietary prefabrication primer;
(*d*) what is the workmen's (trade unions) attitude towards the working of prefabrication primed metals, especially welding;
(*e*) will the removal of primer prior to flame cutting or welding be necessary (critically loaded structures), or can arrangements be made to mask the critical welding surfaces prior to priming;
(*f*) will it be necessary to remove the proprietary primer overall or partially to further coating;
(*g*) what will be the effect of weathering (in stock and in work) on prefabrication primed metal and what precautions will be necessary prior to application of further coatings?

(29) Suitably precoated metals (fabrication process) are preferred to complete or partial post-fabrication treatment where the degree of required protection, the construction and the joining will permit.

(30) The method and technique of application of protective coatings should be accurately specified and adequate inspection instructions issued.

(31) Depending on environmental conditions and purpose simple applications of plastic or combinations of metallising and plastic coatings should

be used where possible. The plastic coatings can be applied by a suitable method (dipping, fluidised bed, electrostatic fluidised bed, electrostatic coating, flocking, flow cladding, etc.).

(32) Typical detail appreciation of plastic coatings (*note*: general appreciation of plastic coatings will be similar to the appreciation of paint coatings):

(*a*) will the plastic coating lend itself to application by available facilities;
(*b*) will the process be rapid and economic enough;
(*c*) will the plastic coating withstand atmospheric weathering conditions;
(*d*) will the plastic coating be tough enough to endure the abrasion and impact of handling, loading and unloading of storage and transport facilities, and stringing equipment;
(*e*) will the plastic coating have sufficient flexibility to withstand the maximum bends utilised at temperatures from 20°F to 140°F (−6.7°C to 60°C);
(*f*) will it not melt or burn back within $\frac{1}{2}$ in (1.2 cm) of the weld and be compatible with a joint system subsequently applied to protect the weld area;
(*g*) will it resist the impact of rocks and soil during backfill operation; also, will it resist the wear and tear of fitting and normal operation;
(*h*) will it not crack or disbond during hydrostatic and other testing;
(*i*) will it not soften at temperatures below 200°F (93°C) when used on hot line service;
(*j*) will it resist penetration of subsurface waters or liquid contents;
(*k*) will it resist the chemical attack from outside (e.g. natural soil chemicals, fertilisers) or inside;
(*l*) will it not be attacked by bacteria and fungus in the soil;
(*m*) will it resist the solvent action of products in permanent contact or occasional contact in the event of overflow, spillage or breakage (e.g. aviation gasolene, jet fuel, crude oil, etc.);
(*n*) will it possess the adhesive forces and chemical inertness which will resist, within an economical lifetime, the effects of cathodic protection systems in soils or sea water of low resistivity?

(33) The designer should decide initially whether the polarisation of materials in conductive media shall be secured:

(*a*) by cathodic protection—ships' hulls and appendages, cargo and ballast compartments, bilges, sea inlets and discharges, off-shore structures, jetties and navigational aids, off-shore pipelines, harbour structures, heat exchangers, box coolers, large seawater storage tanks, buried pipelines, well casings and gathering lines, public utilities, lines and cables, buried feet of overhead power pylons and metallic telephone posts, industrial storage tanks, gas holders, bottle washing machines and other industrial plant, reinforcing rods and wires in prestressed concrete and other structures or equipment immersed in aqueous solutions of electrolyte (pure water, river water, potable water, sea water, wet soils and weak acids) and in weak to medium corrosive environments where proportionally higher consumption of protective currents is allowed;

(*b*) by anodic polarisation of active/passive metals—alloys of nickel, iron, chromium, titanium and stainless steel in weak to extremely corrosive environments, where economy in consumption of protective currents is required;

(*c*) by coating with anodic metals (zinc, aluminium, cadmium), which may be appreciated either as a part of surface separation or a part of cathodic protection.

(34) When the initial decision to use cathodic protection has been made it must be decided upon whether to use impressed currents or sacrificial anodes:

(*a*) size and geometry of the project (impressed currents method is usually used for large projects);
(*b*) availability of the power supply;
(*c*) possibility of the interface problems;
(*d*) necessity for safety from spark hazards and accumulation of hydrogen in enclosed spaces;
(*e*) replaceability of sacrificial anodes;
(*f*) expected economic life of the system.

(35) Typical basic appreciation of cathodic protection by sacrificial anodes:

(*a*) estimate of total current requirements (current densities allowed, spare capacity, allowance for protective coatings and linings, assessment of environmental media);
(*b*) resistivity of water, soil or other electrolyte solution;
(*c*) requirements for insulating flanges and bonding to foreign structures and assessment of extra current allowances;
(*d*) selection of suitable anode metal (zinc, magnesium, aluminium, iron, mild steel or other metals anodic to the protected structures or equipment) and its alloying composition;
(*e*) requirements for introduction of current control to limit output within the optimum parameters;
(*f*) selection of the size of anodes to provide optimum life;
(*g*) selection of the suitable shape of anodes to secure optimum spread;
(*h*) determination of the total number of anodes required;
(*i*) anode spacing to give uniform current distribution;
(*j*) selection of test-point localities;

(*k*) attachment of anodes (*note*: sacrificial anodes should be conductively attached to the protected metal but their sacrificial mass should preferably be separated from the protected surfaces).

(36) Typical basic appreciation of cathodic protection by impressed current/cathodic control:

(*a*) estimate of total current requirements;
(*b*) resistivity of water, soil or electrolyte solutions;
(*c*) requirements for insulating flanges and bonding to foreign structures and equipment and assessment of extra current allowances;
(*d*) selection of suitable groundbed locations (in low resistivity soils or media, reasonably near power supply, at points where there are no interference problems, where beds and cables are reasonably secure from interference or disturbance);
(*e*) decision on the type of anodes and the design of their attachment;
(*f*) decision on whether the anodes (if elongated ones selected) should be installed vertically or horizontally;
(*g*) decision on the voltage to be used;
(*h*) determination of the optimum anode material;
(*i*) optimum number and size of the anodes;
(*j*) decision on anode spacing;
(*k*) type and location of reference electrodes;
(*l*) requirements and design of grounding of propeller shaft, rudder and other attached substructures and equipment within the protected complex—materials and systems;
(*m*) location of controllers, power supply and transmission (cabling and installation);
(*n*) potential hazards of marine and surface traffic;
(*o*) wave action and soil instability;
(*p*) bottom involvement;
(*q*) weed fouling and microbiological effect;
(*r*) malicious damage.

(37) Where cathodic protection is to be used the alkali resistance of the protective paint coatings should be evaluated.

(38) Where possible, cathodically protected surfaces should be preserved by suitable surface coatings or linings.

(39) All precautions should be taken to prevent hydrogen embrittlement of high strength metals arising from their cathodic protection.

(40) Detail design of cathodic protection systems is a highly specialised field of expertise and should be left primarily to a corrosion specialist. However, it will be the designer's task to accommodate, eventually, the diagrammatic detail design rendered by the corrosion specialist in the functional design of the utility to their mutual satisfaction.

(41) Use of zinc-rich primers on cathodically protected structures or equipment in conductive environment is not generally recommended.

(42) Typical basic appreciation of anodic polarisation by impressed currents/anodic control:

(*a*) estimate of total current requirements;
(*b*) is the used chemical/metal system suitable for anodic polarisation (e.g. oleum and carbon steel, cold concentrated sulphuric acid and carbon steel, hot concentrated sulphuric acid and stainless steel, dilute sulphuric acid and stainless steel, etc.)?
(*c*) conductivity of liquid, its temperature, pH, pressure and velocity;
(*d*) minimum, normal and maximum concentration of the liquid;
(*e*) is any substance present which might coat, abrade or coagulate?
(*f*) decision on the type of the cathodes and design of their attachment;
(*g*) decision on the voltage to be used;
(*h*) selection of the optimum cathode material;
(*i*) optimum number, size and spacing of cathodes;
(*j*) type and location of reference electrodes;
(*k*) location of controllers, power supply and transmission;
(*l*) potential fouling of cathodes and reference electrodes.

(43) Reduction of corrosion by a change of environment is to be considered, provided the design is suitable and this can be achieved without excessive cost by any one, or several, of the following methods:

(*a*) lowering the corrosiveness of the atmosphere or other corrosive media by ventilation, dehumidification, air conditioning, reduction of acid strength, sacrificing chemical efficiency for the sake of lower corrosion costs, continuous venting of steam from the unit, reduction of concentration of CO_2 and oxygen in condensate, etc.;
(*b*) adjusting the thermal efficiency of the components by raising or lowering the temperature by reduction of thermal efficiency of preheaters and boilers, by making heat exchangers co-current instead of counter-current, by reduction of peak metal temperature, etc.;
(*c*) using the inhibitors in critical media, e.g. fuels, process liquids, cooling waters, paints, elastomers, etc.

(44) Typical basic appreciation of ventilation, dehumidification and air conditioning for change of environment:

(*a*) requirements for habitability;
(*b*) adjustments of environment to improve protection through control of corrosiveness;
(*c*) corrosion rating of particular design complex;

(*d*) direction of air flow and distribution details;
(*e*) prevention of access of corrosive atmospheres, fumes and liquid media to vital equipment and structures;
(*f*) reduction of condensation precipitation on vital surfaces;
(*g*) facilities for filtration of contaminants and solids from atmospheres; prevention of dusting;
(*h*) upkeep of relative humidity within the utility below 60% and constant;
(*i*) acceleration of drying of the corrosion-prone surfaces;
(*j*) introduction of forced drafts.

(45) Typical basic appreciation of adjustment of thermal efficiency for change of environment:

(*a*) local thermodynamic properties;
(*b*) boundary layer edge mass flux;
(*c*) flow direction;
(*d*) flow separation and reattachment regions;
(*e*) vortices or wake regions;
(*f*) condensed phase impingement;
(*g*) boundary layer heat and mass transfer;
(*h*) environment thermodynamic properties and chemical composition;
(*i*) component configuration and flow field;
(*j*) component material radiation emissivity;
(*k*) component material;
(*l*) interface;
(*m*) surface deposit factor.

(46) Desiccating agents used in corrosion prevention must be cheap, easy to handle and non-corrosive. An easy access for inspection and replacement must be provided and eventually a provision for regeneration *in situ* should be made.

(47) A typical list of inhibitors is given in *Table 10.3*.

(48) Typical basic appreciation of inhibitors for the purpose of change of environment:

(*a*) what is the effect of inhibitor concentration on corrosion rate?
(*b*) minimum concentration needed;
(*c*) tendency to favour pitting—effects at water line;
(*d*) relation to surface area of metal—initial consumption (in coating surface, in reacting with existing corrosion scale);
(*e*) effectiveness as a function of time;
(*f*) tendency to be consumed by reaction with ingredients of corrosive medium;
(*g*) effectiveness under varied conditions that may be found in plant

Table 10.3

	Chromates	*Nitride*	*Phosphates*	*Silicates*
Composition	pH ca 8.5 = Na_2CrO_4 pH *ca* 3–4 = $Na_2Cr_2O_3$	pH *ca* 7.2 = $NaNO_2$ less at pH > 8	Na_2HPO_4, galgon, $(NaPO_3)_6$, etc.	$Na_2O_3.2SiO_2$
Group	Inorg.	Inorg.	Inorg.	Inorg.
Mechanism	Anod. polar.; adsorbed; metal more cathodic; film forming	Anod. polar.; metal more cath.; no visible deposits	Usually visible; insoluble metallic phos. scale	Silica deposit & insoluble met. silicates; not passivator
Toxicity	Toxic	Non-tox.	Non-tox.	
Complex	Polyphosphates;	Phosphates	Chrom, Nitr.	
Consumption	v. slow—Fe by $Fe(OH)_2$, H_2S	By $Fe(OH)_2$, H_2S	At high temp.; galgon to ortho	
Fresh water	pH > 6 Fe, Al, Zn, Cd, Pb, Mg, Cu–Zn	pH > 6 Fe, Al, Ni, SS	pH > 6 Fe, Al, Zn, Cd; Pb, Cu–Zn	pH > 7 Fe, Zn, Cd, Al = (no Cl^-) Pb, Cu–Zn
Brine (non-aerated)	Low temp., *ca* 0.3% in NaCl 0.2% in $CaCl_2$ Fe, Zn, Cu–Zn			Cold, *ca* 50 ppm = Fe; oilfield brines = Al
Alkaline sol.	Zn, Cd, Pb, Al, Mg	Hot strong caustic (no O) = Fe		Weak solutions = Al, Zn, Sn; cleaning = Al
Acids	Strong = border-line maybe pass. SS (> 12 Cr)			
Galvanic couples	Al–Cu, Al–brass, Zn–Cu difficult			
Scale control			Yes	
Fresh water use			1–20 ppm galgon = Fe	1–50 ppm hot or cold
Process cooling water	100–500 ppm	Aerated 50–500 ppm	1–10 ppm galgon = Fe (Ca^{2+} present)	Add other inhib.
Engine cooling water	Closed syst. 1000–2000 ppm	Yes + other inhib. for non-f.		
A/CN water	200–400 ppm			
Refrig. brines	Yes			
Gasoline pipe w.	Yes	Yes		
Gasoline condensate				
Sour crude tanks				
Hydrocarbons				
Sour well flooding	13 ppm Zn Crom			

Table 10.3 (*Contd.*)

	Borates	*Iodates, Molybdates, Tungst. Vanad., Ferrocyanides*	*Calcium*	*Zinc Magnesium Manganous Chromic*
Composition	Borax, $Na_2B_4O_7$	Various	Cao	$ZnSO_4$ $MgSO_4$, $MnSO_4Cr_2(SO_4)_3$
Group	Inorg.	Inorg	Inorg.	Inorg.
Mechanism	Poor inhibitor; mainly alkal. agent	Probably anodic polarisation	Contr. scale w. treatment (req. Ca^{2+} & HCO_3) dep. on cath.	Precip. insoluble hydroxides at cathode by OH^-
Toxicity			Non-tox	
Complex	Yes	Org.		
Consumption				
Fresh water	pH > 7 = Fe; possibly non-ferrous	With some other inhib. = Fe	pH to barely favour dep. of $CaCO_3$ scale	Poor for Fe
Brine (non-aerated)				
Alkaline sol.				
Acids				
Galv. couples				
Scale control			Yes	
Fresh water use		30 ppm minimum		
Process cooling water			Yes	
Engine cooling water	In anti-freeze with oth. inhib.			
A/CN water				
Refrig. brines				
Gasoline pipe w.				
Gasoline condens.				
Sour crude tanks				
Hydrocarbons				
Sour well flooding				

Table 10.3 (*Contd.*)

	Ferric Cupric	*Arsenic Antimony*	*Tin*	*Ammonia*
Composition	$(Fe_2SO_4)_3$, $CuSO_4$	$NaAs_2O_3$, $SbCl_3$	$SnCl_2$	NH_3
Group	Inorg.	Inorg.	Inorg.	Inorg.
Mechanism	Maint. of passive cond. of SS	Depos. of As or Sb film increases H overvoltage	Action not powerful—useful results	Neutralising agent mostly
Toxicity		Toxic		Toxic
Complex			Yes (org. N comp.)	Yes
Consumption				
Fresh water				
Brine (non-aerated)				
Alkaline sol.				
Acids	Strong H_2SO_4, H_3PO_4, etc. = SS; danger low conc. & Cl^-	Strong min. acids; high temp. & conc. = Fe; hot $AlCl_3$ + HCl + arom. hydroc. = Fe, Ni	Fe	Fe
Galv. couples				
Scale control				
Fresh water use				
Process cooling water				
Engine cooling water				
A/CN water				
Refrig. brines				
Gasoline pipe w.				
Gasoline condens.				Yes (combined)
Sour crude tanks				Yes (combined)
Hydrocarbons				
Sour well flooding				

Table 10.3 (*Contd.*)

	Oxygen (air)	*Hydrogen sulphide*	*Carbon monoxide*	*Sodium benzoate*
Composition	O_2	H_2S	CO	C_6H_5.COONa
Group	Inorg.	Inorg.	Inorg.	Org.
Mechanism	Maintenance of passive cond; formation of films			
Toxicity		Toxic		
Complex				
Consumption				
Fresh water	SS & other active/pass. metals.			pH > 7; *ca* 2% for Fe & other metals
Brine (non-aerated)	SS & other act./pas. metals			
Alkaline sol.	Hot strong caustic = Fe; danger = air + S	Prev. stress corr. cracking; brass in NH_3		
Acids	SS–corr. stop (50%, H_2SO_4, 200°F)		Dilute 18–8 SS	
Galv. couples				
Scale control				
Fresh water use				
Process cooling water				
Engine cooling water				Yes
A/CN water				
Refrig. brines	Yes			
Gasoline pipe w.				
Gasoline condens.				
Sour crude tanks				
Hydrocarbons				
Sour well flooding				

Table 10.3 (*Contd.*)

	Mercaptans	*Soluble oils*	*Nitrogen-base compounds/ Polar organics*	*Formaldehyde and derivatives* (NH_3, H_2S, etc.)
Composition	Various	Oil + soap (chiefly petrol sulphonates or Na salt)	C_5H_5N, $C_{13}H_9N$, CH_4N_2S, $C_3H_6N_2S$, etc.	H. CHO
Group	Org.	Org.	Org.	Org.
Mechanism		Spontaneous dispersion in w. & wetting m; thin oil film	Adsorb. incr. of H overvolt; pref. wetting & displ. water	
Toxicity	Toxic		Toxic	Irritant
Complex			Iodite (?)	
Consumption				
Fresh water		Fe & other met.; danger Fe = long immersion water	Not very effect.; Fe in waterless oil phase	
Brine (non-aerated)				
Alkaline sol.				
Acids	Fe		Pickling acids, Fe	Fe
Galv. couples		Preferential deposit on anodic points		
Scale control				
Fresh water use				
Process cooling water		Yes, but danger of fouling		
Engine cooling water		Yes, but danger of fouling		
A/CN water				
Refrig. brines				
Gasoline pipe w.			Yes	
Gasoline condens.				
Sour crude tanks			Pitting	
Hydrocarbons			Yes	
Sour well flooding			Yes	
Cutting oils			Yes	

(different temperatures, concentrations of corrosive, velocities, aeration, etc.);

(*h*) effectiveness on metal already corroded.

(*i*) Can the cost of maintaining a sufficient quantity of inhibitor in the system, and the cost of testing that this quantity is being maintained at an appropriate level, be kept within reasonable economic boundaries;

(*j*) can the inhibitor contaminate the product/contents;

(*k*) can the inhibited fluid present an effluent problem;

(*l*) can the inhibitor loosen corrosion deposits and thus cause blockages;

(*m*) can the inhibitor precipitate on stream and can the sludge or scale thus formed be acceptable;

(*n*) can the organic inhibitor coat heavily the surfaces to the important and considerable detriment of efficiency of the heat transfer and filtration or may it give undesirable emulsification, ion exchanger, etc.;

(*o*) what effect will it have on other metals that may be present—effect on bimetallic couples;

(*p*) can the inhibitor cause foaming and thus impair the operation;

(*q*) what are the hazards in handling—toxicity;

(*r*) what would be the cost and effect of the fall in the inhibitor concentration?

(49) The combined effect of inhibitors and cathodic protection is far greater than the individual effect of each method separately.

(50) Avoid packaging materials containing soluble salts or acids in significant quantities or emitting corrosive vapours. Prevent entrapment of gaseous contaminants carried by air between the metallic components and the packaging materials.

10.4 Structures

(1) Anodic metallic coatings have proved their economic value for the protection of capital structures (galvanised, metallised, zinc-rich paints).

(2) Where, however, the use of metallic coatings is contemplated for protection of strength structures, attention should be given to the problems of ageing, cracking, diffusion, corrosion and hydrogen embrittlement (this one due both to the methods of preparation of surface and the development of gaseous hydrogen by the cathodic protection process).

(3) Metallic coatings used under insulation should always be well sealed and protected.

(4) Zinc coatings have a good corrosion resistance in most neutral environments, especially if passivated. Zinc coatings without sealer should not be used in corrosive conditions (marine and industrial environments), in totally unventilated spaces and in proximity of electronic equipment subject to phenolic vapours emanating from insulating materials, varnishes or encapsulants.

(5) Average thickness of zinc sprayed on structural steel is normally 3 mils

(76 μm); in corrosive conditions up to 6 mils (153 μm) thickness is used. Average weight of zinc applied by galvanising on structural steel is 2 oz/ft^2 (61 mg/cm^2) surface.

(6) Aluminium coatings (99.5% commercial purity aluminium) have a good corrosion resistance to marine conditions, industrial atmospheres, weak acids, etc.; layer corrosion of heat treated aluminium can be completely stopped by hot sprayed aluminium coating (main impurity must not be copper) of its surfaces. Coupling of so-protected structures to copper, lead or other noble metals is not normally recommended.

(7) Average thickness of hot sprayed aluminium on structural metals (steel, aluminium) is normally 4 mils (102 μm); for immersed conditions up to 8 mils (203 μm) of aluminium spray can be specified.

(8) All provisions must be made in design for application of a uniform thickness of protective metallic coatings (see Chapter 9).

(9) Cadmium metallic coating is superior to zinc coating in stain and tarnish resistance in rural environments. In marine conditions its resistance is uncertain. Chromate post-treatment should be used. Cadmium coating should not be used in totally unventilated spaces and in proximity of electronic equipment subject to phenolic vapours emanating from insulating materials, varnishes or encapsulants.

(10) Lead coatings have good corrosion resistance to sulphuric acid and to industrial atmospheres without chlorides or nitrates.

(11) Cathodic metallic coatings should not be used on submerged and underground structures subject to physical damage and abrasion.

(12) Areas of structural metals affected by cavitation can be surfaced by welding wire or strip, or overlay welding, or coating with dense high-tensile materials that resist damage of cavitation (e.g. chromium stainless steel 18–8).

(13) Thermal-sprayed coatings are not homogeneous isotropic entities, and they do not have the same properties as identical bulk materials:

(*a*) passage through flame or arc causes preferential oxidation;
(*b*) contaminants are picked up;
(*c*) strength is lost and coating may become embrittled;
(*d*) reaction to heat treatment changes;
(*e*) thermal conductivity changes;
(*f*) porosity of coating influences fracture behaviour;
(*g*) bond adherence varies.

(14) Perform detailed analysis of local environmental conditions prior to undertaking activities appertaining to selection of protective systems (*Figure 10.1*).

(15) Weathering, etching, hot phosphating or priming with calcium

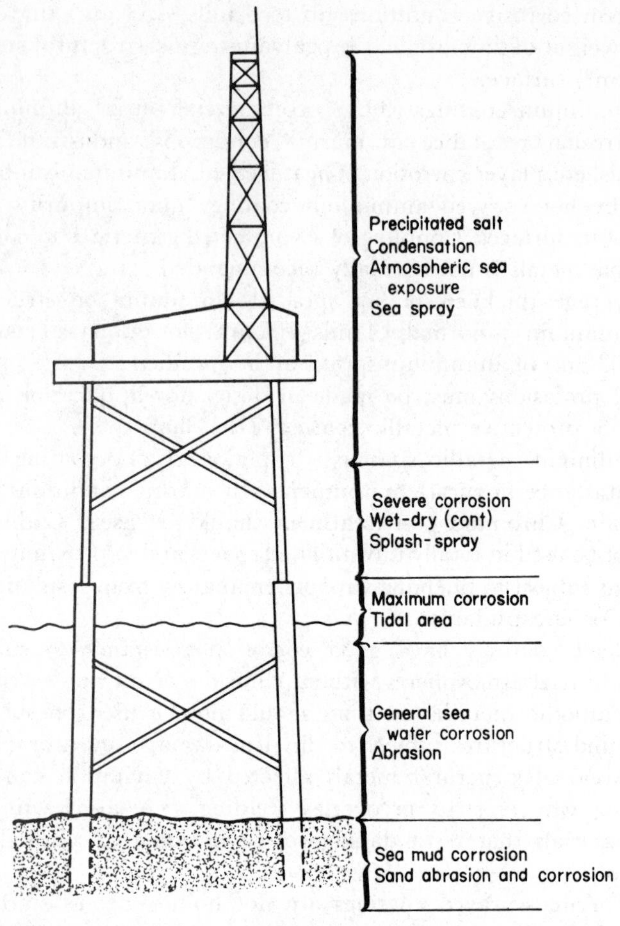

Figure 10.1

plumbate should be specified prior to application of sealer or paint on top of galvanising.

(16) Appropriate cleaning, etching or priming with zinc chromate primer or barrier coat should be specified prior to application of sealer or paint on top of hot sprayed metal or zinc-rich primer.

(17) Due consideration should be given to any adverse effect of the coating on the substrate of metal or metallic coating (e.g. lead- or copper-containing compounds should not be applied on top of solid or coating zinc or aluminium). This applies also to application over zinc-rich primers.

(18) Prefabrication treatment of structural metals, critical strength permitting, is recommended. Fabrication procedures must be fitting to the use of prefabrication-treated metals.

(19) Basic thickness of anti-corrosive coating should fit the surface texture and the demand on initial protection. Prefabrication primer approximately. 0.5–1.0 mil (12–25 μm) thickness; anti-corrosive primer approximately 1.5–2.5 mil (38–64 μm) thickness. Number of coats varies with expected life and the environmental conditions, observing that under ideal conditions the over-coating of metallic or zinc-rich anti-corrosive coatings may not be necessary.

(20) Where corrosive conditions are more severe, such as in marine or industrial environments, then the application of sealer or standard paint coatings forming the coating system over the anti-corrosive layer will be necessary, and the total dry thickness will be progressively increased from the normal minimum thickness of the coating system in atmospheric conditions of 5 mils (127 μm) to whatever maximum thickness is required for its purpose, allowable by the material composition of the coating and economical feasibility of the application. Note difference between standard coatings and high build paints.

(21) To facilitate application and inspection select individual and different colours or tinting of successive coats within a paint system.

(22) In order to obtain satisfactory protection it is necessary to ensure complete coating of the surface with absence of interruptions, pinholes and misses.

(23) High duty paints and compositions should be specified for protection from corrosive fluids, in less accessible spaces and for protection of the cathodic metal in a galvanic couple.

(24) Painting of faying surfaces of crevices before these are mated, preferably with inhibitive paints, should be planned and specified.

(25) Painting of surfaces (complete), which become inaccessible on assembly, should be specified on design drawings prior to assembly.

(26) Additional brush coat on all edges, corners, rivets and other fasteners, welds and in inaccessible spaces should be specified (two-step application).

(27) Post-assembly and post-painting flame cutting and welding should be reduced to a minimum. Specify restoration of damaged coatings to their original integrity.

(28) Provide against any unnecessary damage to coatings applied at the preassembly stage.

(29) Aluminium or aluminium coatings should not be anodised if electrical conductivity is required.

(30) Fully heat treated aluminium alloys, prior to painting, should be primed with chromate primer containing not less than 20% zinc chromate pigment.

(31) Comprehensive and technically accurate coating specifications should be an inherent part of all structural designs; reference to approved standards of coatings and application methods is to be made.

(32) Limitations to free application of coating in adverse conditions should be an inherent part of specifications and design.

(33) Instructions for expert inspection of protective coatings should be given for the particular design; this throughout the application of the whole system, operation after operation.

(34) Cathodic protection dielectric shields should have good insulating qualities, low permeability, good adhesion and good alkaline resistance. The shields should be of sufficient size to prevent damage to the adjacent

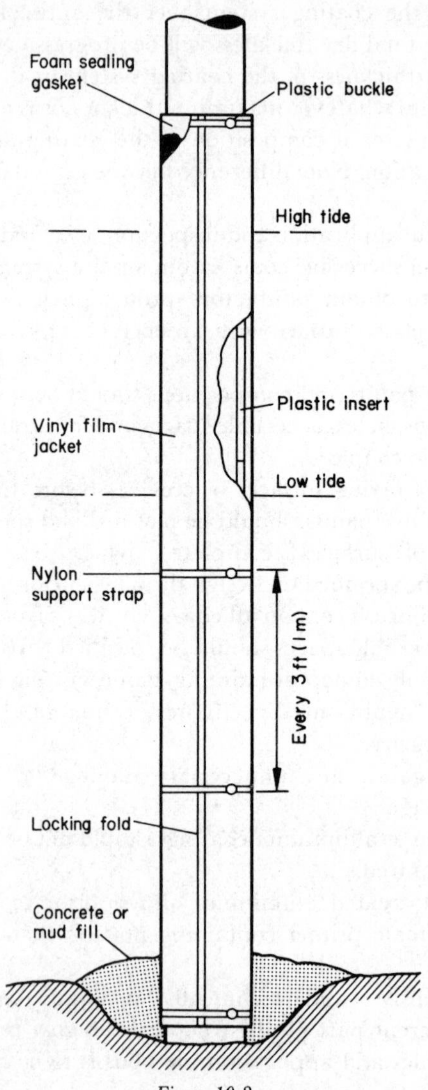

Figure 10.2

coating system and ensure good current distribution. It is recommended to increase the coating thickness of the adjacent paint coating in the immediate periphery of the shield.

(35) The limit on polarisation level below 1 V negative (Ag/AgCl) is valid for marine coatings, including zinc primers used together with cathodic protection systems.

(36) Environmental anti-pollution regulations and health precautions should be incorporated into specifications and design:

(*a*) cleaning of materials—in-shop cleaning, vacublast, wet blasting;
(*b*) supply of raw materials (paints, solvents)—non-toxic or reduced toxic contents.

(37) Design parameters for selection of coatings for hydrodynamic structures:

(*a*) frictional resistance (relative speed, wetted surface area, surface roughness);
(*b*) wave and eddy resistance (from total resistance obtained by tests in hydrodynamic tanks subtract frictional resistance).

(38) Concrete structures lying in waterlogged ground should be protected with sealing membranes (e.g. a high build bituminous composition on top of primer).

(39) Piles and structures to be enveloped or jacketed should first be cleaned and freed of all contamination and fouling. Surfaces to be jacketed should also be by preference and, if possible, primed with anti-corrosion composition prior to jacketing (see also Chapters 6 and 9) (*Figure 10.2*).

(40) Prior to application of corrosion inhibiting or insulating wrapping tapes to structural steel, the steel should be thoroughly cleaned and primed. Tubular structures should be wrapped as described in Section 10.6. Structural shapes should be taped longitudinally. The tape should be well pressed down and smooth, and the tension should not be excessive. Folds and air pockets should be avoided; the tape over protruding nuts, bolts, etc., should be cut in the form of cross, with the tape pressed firmly to the metal and the exposed surface patched up with a piece of tape (*Figure 10.3*).

(41) Surfaces exposed to serious damage by abrasion or repeated impact in corrosive conditions may be protected by loosely hung or bonded rubber liners in required thickness, $\frac{1}{4}$ in (6 mm) thick and up. Edges and metal surfaces covered by loose lining should be protected against corrosion (*Figure 10.4*).

(42) Use of precoated, in-factory or *in situ* plastic clads and simple or complex plastic laminates (e.g. fibre-reinforced plastic laminate, polypropylene sheet with glass fibre cloth, etc.), for the fabrication of suitably designed

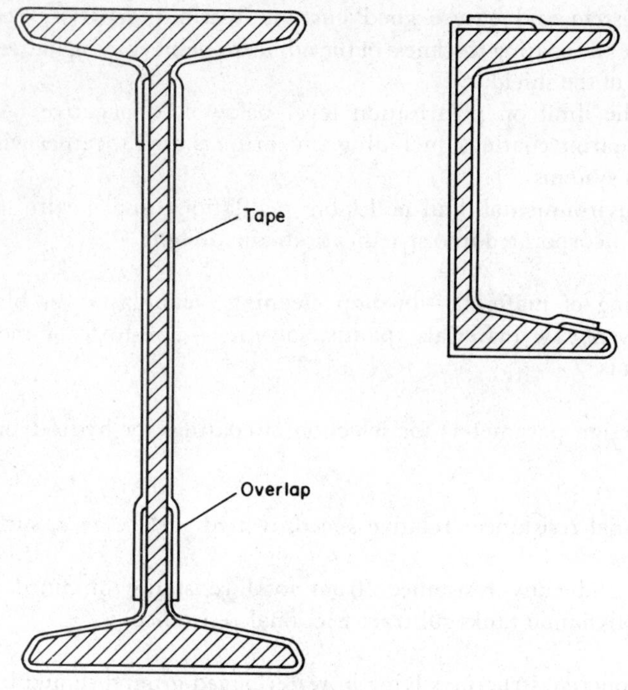

Figure 10.3

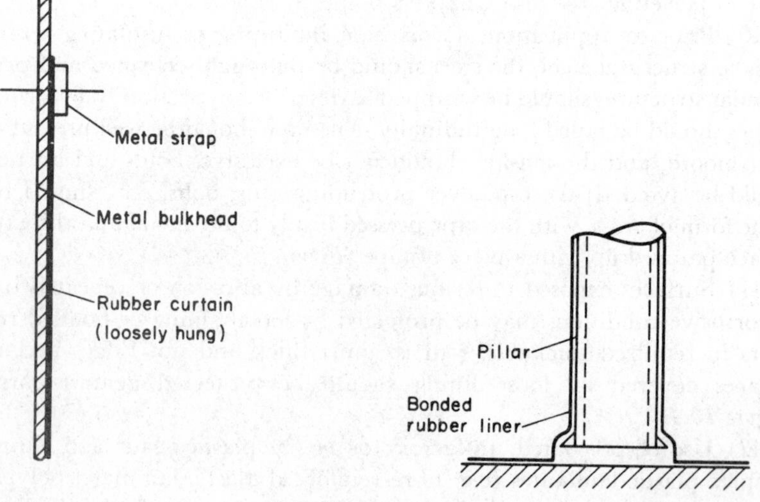

Figure 10.4

corrosion-resistant structures, should be evaluated (*Figure 10.5*): (*a*) precoated metals; (*b*) in factory (preferred) or *in situ* plastic-clad metals.

(43) Design changes from standard on prestressed concrete water reservoirs:

(*a*) cable-stressed reservoirs—use airtight flexible metallic conduit for horizontal encased cables;
(*b*) bar-stressed reservoirs—fill the vertical coupling beams with cement grout on construction; apply minimum 2 in (5 cm) cover of cement mortar over bars and beams (*note*: corrosiveness of water);

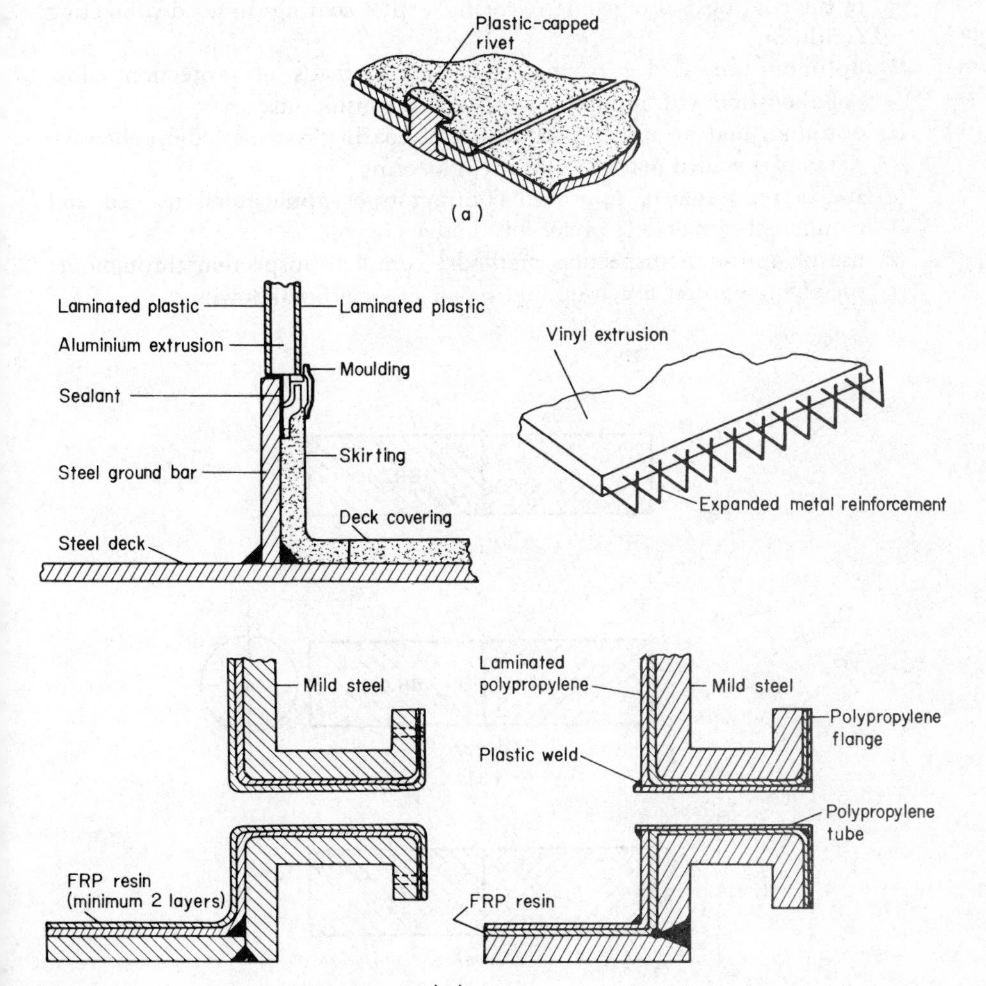

Figure 10.5

(*c*) wire-stressed reservoirs—no cavities round wires caused by their bunching; apply cement slurry coating just before and after wrapping operation; mortar uniform in density, minimum 2 in (5 cm) for unpainted surfaces and 3/4 in (2 cm) for painted surfaces; mortar thoroughly moist during curing period; sealing coat applied as soon as possible after curing; if backfilled exterior wall to be sealed.

(44) Basic requirements for obtaining optimum result from protective coatings:

(*a*) optimum geometry for cleaning, application, inspection and maintenance of the coatings; also geometry for upkeep of coatings in good protecting condition;
(*b*) optimum knowledge of materials and methods of protection, close collaboration with reputable suppliers or consultants;
(*c*) optimum and accurate specification of coating systems; comprehensive detail of specified matter, coating engineering;
(*d*) use of reputable or approved contractors or applicators; trained and competent personnel; preferably under cover;
(*e*) use of optimum inspection methods; complete inspection throughout;
(*f*) specifying earliest touch-up and repair of local breakdowns.

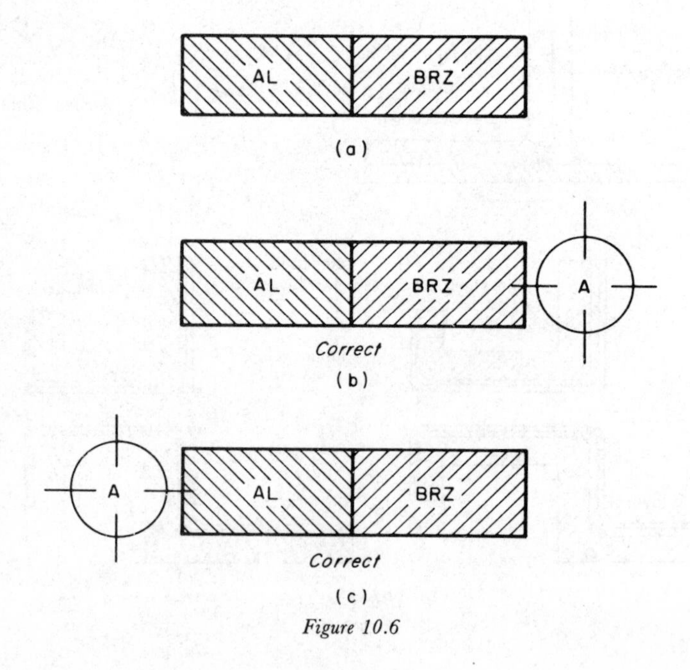

Figure 10.6

(45) Attachment of sacrificial anodes to galvanic couples (*Figure 10.6*): (*a*) galvanic couple; (*b*) brings potential of cathode to the level of anode and then reduces the whole to potential of the couple (*danger*: excessive formation of zinc oxide); (*c*) to be used when excessive formation of zinc oxide is to be avoided (problem of space and operation) or in closed pipe systems.

(46) Alternative protection of fasteners in design by sacrificial action of dissimilar metal (*Figure 10.7*): (*a*) structural carbon steel is sacrificial and protects the fasteners—this design can be used where the excess weight

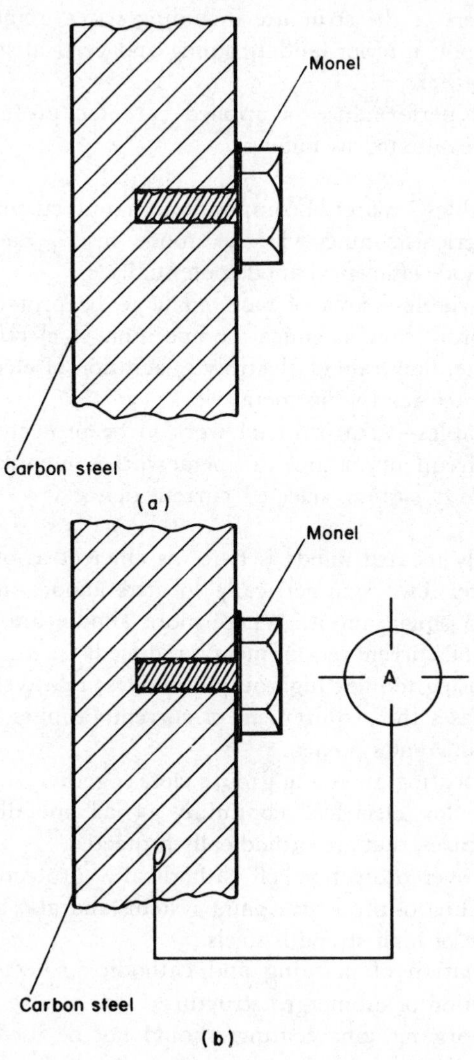

Figure 10.7

can be added to the established design requirements, corrosion and pitting of the steel will not be detrimental to the function and the structure is not highly stress loaded; (*b*) sacrificial anode is the sacrificial metal which protects both the fastener and the structural steel—this design should be used on structures in conductive environments which are subject to weight limits, where corrosion and pitting would interfere with movement of operations and where the corrosion or pits could form stress-raisers on structures under critical stress loading.

(47) In design of prestressed concrete water reservoirs or underwater reinforced concrete tunnels for future installation of cathodic protection, all metallic members of the structure including wires, reinforcing bars in the walls and the floor, interior ladder, piping and vertical stressing rods should be bonded together.

(48) General performance of applied cathodic protection by sacrificial anodes can be evaluated as follows:

(*a*) anode variables—material composition, manufacturing method, physical shape, electrical contact via the anode insert, anode output, anode capacity, anode efficiency, anode potential.
(*b*) external variables—area of bare metal to be protected, system life or length of protection, chemical composition of electrolyte, temperature of electrolyte, flow rate of electrolyte, aeration of electrolyte, position of anode in space relative to metal work;
(*c*) design variables—area of metal work to be protected, type of coating, length and frequency of time the metal work is in contact with electrolyte, required life of system, selected current density.

(49) Centrally located anode is twice as efficient as one mounted at one end of structure. Two symmetrically located anodes are about 1.6 times as effective as a single anode. Where more anodes are used the efficiency increases and total current requirement is reduced.

(50) Avoid using too few high output anodes; relatively poor distribution efficiency increases the requirement of current (applies to sacrificial anode and impressed current systems).

(51) Avoid locating anodes in groups close together.

(52) Provide for electrical continuity of all metallic components, in immersed structures, that are cathodically protected.

(53) Avoid over-protection of cathodically protected structures; this could cause peeling of protective paint systems and also hydrogen embrittlement, especially of high strength steels.

(54) Combination of painting and cathodic protection is required for effective protection of submerged structures.

(55) Bare inorganic zinc coatings should not be used together with the cathodic protection systems in submerged conditions (sea or brackish water).

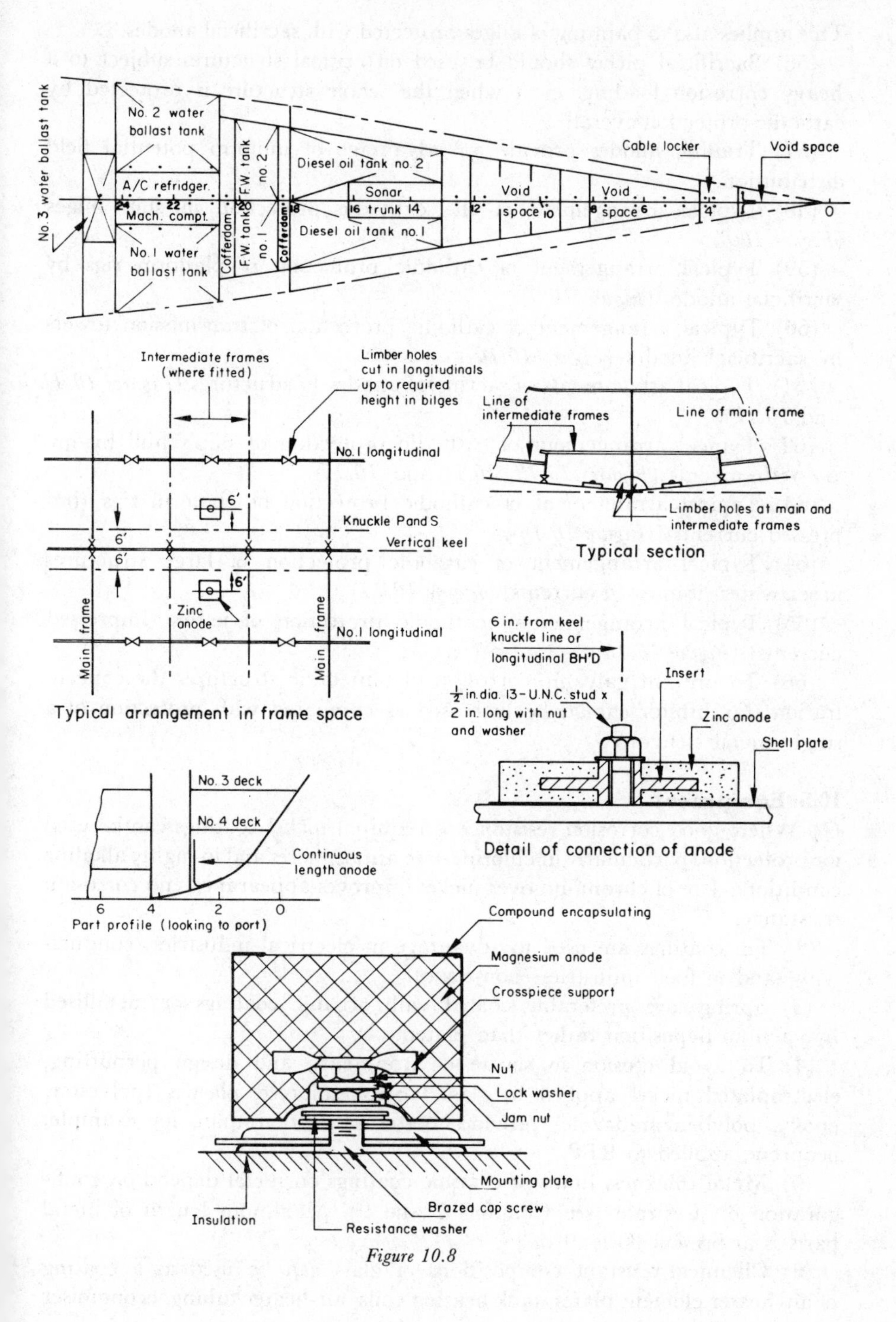
No. 3 water ballast tank
No. 2 water ballast tank
A/C refridger.
Mach. compt.
No. 1 water ballast tank
Cofferdam
F.W. tank no. 2
F.W. tank no. 1
Cofferdam
Diesel oil tank no. 2
Sonar trunk
Diesel oil tank no. 1
Void space
Void space
Cable locker
Void space
24
22
20
18
16
14
12
10
8
6
4
0
Intermediate frames (where fitted)
Limber holes cut in longitudinals up to required height in bilges
No. 1 longitudinal
6'
Knuckle P and S
Vertical keel
Zinc anode
Main frame
Main frame
No. 1 longitudinal
Typical arrangement in frame space
Line of intermediate frames
Line of main frame
Limber holes at main and intermediate frames
Typical section
6 in. from keel knuckle line or longitudinal BH'D
Insert
½ in. dia. 13-U.N.C. stud x 2 in. long with nut and washer
Zinc anode
Shell plate
Detail of connection of anode
No. 3 deck
No. 4 deck
Continuous length anode
6
4
2
0
Part profile (looking to port)
Compound encapsulating
Magnesium anode
Crosspiece support
Nut
Lock washer
Jam nut
Mounting plate
Brazed cap screw
Insulation
Resistance washer

Figure 10.8

This applies also to painting of bilges protected with sacrificial anodes.

(56) Sacrificial pieces should be used on critical structures subject to a heavy corrosion loading, even when the whole structure is protected by cathodic protection overall.

(57) Trailing anodes provide an advantage of uniform potential field distribution.

(58) Typical arrangement of the cathodic protection in ship bilges (*Figure 10.8*).

(59) Typical arrangement of cathodic protection of platform rigs by sacrificial anodes (*Figure 10.9*).

(60) Typical arrangement of cathodic protection of transmission towers by sacrificial anodes (*Figure 10.10*).

(61) Typical attachments of sacrificial anodes to structures (*Figures 10.11* and *10.12*).

(62) Typical arrangement of cathodic protection of ship's hull by impressed currents (*Figures 10.13, 10.14* and *10.15*).

(63) Typical arrangement of cathodic protection of platform rigs (impressed currents) (*Figure 10.16*).

(64) Typical arrangement of cathodic protection of large structures in sea water (impressed currents) (*Figure 10.17*).

(65) Typical arrangement of cathodic protection of jetties (impressed currents) (*Figure 10.18*).

(66) To prevent galvanic corrosion of bimetallic structures the concentration of inhibitor should be increased as compared with protection of a single metal structure.

10.5 Equipment

(1) Where good corrosion resistance is required nickel coatings can be used for protection, particularly in chloride-free atmospheres and in highly alkaline conditions. Use of chromium over nickel improves appearance and corrosion resistance.

(2) Tin coatings are used to advantage in electrical industries (conductivity) and in food industries (non-toxic).

(3) Springs are preferably coated with organic coatings or metallised by vacuum deposition rather than plated.

(4) To avoid erosion in severe environments, and design permitting, electroplated nickel applied to glass fabric-reinforced plastics (polyester, epoxy, polybenzimidazole) provides better protection than, for example, neoprene applied to RFP.

(5) Metal thickness limits of ceramic coatings on metal depend on configuration of substrate (see Chapters 7 and 9). Maximum length of metal parts is at present 30 ft (9 m).

(6) Chemical-resistant compositions of glass can be used as a coating of air-heater element plates, tank heating coils, air-heater tubing, economiser tubing, economisers, etc.

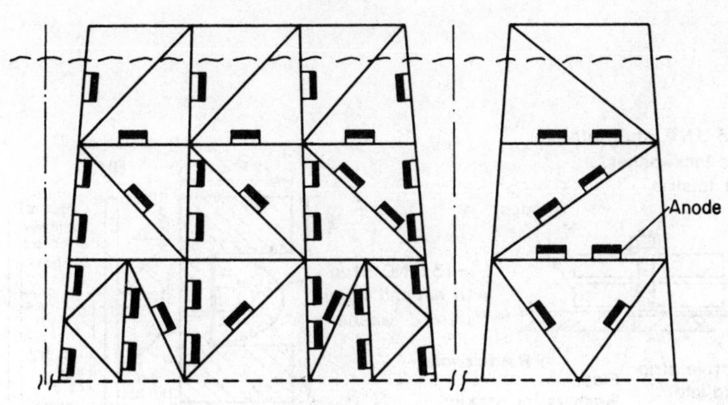

Figure 10.9

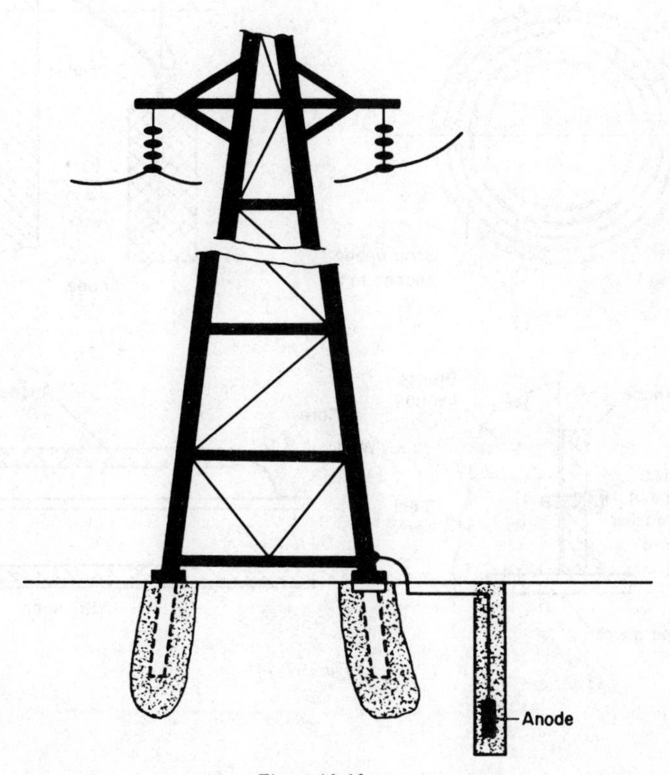

Figure 10.10

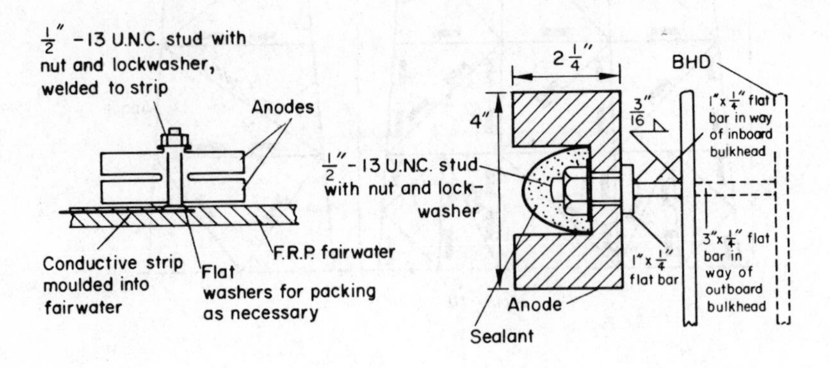

$\frac{1}{2}$" - 13 U.N.C. stud with nut and lockwasher, welded to strip
Anodes
Conductive strip moulded into fairwater
Flat washers for packing as necessary
F.R.P. fairwater
$2\frac{1}{4}$"
4"
$\frac{3}{16}$"
$\frac{1}{2}$" - 13 U.N.C. stud with nut and lock-washer
BHD
1" x $\frac{1}{4}$" flat bar in way of inboard bulkhead
3" x $\frac{1}{4}$" flat bar in way of outboard bulkhead
1" x $\frac{1}{4}$" flat bar
Anode
Sealant

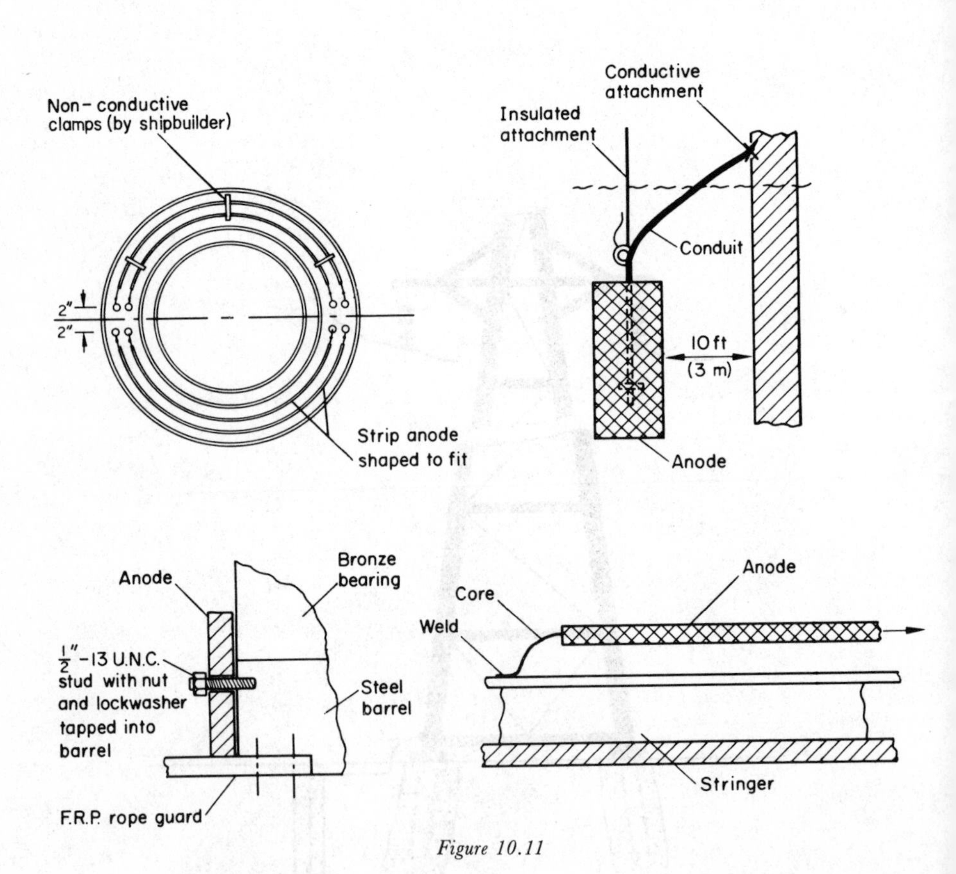

Non-conductive clamps (by shipbuilder)
2"
2"
Strip anode shaped to fit
Conductive attachment
Insulated attachment
Conduit
10 ft (3 m)
Anode
Anode
Bronze bearing
$\frac{1}{2}$" -13 U.N.C. stud with nut and lockwasher tapped into barrel
Steel barrel
F.R.P. rope guard
Anode
Core
Weld
Stringer

Figure 10.11

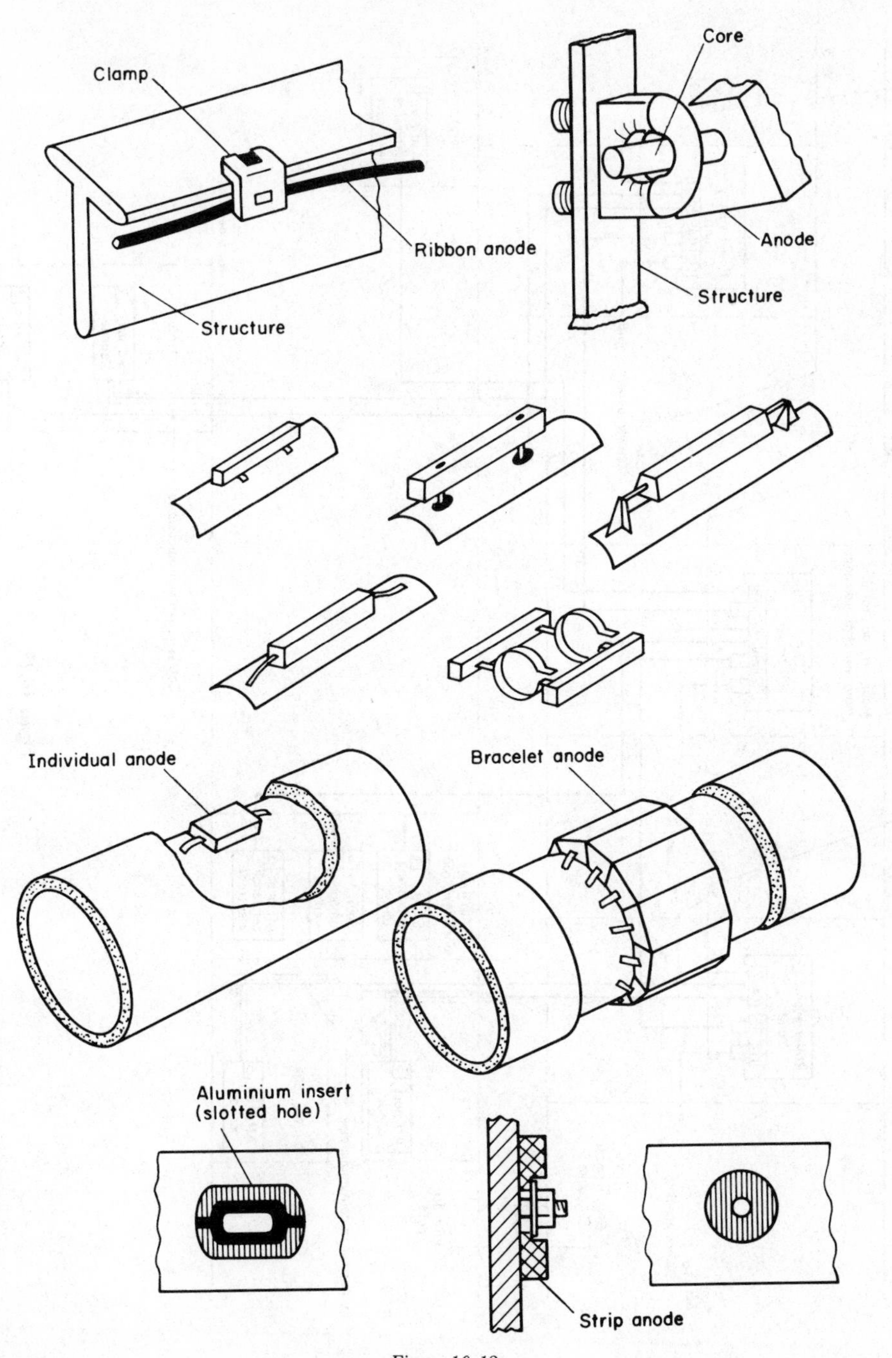
Clamp
Ribbon anode
Structure
Core
Anode
Structure
Individual anode
Bracelet anode
Aluminium insert
(slotted hole)
Strip anode

Figure 10.12

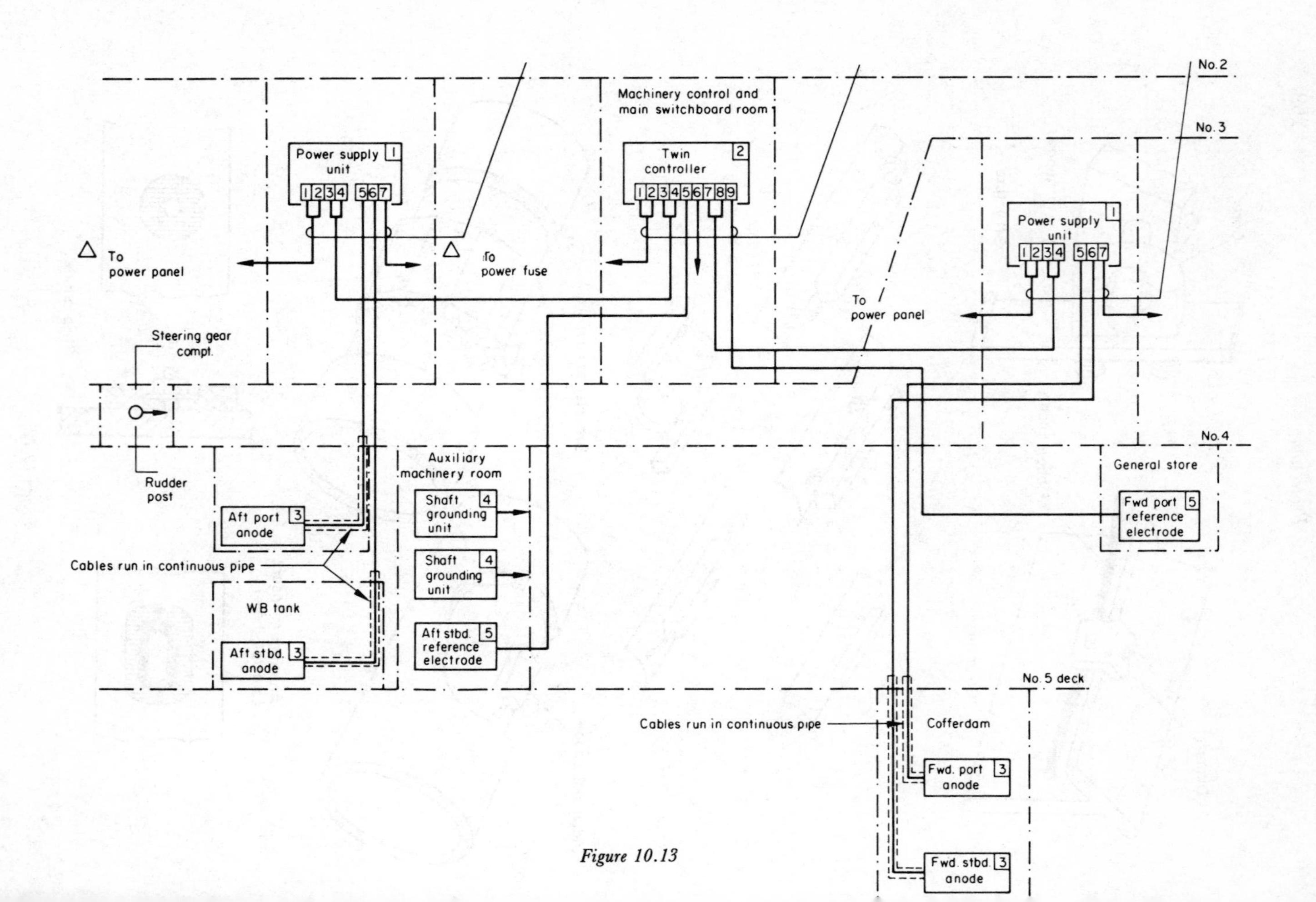
No. 2
No. 3
No. 4
No. 5 deck
Machinery control and main switchboard room
Power supply unit
1
1 2 3 4 5 6 7
Twin controller
2
1 2 3 4 5 6 7 8 9
To power panel
To power fuse
To power panel
Steering gear compt.
Rudder post
Aft port anode
3
Cables run in continuous pipe
WB tank
Aft stbd. anode
Auxiliary machinery room
Shaft grounding unit
4
Aft stbd. reference electrode
5
General store
Fwd port reference electrode
Cables run in continuous pipe
Cofferdam
Fwd. port anode
Fwd. stbd. anode

Figure 10.13

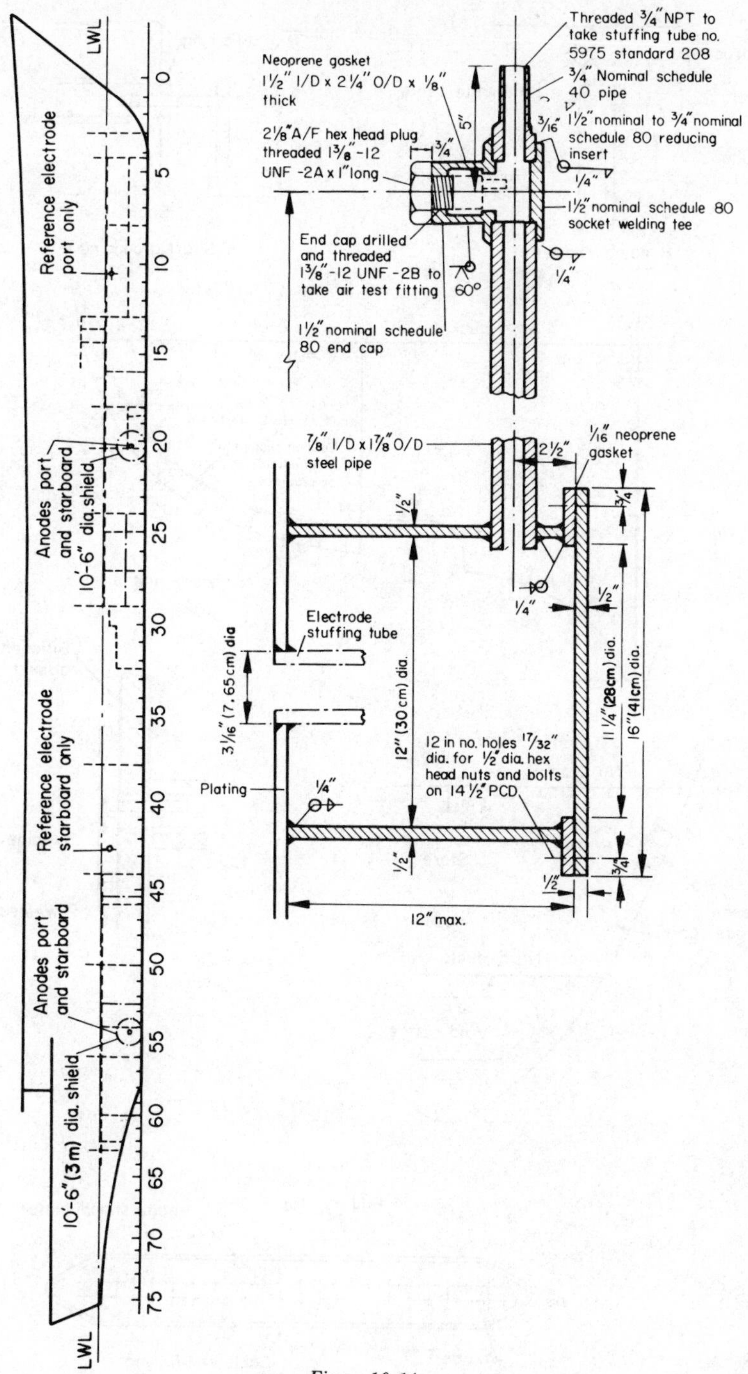
LWL
Reference electrode port only
Anodes port and starboard 10'-6" dia. shield
Reference electrode starboard only
Anodes port and starboard 10'-6" (3m) dia. shield
LWL
Threaded 3/4" NPT to take stuffing tube no. 5975 standard 208
Neoprene gasket 1 1/2" I/D x 2 1/4" O/D x 1/8" thick
3/4" Nominal schedule 40 pipe
1 1/2" nominal to 3/4" nominal schedule 80 reducing insert
2 1/8" A/F hex head plug threaded 1 3/8"-12 UNF -2A x 1" long
1 1/2" nominal schedule 80 socket welding tee
End cap drilled and threaded 1 3/8"-12 UNF -2B to take air test fitting
1 1/2" nominal schedule 80 end cap
7/8" I/D x 1 7/8" O/D steel pipe
1/16" neoprene gasket
Electrode stuffing tube
3 1/16" (7.65cm) dia
12" (30 cm) dia.
11 1/4" (28cm) dia.
16" (41cm) dia.
12 in no. holes 17/32" dia. for 1/2" dia. hex head nuts and bolts on 14 1/2" PCD
Plating
12" max.

Figure 10.14

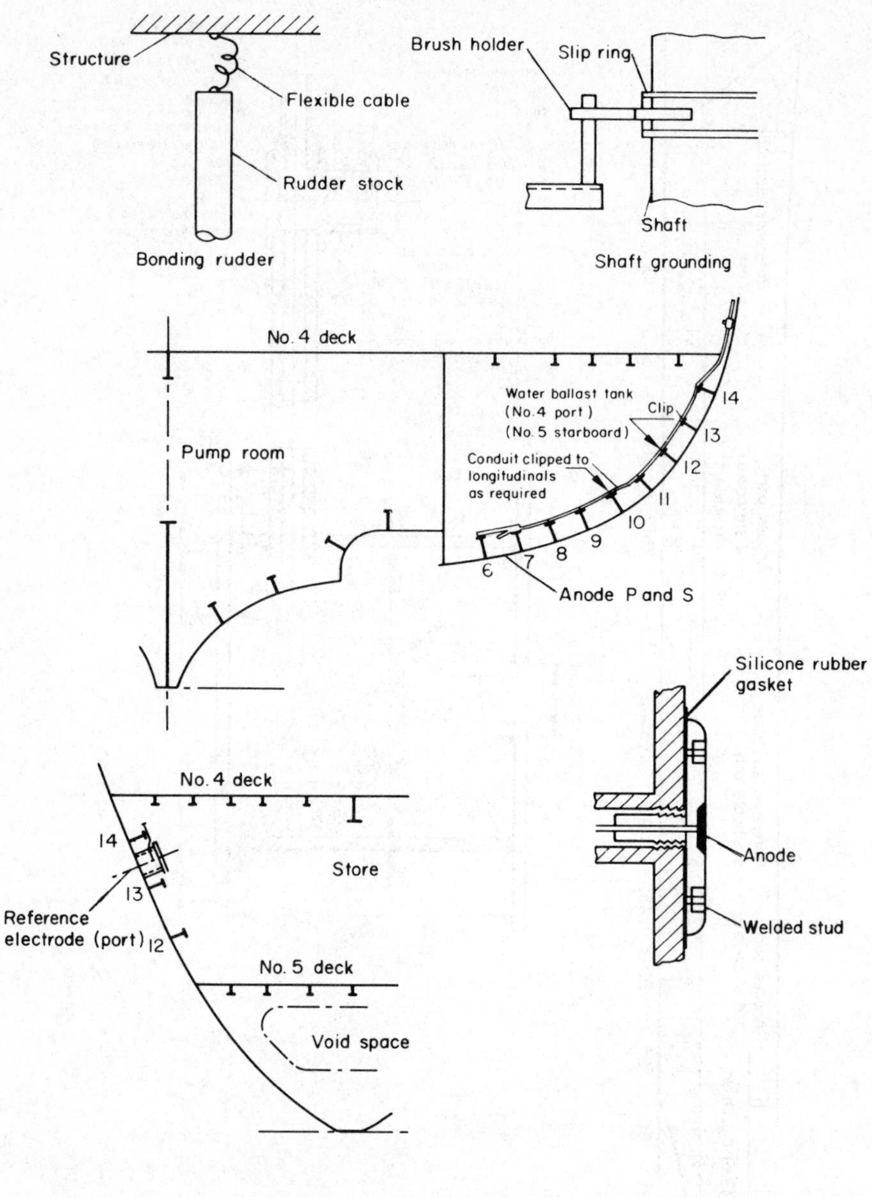
Structure
Flexible cable
Rudder stock
Bonding rudder
Brush holder
Slip ring
Shaft
Shaft grounding
No. 4 deck
Pump room
Water ballast tank
(No. 4 port)
(No. 5 starboard)
Clip
Conduit clipped to
longitudinals
as required
14
13
12
11
10
9
8
7
6
Anode P and S
No. 4 deck
Store
14
13
Reference
electrode (port)
12
No. 5 deck
Void space
Silicone rubber
gasket
Anode
Welded stud

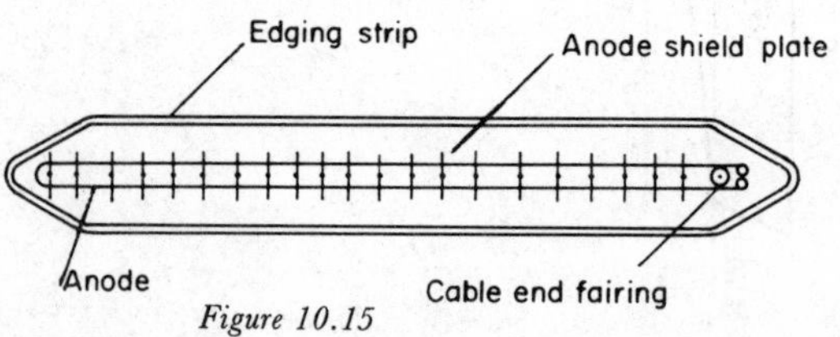
Edging strip
Anode shield plate
Anode
Cable end fairing

Figure 10.15

Reference electrode
Anode

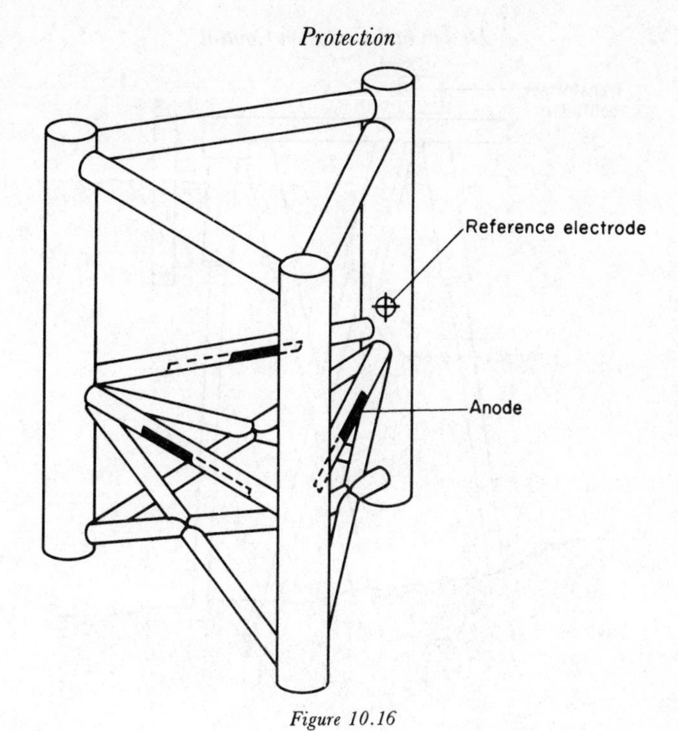

Figure 10.16

Anode
Anode cable
Ballast
Sand
Heavy gravel
U/W tunnel
Heavy gravel
Gravel bed
Anode
600 ft (183 m)

Figure 10.17

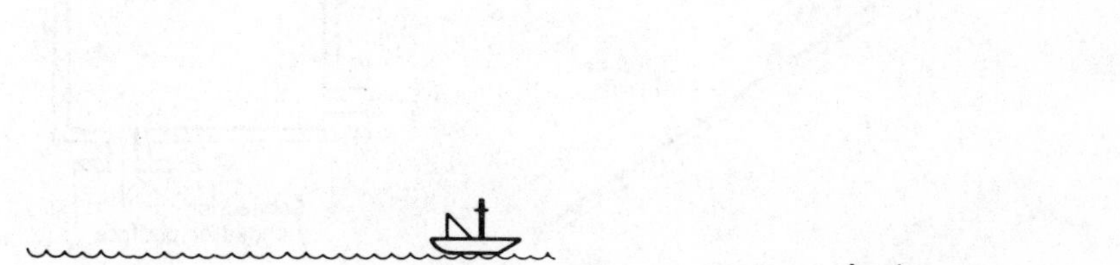

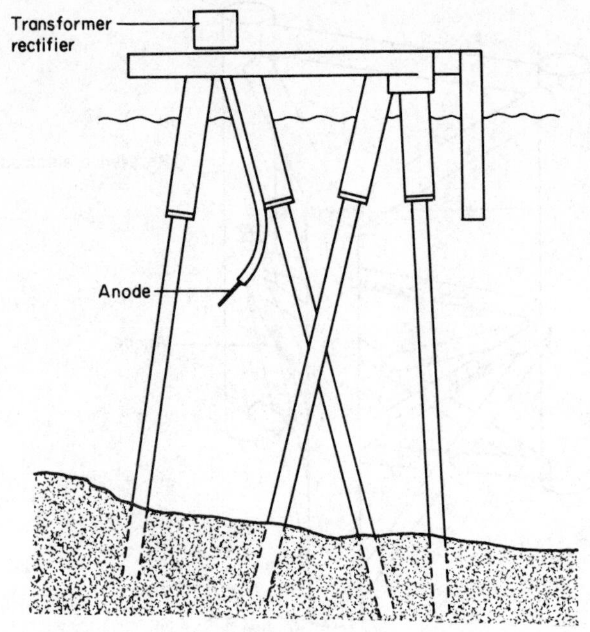

Figure 10.18

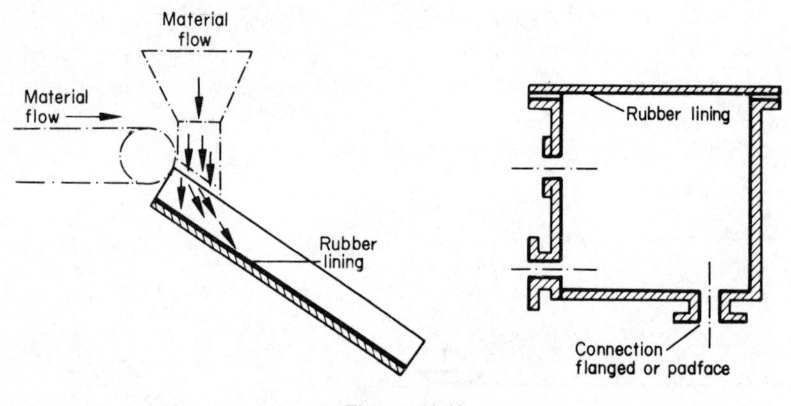

Figure 10.19

(7) Avoid specifying and designing for dry working surfaces of exposed fittings (e.g. door clips)—specify suitable lubrication.

(8) Parts which are totally and continuously immersed in oil or grease (preferably inhibited), and those embedded in encapsulant or moisture-proof compound, need not necessarily be given any further protection. Local application of oil or grease, however, does not automatically provide good corrosion protection.

(9) Dip-coating in suitable plastics may be used for submerged mild steel equipment exposed to high pressures and relatively low temperatures.

(10) Use of precoated and plastic-clad metals may be considered for production of equipment cabinets.

(11) When impingement of surfaces by vapour bubbles or erosion by abrasion action is possible, a lining of surfaces with loose or bonded elastomers may be considered (*Figure 10.19*). The surfaces of equipment to be so treated must be clean, smooth and preferably inhibited or coated with anti-corrosion coatings. The thickness of lining depends upon flow and impact energy. Screwed joints should be avoided.

(12) Tanks to contain corrosive liquids may be overall or partially lined with bonded or independent flexible laminated plastic liners (*Figure 10.20*). Ingress of corrosive liquids to the interface between the elastomer liner and the metal must be prevented.

(13) Under optimal conditions some surface coatings can serve several purposes, i.e. corrosion prevention, electrical neutrality and chemical inertness, self-cleaning and lubricity (see TFE-impregnated coating, page 96). Balanced evaluation for the particular design is necessary.

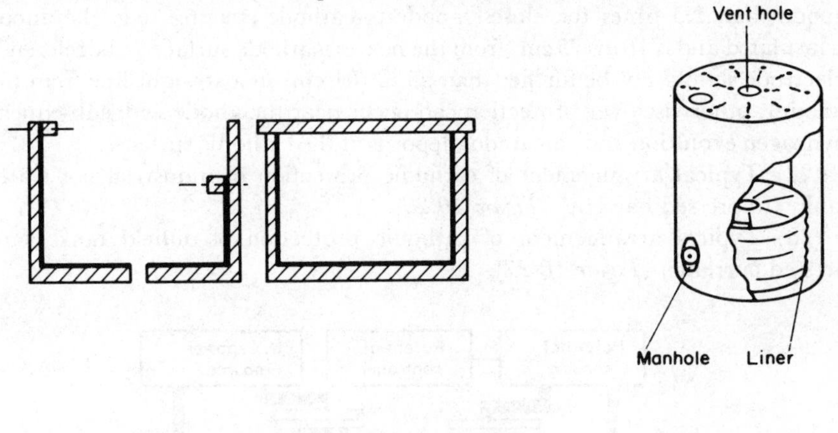

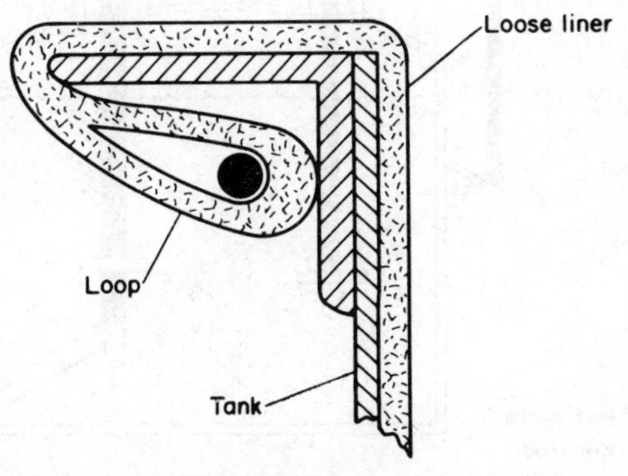

Figure 10.20

(14) Use of coatings and linings for absorption or attenuation of noise should be considered.

(15) Typical arrangement of anodic protection system (*Figure 10.21*).

(16) Typical arrangement of cathodic protection of oil storage tanks by impressed currents (*Figure 10.22*). *Note*: for completely submerged or buried tanks see marine structures and underground pipelines, pages 273 and 294.

(17) Typical arrangement of cathodic protection of deep well pump by impressed currents (*Figure 10.23*).

(18) Typical arrangement of cathodic protection of freshwater storage tanks (impressed currents) (*Figure 10.24*). *Note*: arrangements for venting of hydrogen.

(19) Typical arrangement of cathodic protection of buried storage tanks (impressed currents) containing flammable liquids (*Figure 10.25*).

(20) Rough surfaces of steel- and copper-base alloys require substantially less current density than smooth surfaces.

(21) In applying cathodic protection to a closed vessel, such as a tank or water box, the reference electrode should be positioned no further from the anode than 2.5 times the closest anode-to-cathode spacing (e.g. the anode is insulated and is 10 in (25 cm) from the nearest cathode surface—the reference electrode should not be further than 25 in (64 cm) in a straight line from the anode); otherwise over-protection can occur near the anode with subsequent hydrogen evolution and calcareous deposits at the cathodic surfaces.

(22) Typical arrangement of cathodic protection of industrial hot water tanks (impressed current) (*Figure 10.26*).

(23) Typical arrangement of cathodic protection of oilfield tanks (impressed currents) (*Figure 10.27*).

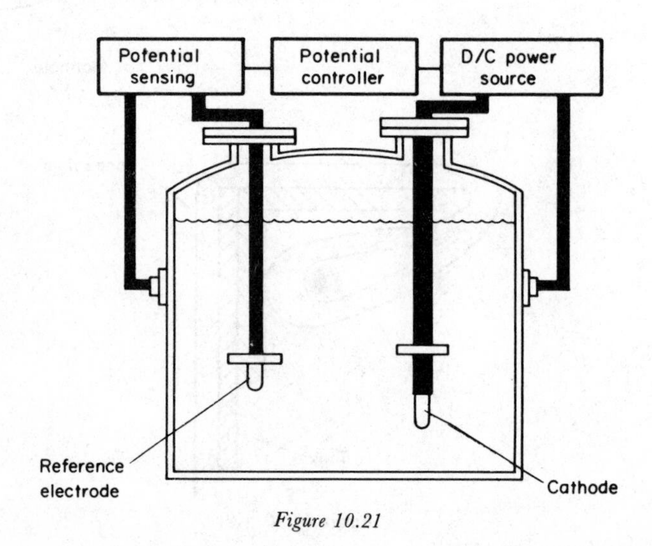

Figure 10.21

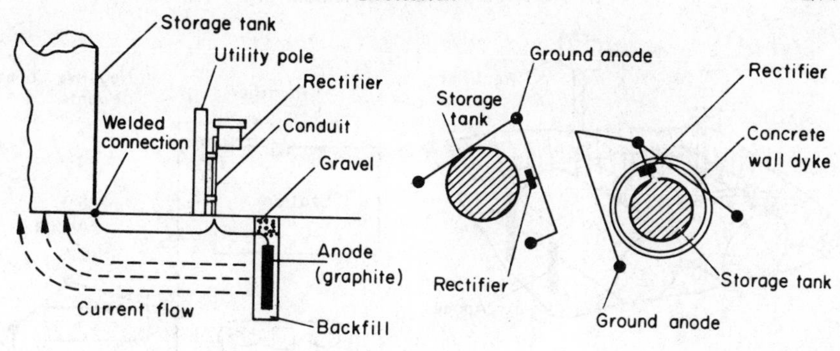

Figure 10.22

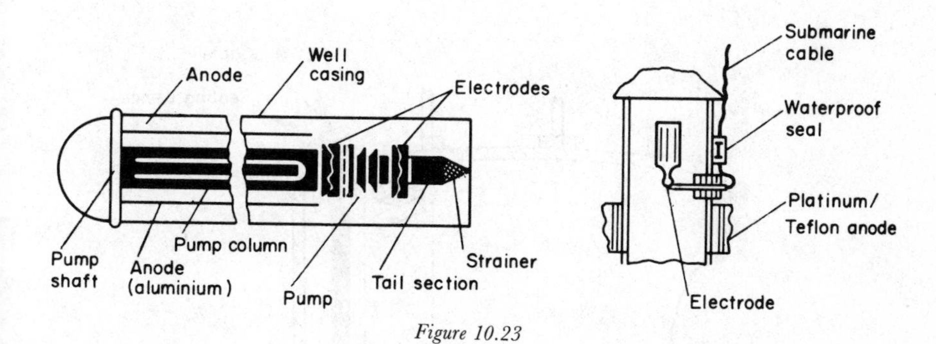

Figure 10.23

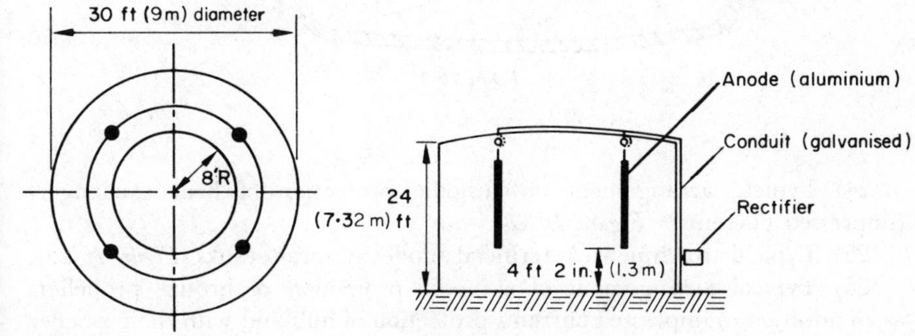

Figure 10.24

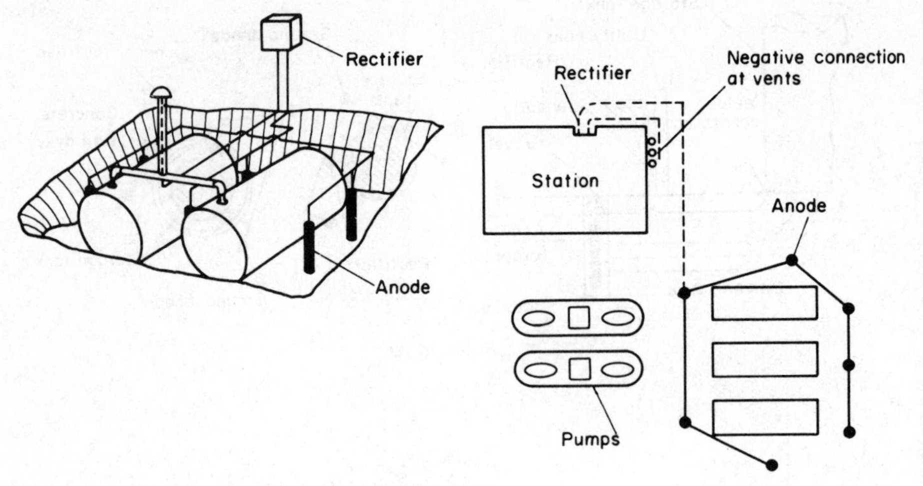

Figure 10.25

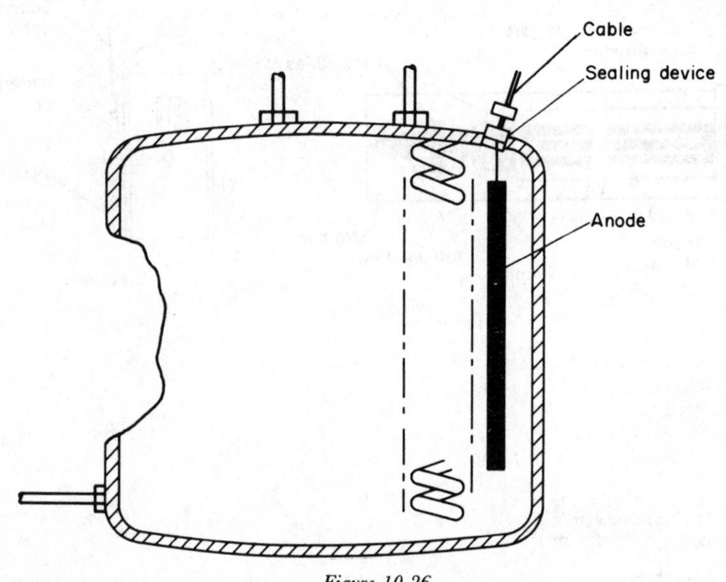

Figure 10.26

(24) Typical arrangement of cathodic protection of heat exchangers (impressed currents) (*Figure 10.28*).

(25) Typical attachment of sacrificial anodes in storage tanks (*Figure 10.29*).

(26) Typical arrangement of cathodic protection of bronze propellers as an addition to impressed currents protection of hull and with the propeller shaft efficiently grounded (*Figure 10.30*).

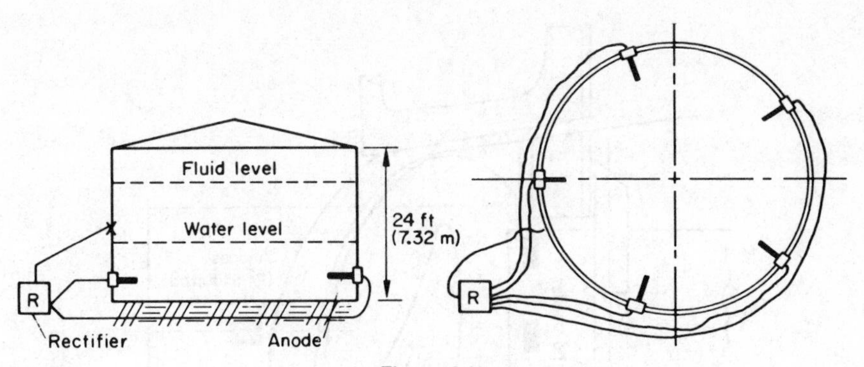

Figure 10.27

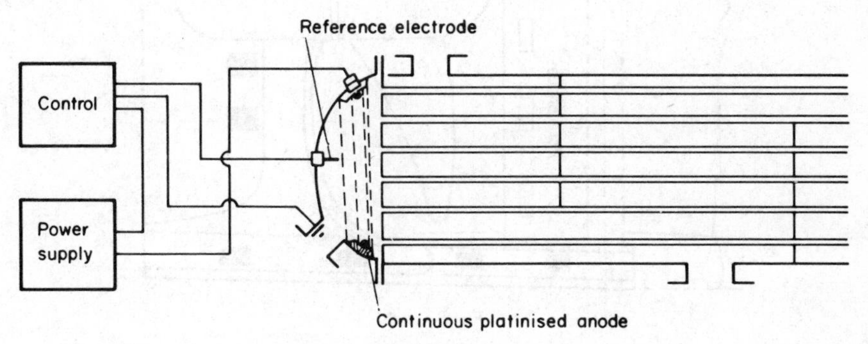

Figure 10.28

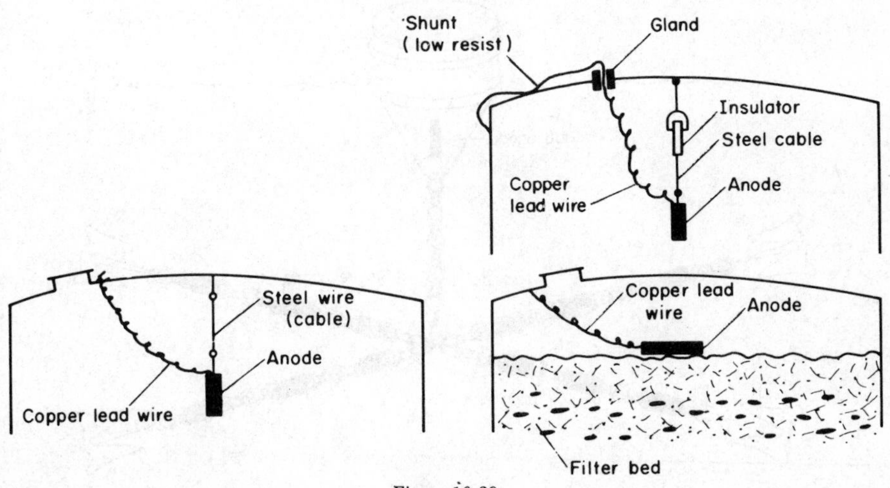

Figure 10.29

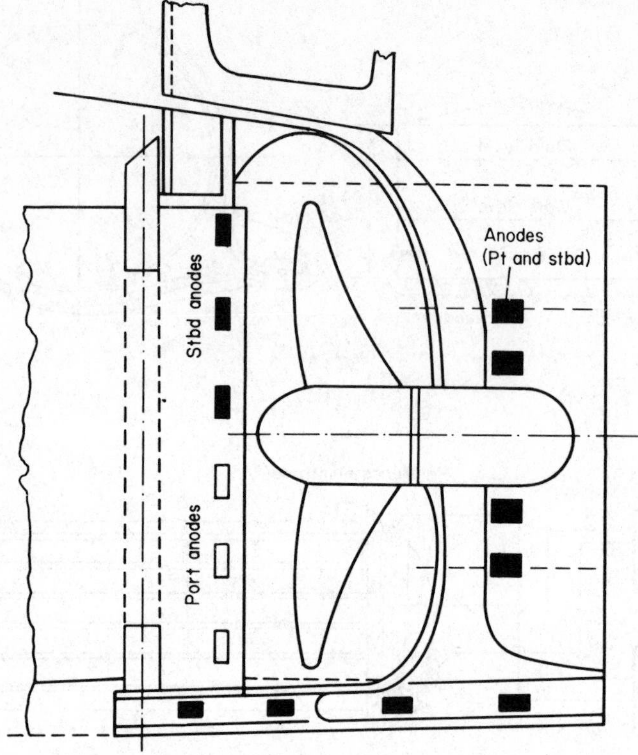

Figure 10.30

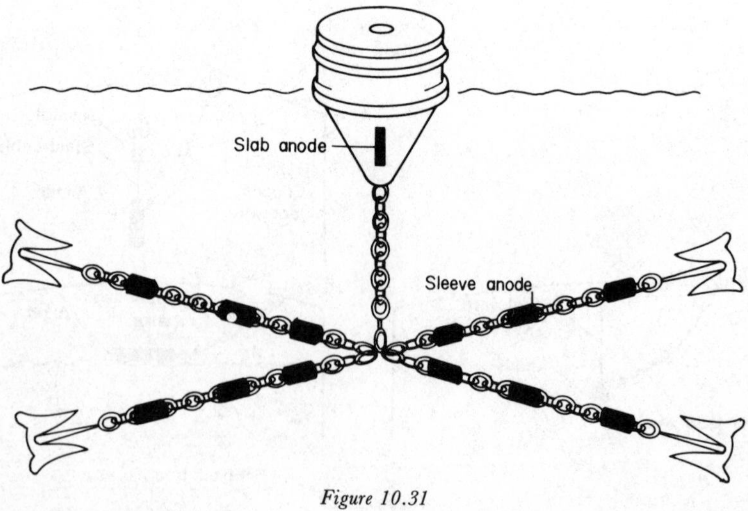

Figure 10.31

(27) Typical arrangement of cathodic protection of mooring buoys and chain (*Figure 10.31*).

(28) Cathodically protected equipment should not be over-protected; over-protection may cause peeling of paint and hydrogen embrittlement of high strength steels.

(29) Both painting and cathodic protection are mostly required for protection of submerged equipment.

(30) Critical equipment should be protected locally by sacrificial pieces where exposed to heavy corrosion loading.

(31) Cathodic protection of condenser boxes and other heat exchangers using natural waters is a necessity (*Figure 10.32*).

(32) Typical arrangement of cathodic protection of oilfield tanks by sacrificial anodes (*Figure 10.33*).

(33) Cathodic protection of inhibited recirculating cooling water systems (copper tubes, steel tubesheet) can be effective if:

(*a*) sodium–zinc molecularly dehydrated glassy phosphate inhibitor is used;
(*b*) overhang of the copper tubes in the sheet is below $\frac{1}{4}$ in (6 mm) for 3/4 in (19 mm) diameter tubes on 1 in (25 mm) centres;
(*c*) magnesium anodes located within 12 in (30 cm) from tubesheets are used and their shape is optimal.

(34) Typical arrangement of cathodic protection of heater treaters (*Figure 10.34*).

(35) Typical attachments of various anodes to equipment (*Figure 10.35*).

(36) Provide for admittance of sufficient air to the flowing fluid to relieve the local or general low pressure areas and so eliminate the cause of cavitation.

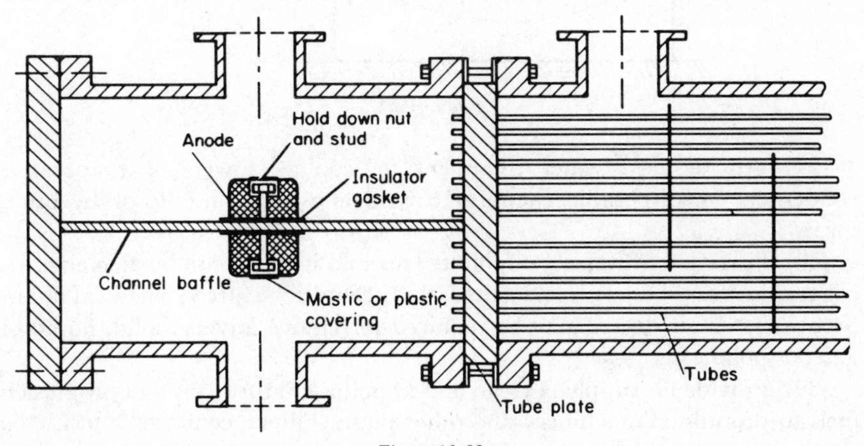

Figure 10.32

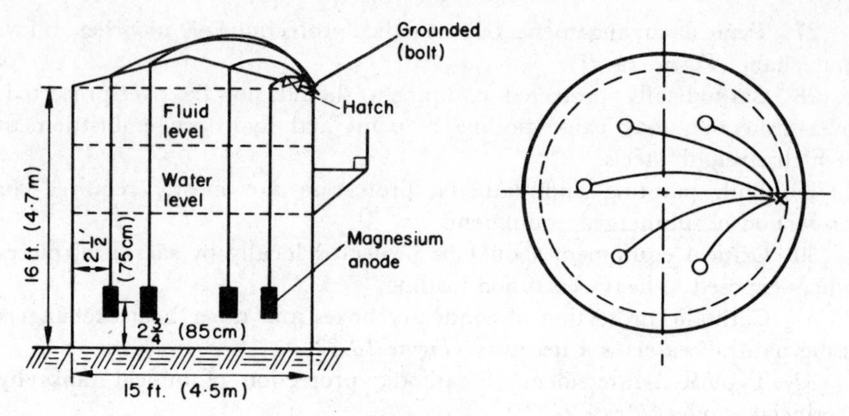

Figure 10.33

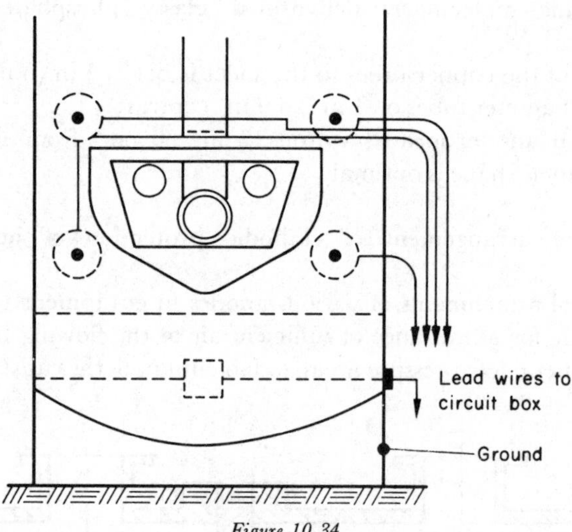

Figure 10.34

(37) Provide feed water treatment on cooling towers or evaporative condensers by disposable chemical cartridges where suitable or by other means.

(38) Provide for intake of unpolluted air into internal combustion engines, gas turbines and boilers (see Section 10.7). Demisters, filters, chemical filters, preheaters or air dryers may be required to remove adverse solid, liquid or gaseous pollutants.

(39) Provide for supply of clean and unpolluted lubricants and propulsion fuels to propulsion machinery and other plant. Filters, coalescer-filters, etc., may be necessary to remove solid, liquid or gaseous pollutants.

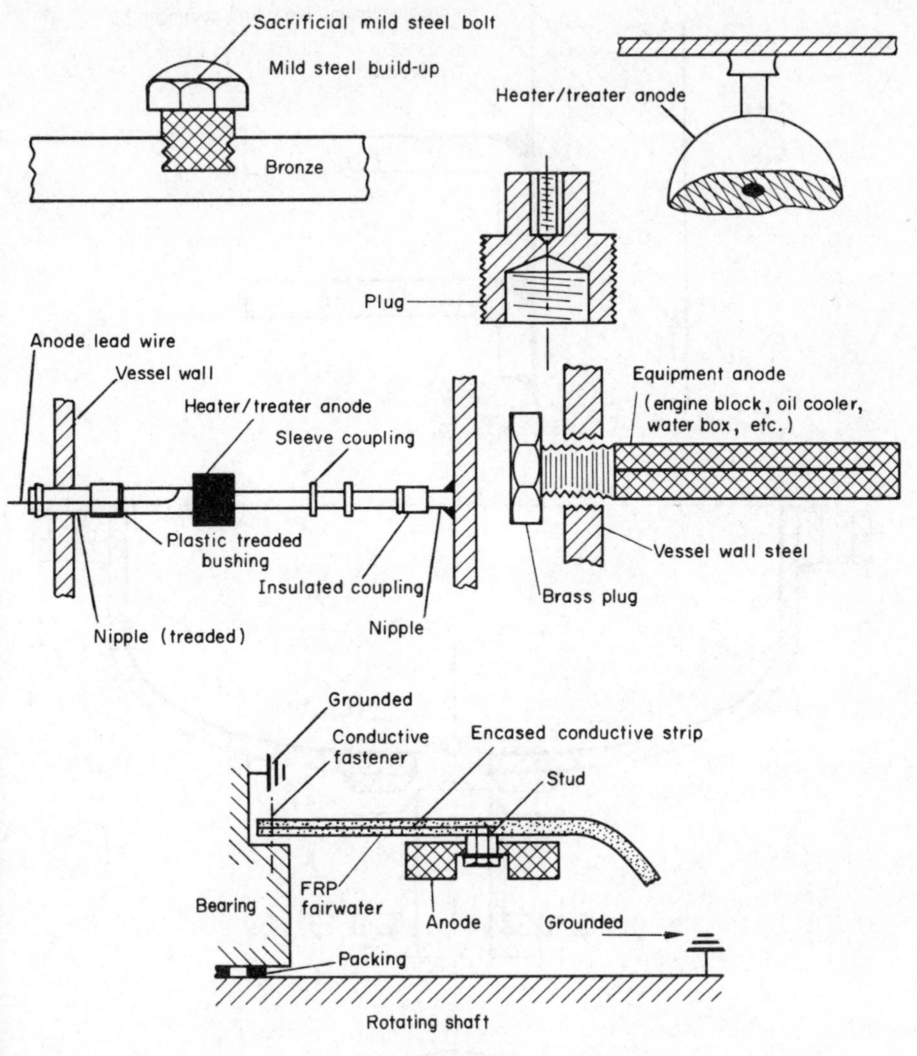

Figure 10.35

10.6 Pipe Systems

(1) Metallic coatings can be used not only for protection but also for simplification of surfaces, electrical or thermal conductivity, avoidance of crevices, etc. (*Figure 10.36*).

(2) Selection of buried pipelines coatings is based on the following requirements:

(*a*) highest insulating resistance over the entire operating service life;
(*b*) resistance to deformation stresses along the weight-bearing surfaces;

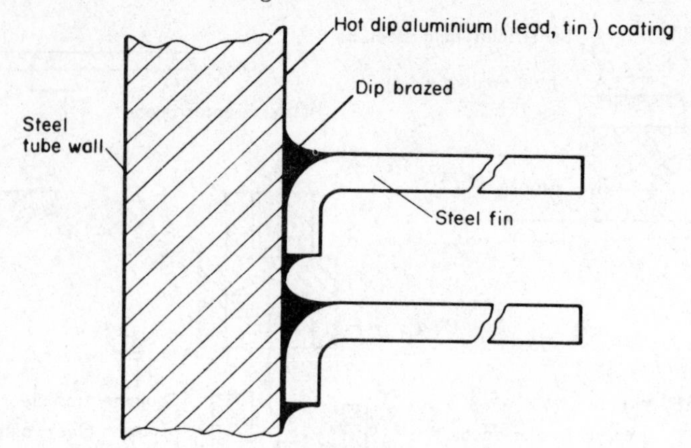

Figure 10.36

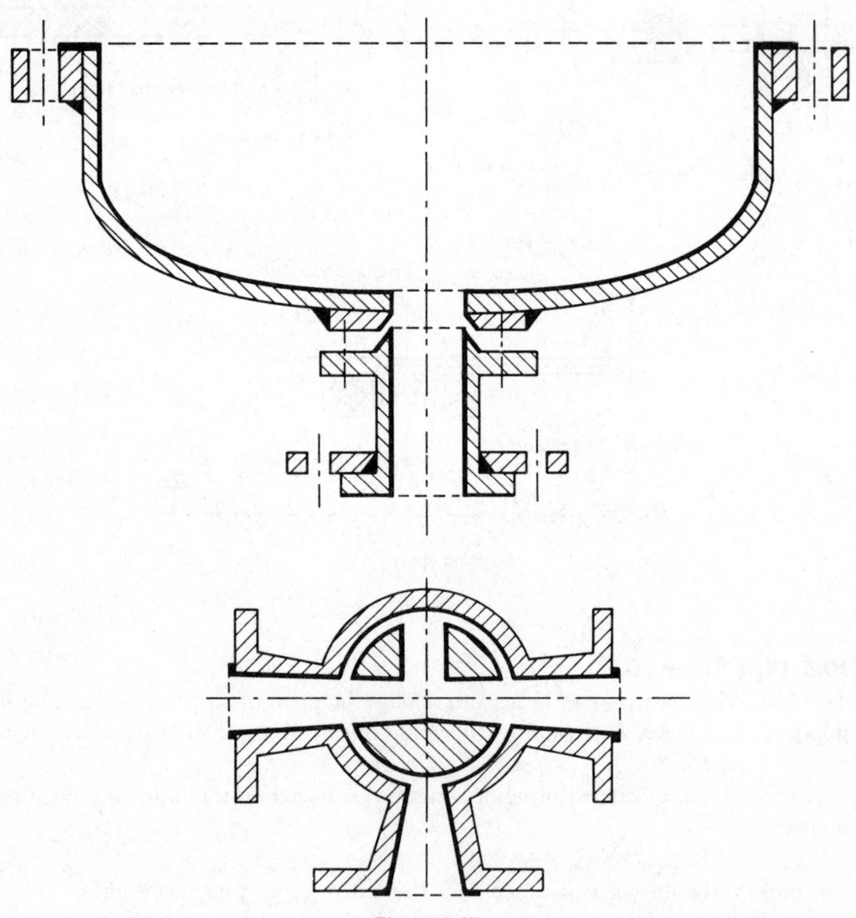

Figure 10.37

(*c*) resistance to temperature (high and low) deformation forces;
(*d*) chemical inertness to water-soluble electrolytes;
(*e*) insolubility in petroleum crudes, derivatives and solvents;
(*f*) resistance to microbiological attack;
(*g*) optimum economy;
(*h*) easy maintainability.

(3) Residual stresses input in pipe weldments by fabrication processes are aggravated by hot dip galvanising.

(4) Plastic- or rubber-lined pipe systems should be designed for the least exposure of unprotected edges to exposure of corrosive environments (submerged or spillage) (*Figure 10.37*).

(5) Pipes up to a 2 in (5 cm) bore can be lined, normally in lengths up to 10 ft (3 m), with rubber where required; larger bores can be lined in lengths up to 20 ft (6 m) (*Figure 10.38*). Standard flanged-bolted type pipework can be used for rubber lining; screwed joints are generally not satisfactory. Elbows, bends and tees can also be lined. Normally the substrate surfaces should be clean, metallised or primed prior to application of rubber lining.

(6) Where high pressure and relatively low temperature are involved, underground pipes can be dip-coated in suitable plastics for protection.

(7) Prior to application of corrosion-inhibiting or insulating wrapping tapes to pipes and tubes, these should be thoroughly cleaned and primed. Tape should be applied spirally with minimum $\frac{1}{2}$ in (13 mm) overlap per spiral; for better protection 55% overlap is required (*Figure 10.39*). In limited space, pipes can be wrapped in the same way as sectional steel (see structures page 260). Insulation and overwrap to follow.

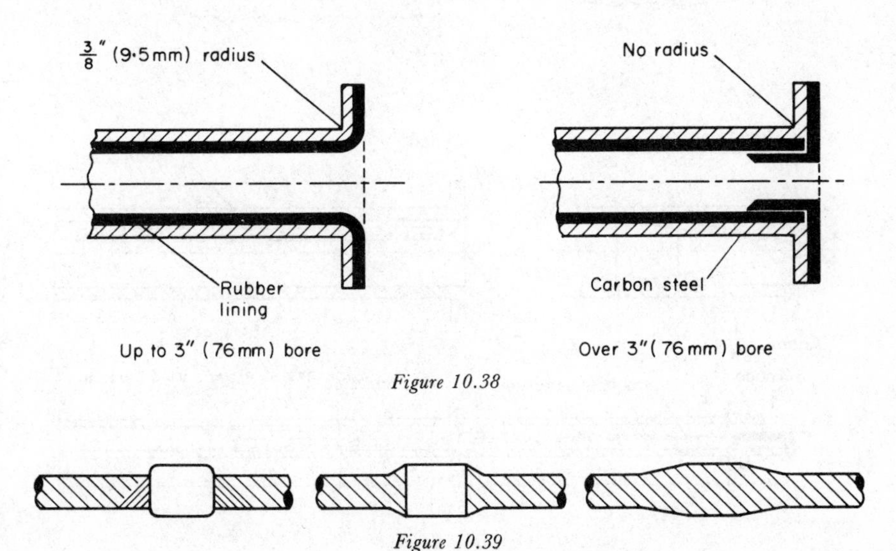

Figure 10.38

Figure 10.39

(8) Pipelines encased in or coated with concrete should be separated from pipelines buried in soil, whether these are bare, coated with insulating-type materials or insulated (*Figure 10.40*). Appropriate insulating devices should be used for separation. If such precautions are not possible, compensating cathodic protection should be used.

(9) Design of mortar-lined-and-coated steel pipes (*Figure 10.41*).

(10) Where no coatings are used on the cathode surface of a closed system,

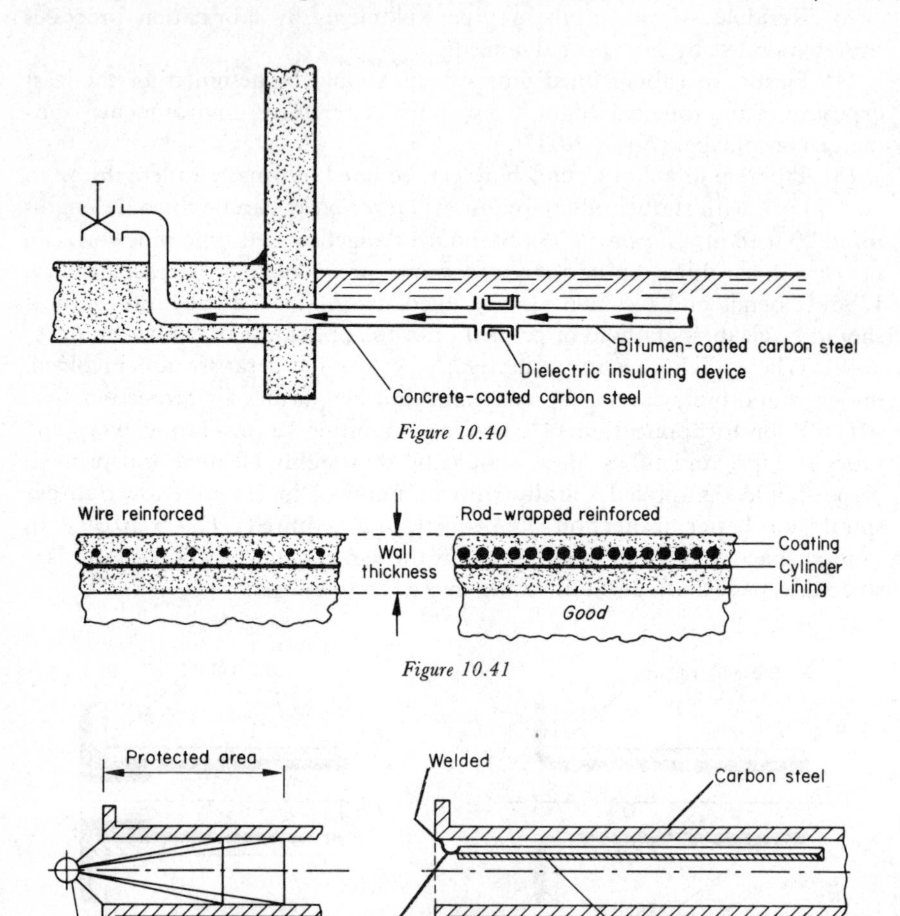

Figure 10.40

Figure 10.41

Figure 10.42

the area of optimum protection will cover a 2.5:1 spacing. If the closest anode to cathode spacing is 1 ft (0.3 m) and a potential of 1000 mV is obtained, a potential of at least 850 mV will be achieved at 2.5 ft (0.75 m) distance from this anode.

(11) At a surface velocity of 1 ft/s (0.3 m/s) over a smooth metal surface, a current density of 200 mA/ft^2 (2.2 A/m^2) will protect steel, stainless steel and copper-based alloys.

(12) A single sacrificial anode attached to one end of a pipe can normally, and in freshly piped sea water, protect the internal surfaces of a pipe only up to 2–3 diameters in depth and an installation of continuous strip anode may be required (*Figure 10.42*). In stagnant sea water, however, even 0.5–2 in (12–50 mm) nominal diameter stainless steel (e.g. type 304) and copper pipes can obtain effective cathodic protection after an initial polarisation period, which can vary between 4 days for 2 in (5 cm) dia SS pipe and 186 days for 0.5 in (12 mm) dia copper pipe, by a single sacrificial anode fitted to one end to a depth of approximately 20 ft (6 m).

(13) A sacrificial anode is not normally effective in the crevice between two pipe flanges (*Figure 10.43*). Either one of the flanges should be made sacrificial and the dissimilar metals left in contact, or only the apparent surface of the crevice should be effectively protected and the supply of protective current into crevice may not be necessary (see above and Chapter 6).

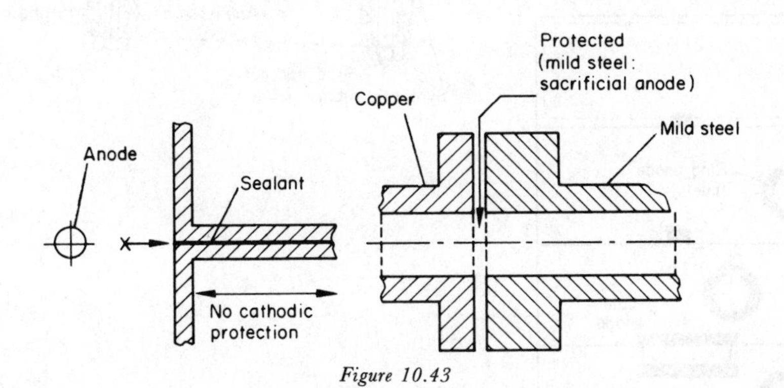

Figure 10.43

(14) Typical arrangements of cathodic protection of sea inlets by sacrificial anodes: main bay (*Figure 10.44*); sea box (*Figure 10.45*); sea tube (*Figure 10.46*).

(15) Typical attachment of sacrificial anodes to pipes (interior) (*Figure 10.47*). See also (6)–(12) above.

(16) Typical sacrificial anode for protection of service riser pipes (*Figure 10.48*).

(17) Typical pipeline installation of sacrificial anodes (*Figure 10.49*).

(18) Typical arrangements for cathodic protection of buried pipelines by impressed currents (bonded connections) (*Figure 10.50*).

(19) Selection of rectifiers:

(*a*) direct current output rating of rectifier necessary to secure effective polarisation of buried structures, pipelines and cables;
(*b*) type of power input to the unit (single, three-phase);
(*c*) type of output regulation (constant voltage, constant current, non-regulated);
(*d*) type of adjustment (manual, automatic);
(*e*) ease of installation and sturdiness of construction;

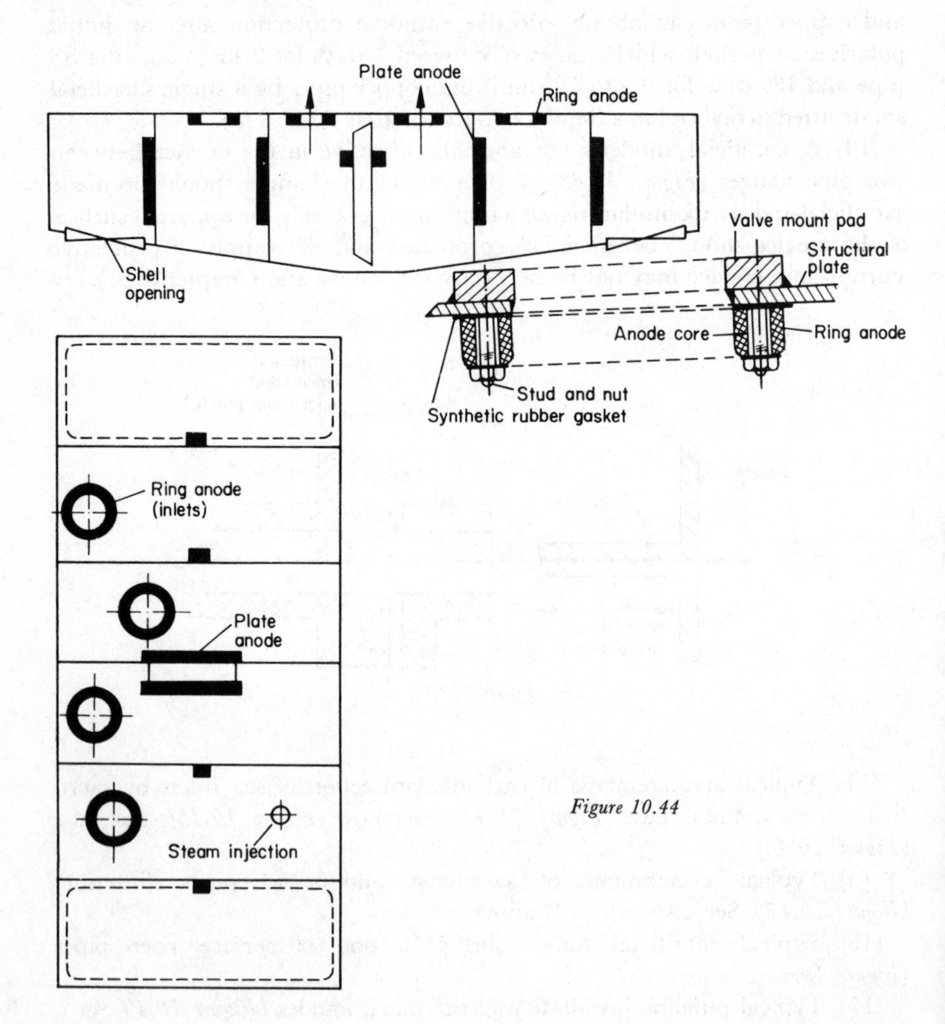

Figure 10.44

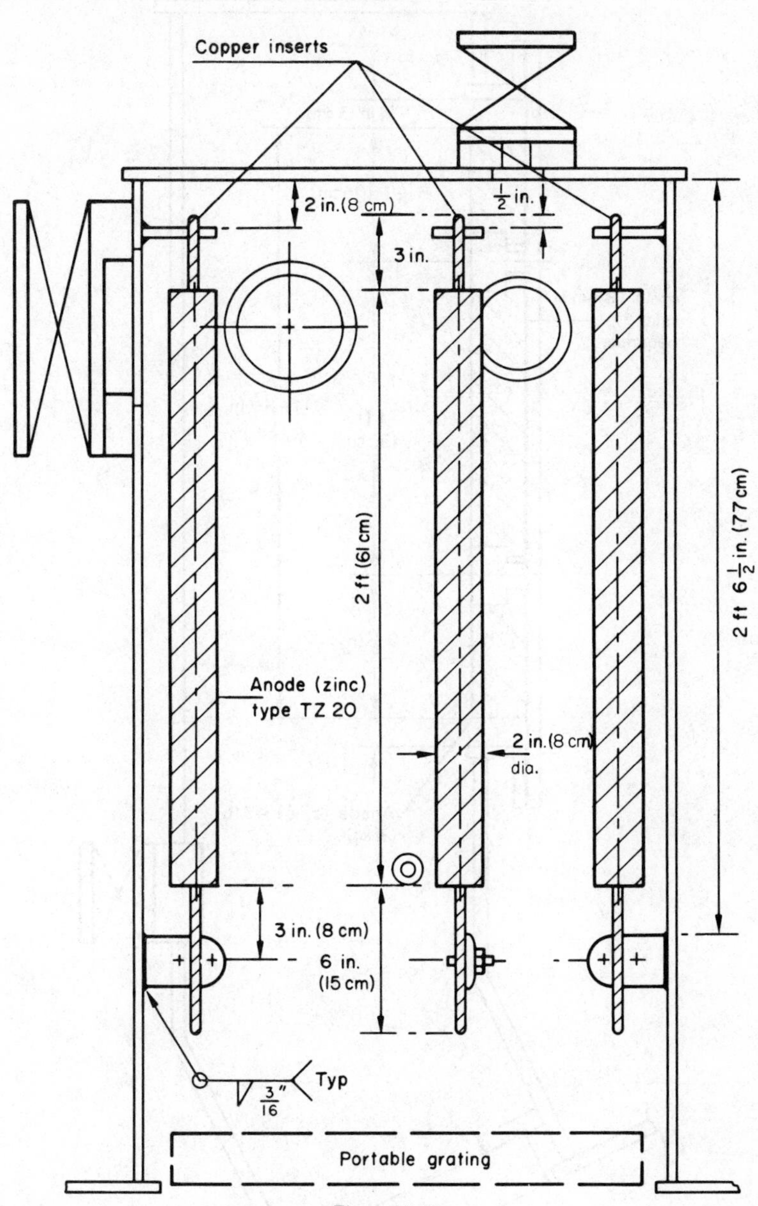
Copper inserts
2 in.(8 cm)
1/2 in.
3 in.
2 ft (61 cm)
2 ft 6 1/2 in. (77 cm)
Anode (zinc)
type TZ 20
2 in.(8 cm)
dia.
3 in. (8 cm)
6 in.
(15 cm)
3/16"
Typ
Portable grating

Figure 10.45

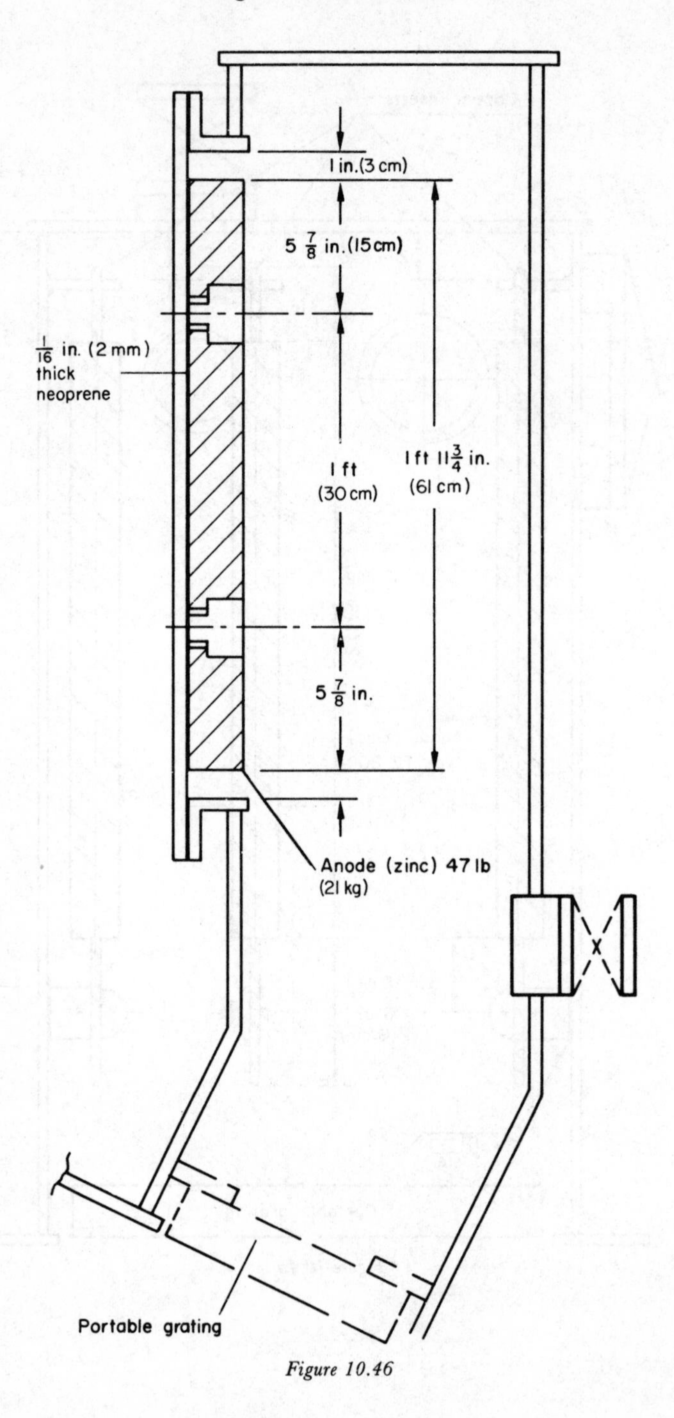
1 in.(3 cm)
5 7/8 in.(15cm)
1/16 in. (2 mm) thick neoprene
1 ft (30 cm)
1 ft 11 3/4 in. (61 cm)
5 7/8 in.
Anode (zinc) 47 lb (21 kg)
Portable grating

Figure 10.46

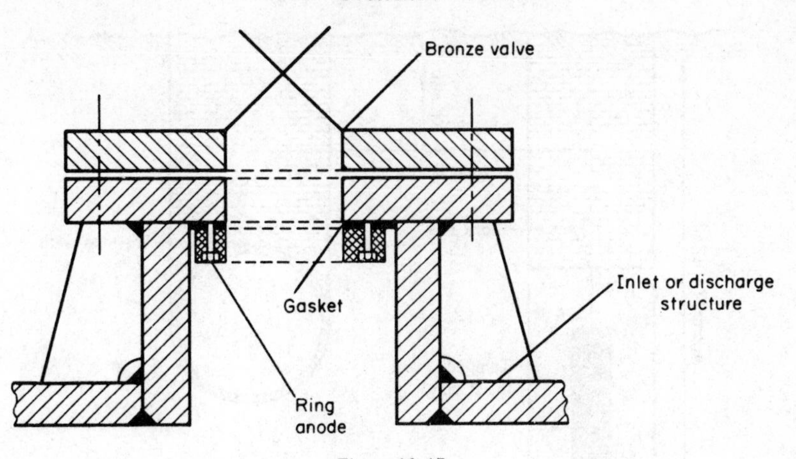

Figure 10.47

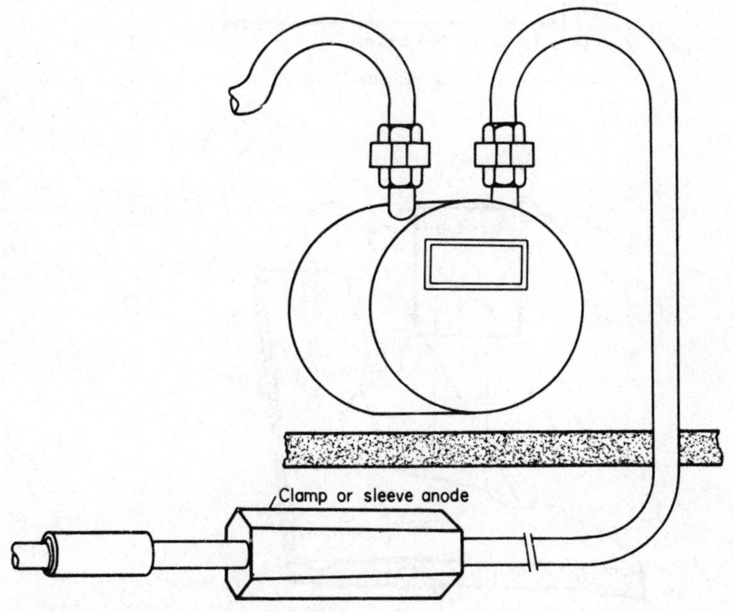

Figure 10.48

(*f*) electrical construction to conform to Standards;
(*g*) electrical construction to operate at specified ambient temperature in direct sunlight;
(*h*) size of output terminal to suit required cables;
(*i*) good ventilation and adequate insect protection;
(*j*) readable meters, standard accuracy; shunt accessible, well marked.

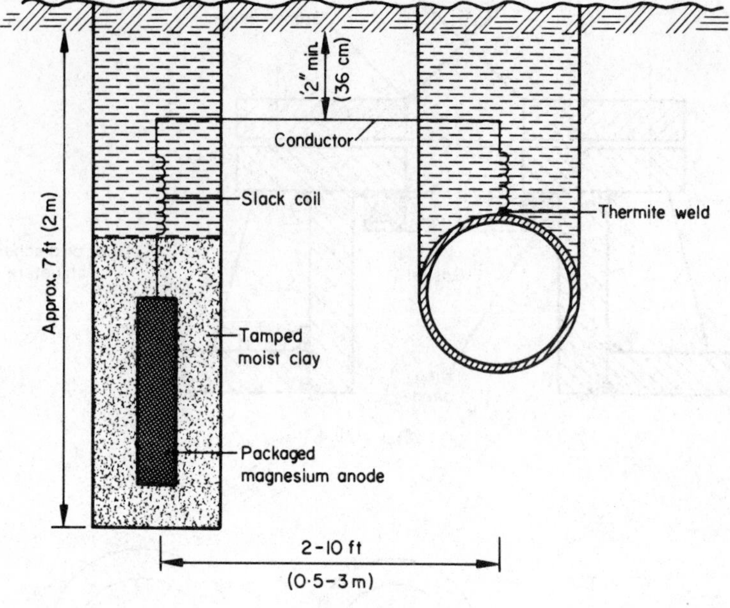

Figure 10.49

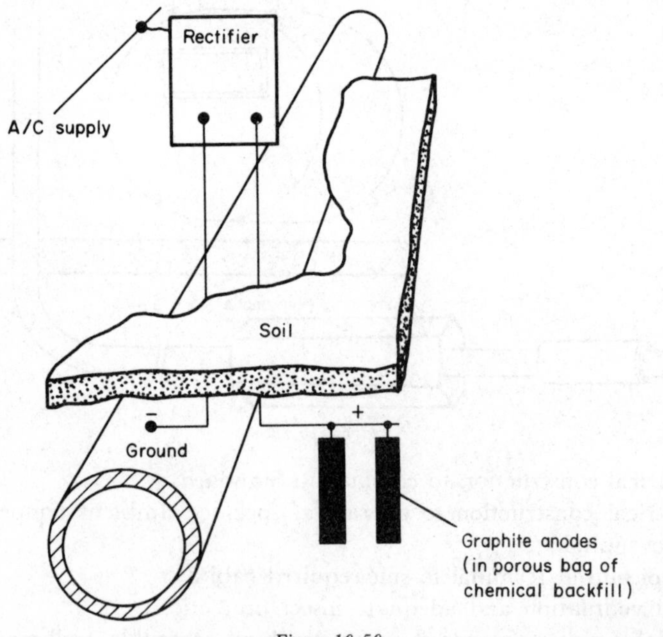

Figure 10.50

(20) Placement of reference electrode (*Figure 10.51*).

(21) Specify, where necessary, sand padding or use of special backfills to prevent adverse effect of aggressive soils on expensive or critical structures or equipment and pipelines buried in aggressive soils or old household or industrial dumps.

(22) Secure in all cases of important buried pipeline, in any type of environment, electrical continuity, preferably at the design stage:

(*a*) install flexible bonding jumpers of adequate size to prevent excessive IR loss through the bonds and connections;
(*b*) install bonding jumpers or test leads for fitting that could cause discontinuities;
(*c*) apply dielectric insulating coating and electrically connect discontinuous fittings to prevent cathodic interference and improve CP;
(*d*) provide dielectric coating at electrical discontinuities in pipelines, particularly on the side under cathodic protection;
(*e*) try to eliminate, as much as possible, many small isolated segments of pipeline requiring separate protection – design large unified cathodic protection systems;
(*f*) See Chapter 6.

(23) In general, buried pipelines separated more than 50 miles from a 2000 A hvdc electrode should be safe when equipped with normal cathodic protection; structures within 50 miles (80 km) may be affected. These should be analysed and appropriate steps taken (numerous small capacity self-modulated cathodic protection stations, special pipeline insulation, elimination of interference of adjacent buried structures, conductive co-ordination with adjacent buried structures, possibly a use of longitudinal pipeline currents adjusted to provide potential profile of the pipeline that is compatible with the earth surface profile).

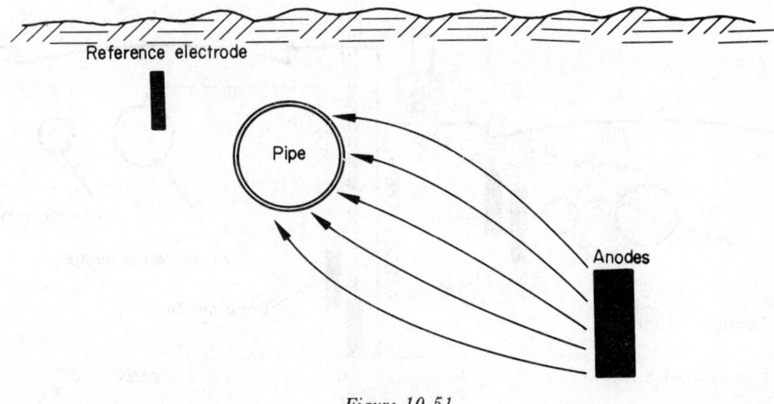

Figure 10.51

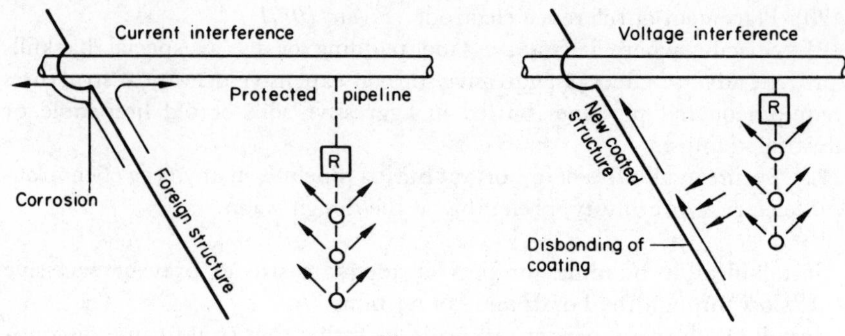

Figure 10.52

(24) Typical methods of avoiding interference of cathodic protection stations on other buried structures:

(*a*) use of zinc or magnesium sacrificial anodes as a current source;
(*b*) avoidance of large current densities (multiple rectifiers with small current outputs, multiple anode system);
(*c*) location of anodes as far away from adjacent structures as possible (*Figure 10.52*);
(*d*) use of deep anode systems (*Figure 10.53*);
(*e*) avoidance of locating shallow anodes so that the foreign buried structure is between the anode and the protected pipeline.

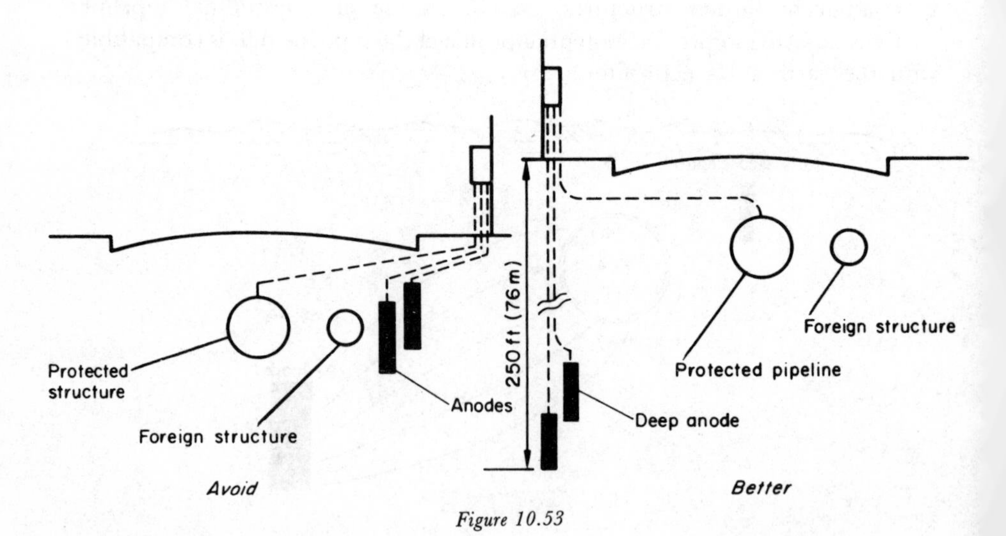

Figure 10.53

(25) Typical arrangement of cathodic protection of natural gas pipelines, using gas flow-powered turbine (impressed currents) (*Figure 10.54*).

(26) Typical arrangement of cathodic protection in ground bed installation (*Figure 10.55*): d = 20–30 ft (6–9 m) depending on soil resistivity and availability of space; 3–21 anodes in ground bed.

(27) Condensation on external surfaces of piping systems should be prevented by the appropriate application of insulation and ventilation (*Figure 10.56*) (see also Chapter 7).

(28) Provide for removal of oxygen from feed water and keep the oxygen within the required limits; this is to eliminate corrosion and pitting in boiler tubes (*Figure 10.57*) (for other types see Chapter 7).

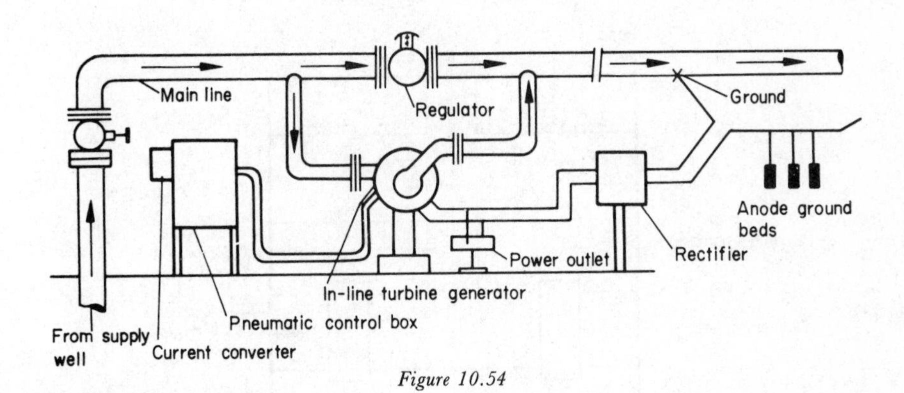

Figure 10.54

Figure 10.55

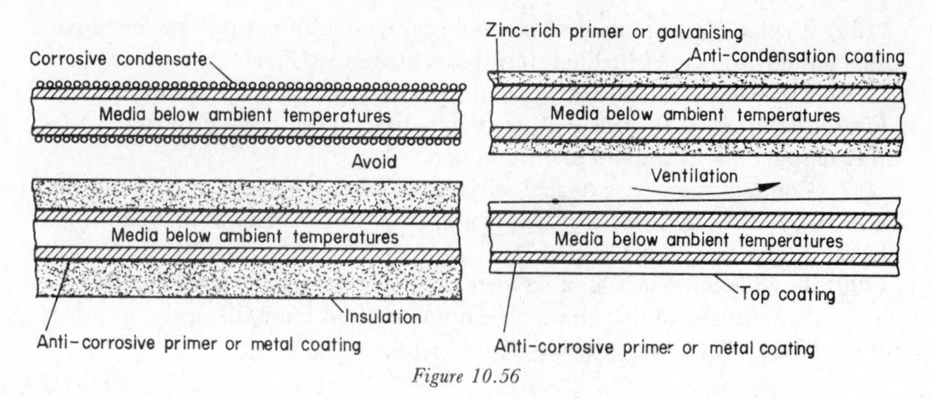

Figure 10.56

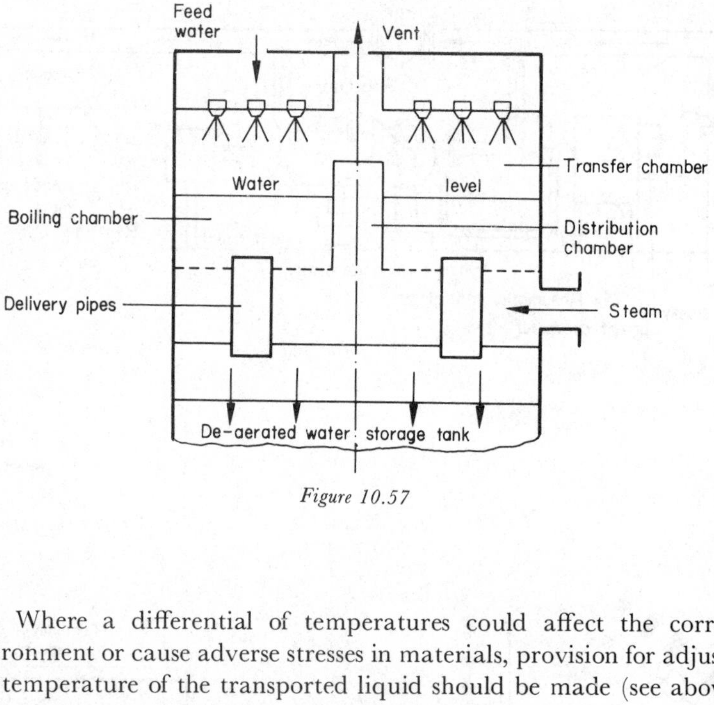

Figure 10.57

(29) Where a differential of temperatures could affect the corrosivity of environment or cause adverse stresses in materials, provision for adjustment of the temperature of the transported liquid should be made (see above and under).

(30) Where liquid is being transported through pipes made in active/passive metals, a natural or forced oxygenation of liquid may be necessary for renewal of the protective films (*Figure 10.58*); also for release of H_2S.

(31) For removing corrosive particulates and soluble gases from ducted exhaust air streams, wet cyclone scrubbers or spray towers or chambers should be introduced into the system (*Figure 10.59*).

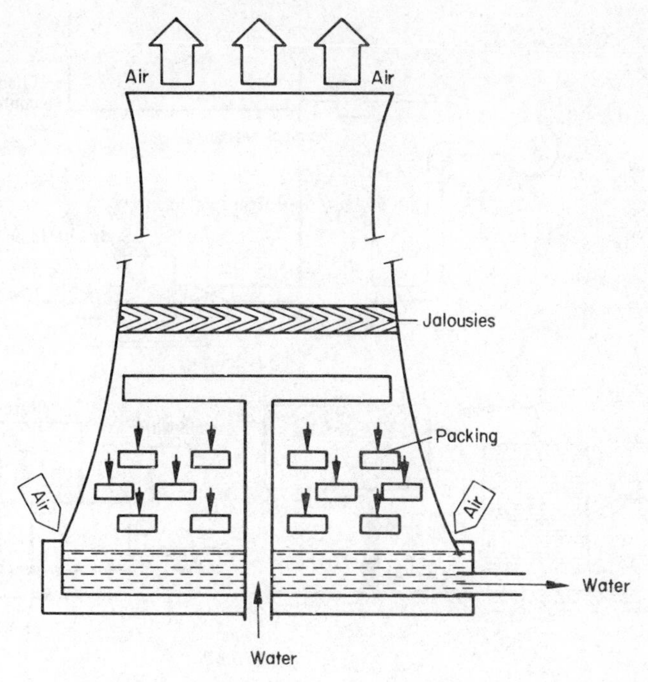

Figure 10.58

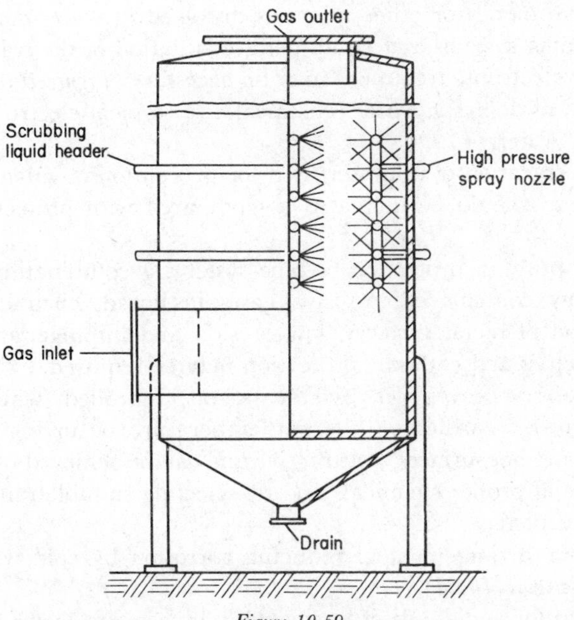

Figure 10.59

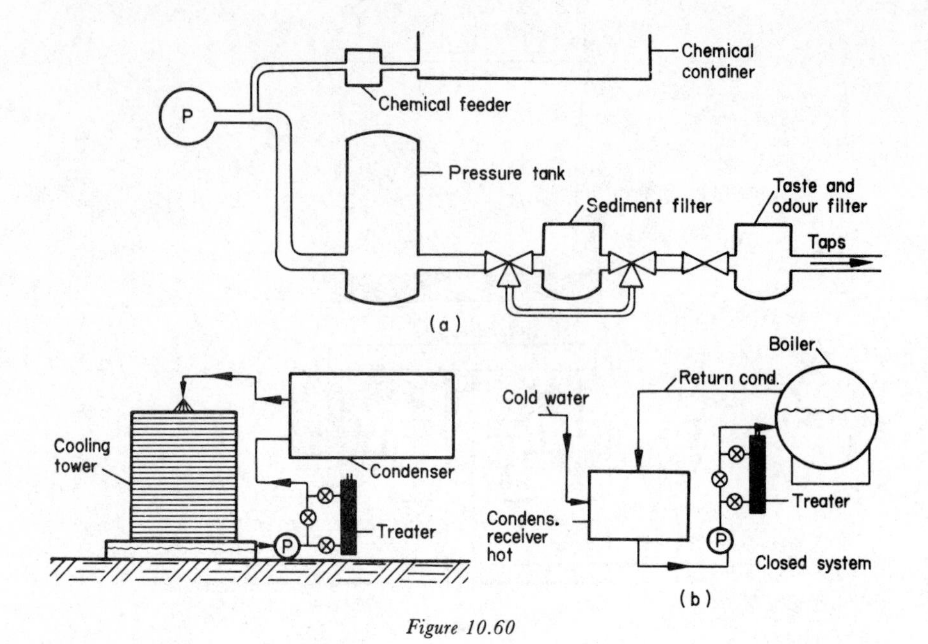

Figure 10.60

(32) Where metals or other minerals dissolved in water could adversely affect the piping systems and equipment, distillation of the water, chemical treatment or electronic treatment may be necessary (*Figure 10.60*): (*a*) typical treatment of domestic water (chemical); (*b*) typical electronic treatment of industrial water.

(33) Concentration of inhibitor used for prevention of galvanic corrosion in pipe systems should be increased as compared with protection of single metal.

(34) For optimum protection of pipe systems a combination of resistant materials, low contents of corrosive gases in liquid, chemical treatment for passivation of metal surfaces, low erosion and impingement geometry, suitable velocities and cathodic protection may be required.

(35) Corrosion prevention by means of controlled scale deposition (uniformly thin), regardless of different temperatures occurring in the system on both cool and hot surfaces, is attractive and can be achieved by eliminating inhibitors, using proper chemicals and by selecting suitable ranges for scale-forming constituents.

(36) Typical arrangement of reducing corrosion by cold water vacuum de-aeration (*Figure 10.61*).

(37) Design for avoidance of biological fouling in sea water systems:

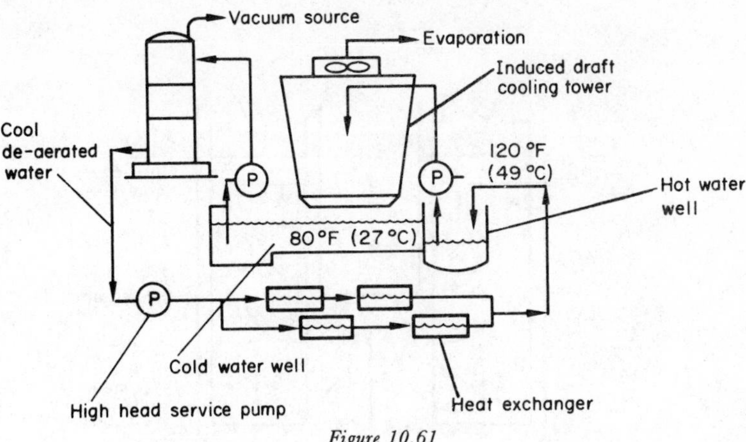

Figure 10.61

(*a*) in intermittent service provide against dead organisms clogging restricted passages;
(*b*) in critical systems design for water velocity of more than 6 ft/s (1.8 m/s) but less than the velocity causing damage to the system;
(*c*) in critical systems provide smooth internal surfaces without porosity and physical damage—line with soft rubber if possible;
(*d*) in critical systems the design of critical surfaces should allow either for access of light or be light coloured if possible;
(*e*) in pipe systems made of high content copper metals no residual deposits of resin, pitch or pollutant should remain; stainless steel should be flushed after each operating period; plastic to be restricted to intermittent service;
(*f*) where the above measures do not help, arrange for a periodic raising of temperature over 100°F (38°C) or introduce massive doses of chlorine.

(38) Fouling of internal surfaces of pipe systems by contained solids can cause heavy corrosion and erosion. Desludging, sedimentation, clarification and filtration arrangements should be made (*Figure 10.62*) (see also Chapters 7 and 11).

(39) The employment of fouling monitors to check the effectiveness of physical and chemical filtration and anti-fouling precautions is recommended.

(40) To prevent re-ingestion of solids and micro-organisms from discharge units into the intakes of pipe systems, only clean and safe effluent should be discharged.

(41) Desludging, cleaning and de-watering of lubrication and fuel oils (including diesel oil) should be undertaken for reduction of corrosion (*Figure 10.63*).

(42) Typical installation of automatic inhibition control (*Figure 10.64*).

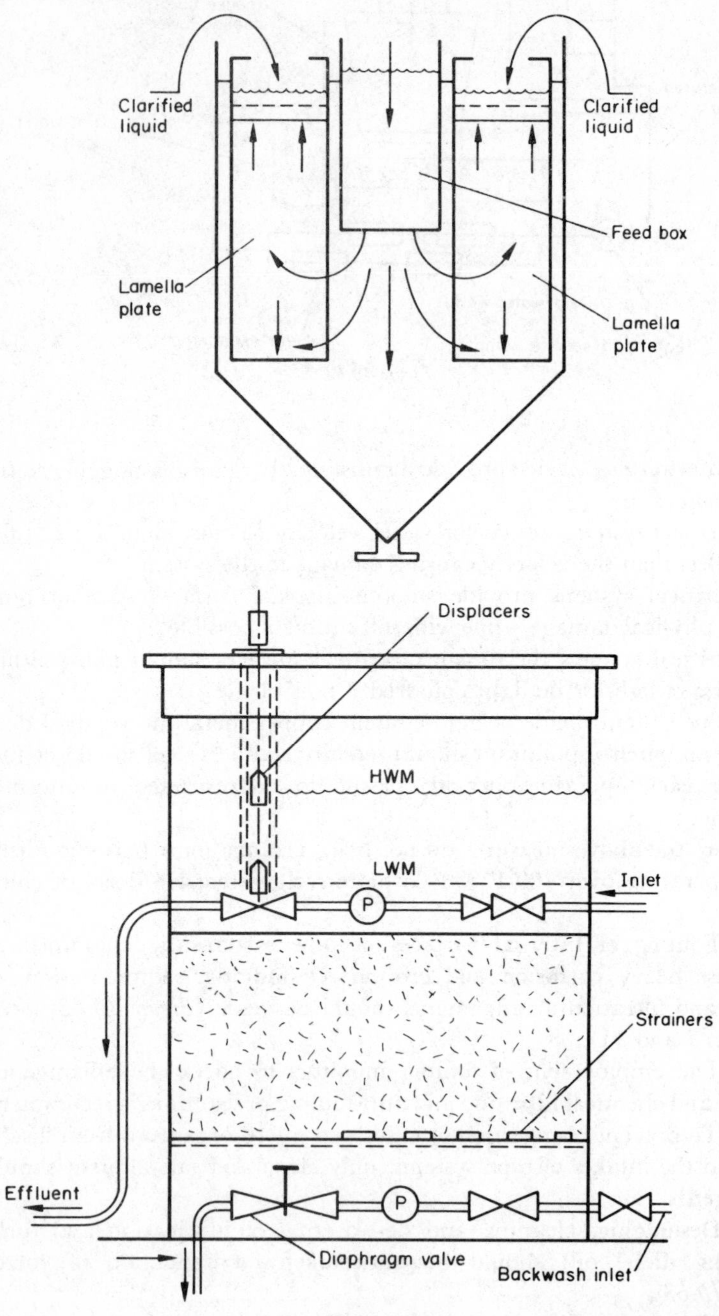
Clarified liquid
Clarified liquid
Feed box
Lamella plate
Lamella plate
Displacers
HWM
LWM
Inlet
P
Strainers
Effluent
P
Diaphragm valve
Backwash inlet

Figure 10.62

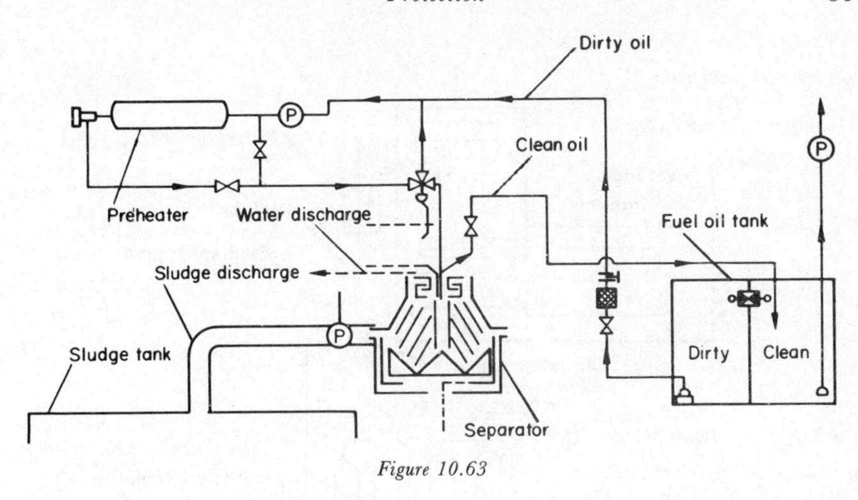

Figure 10.63

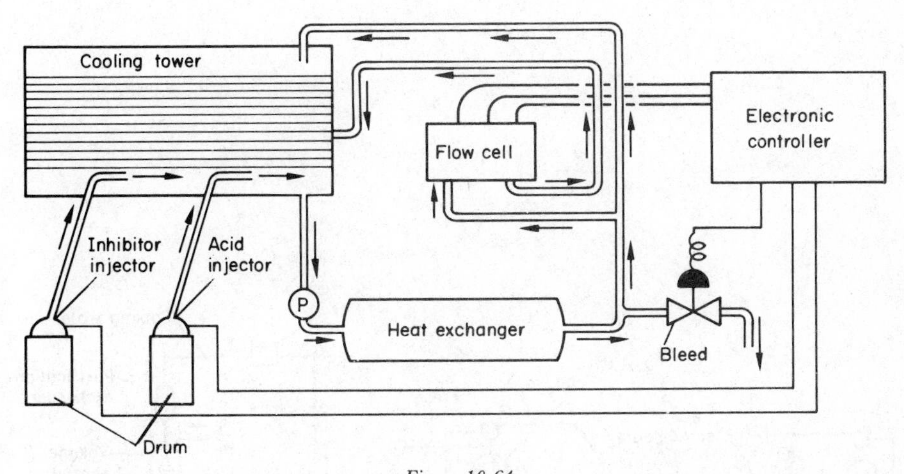

Figure 10.64

(43) Typical arrangement and flow diagram of oilfield steam generation system (*Figure 10.65*).

(44) Typical arrangement of reactor coolant system and chemical and volume-control system schematic (*Figure 10.66*).

(45) Reduction of fouling and corrosion in refinery processes with anti-foulants and inhibitors (*Figure 10.67*).

(46) Typical treatment of boiler feed water in refinery (*Figure 10.68*).

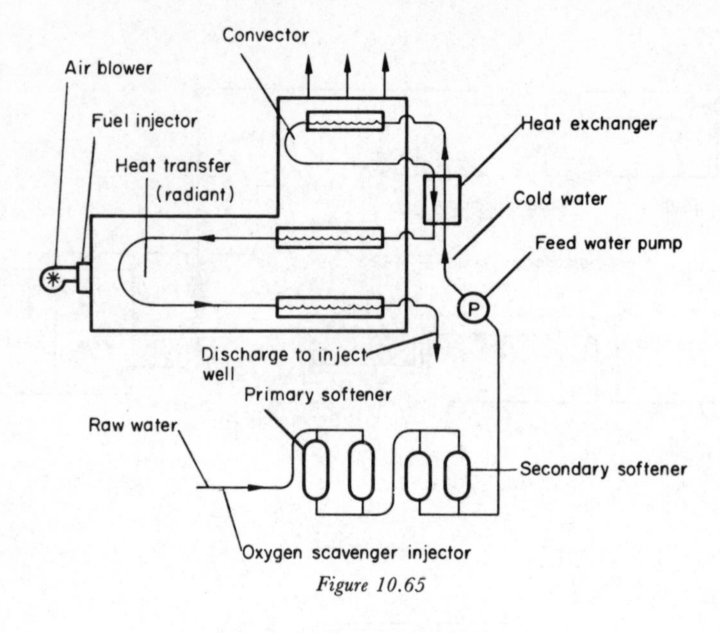

Figure 10.65

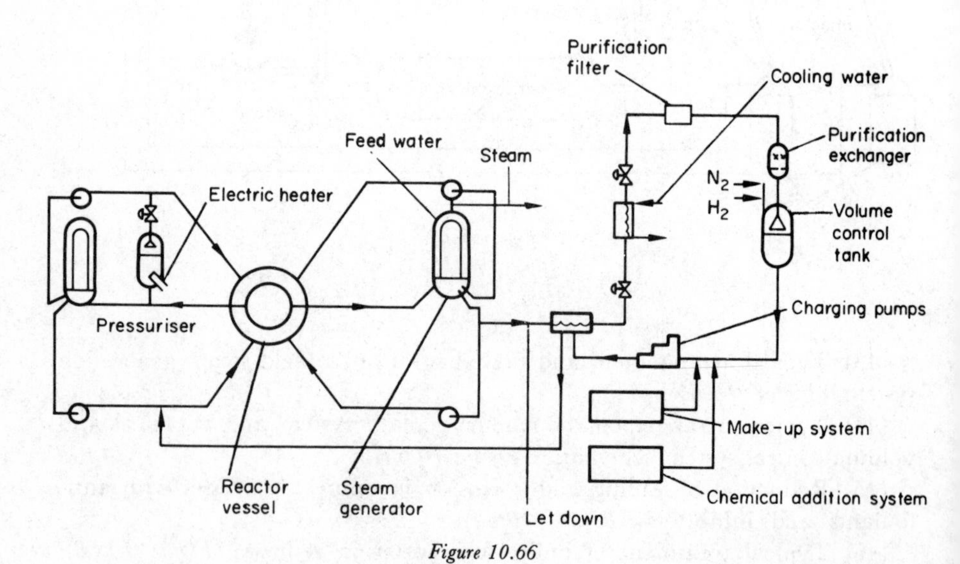

Figure 10.66

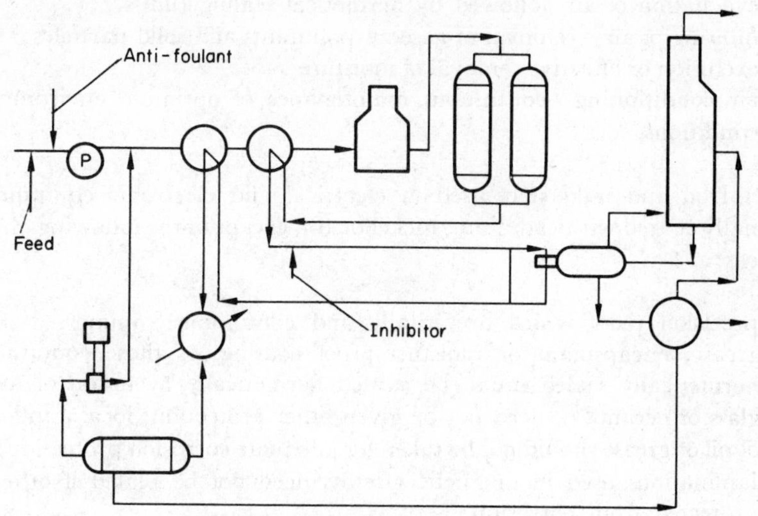

Figure 10.67. Inhibition—hydrodesulphurisation

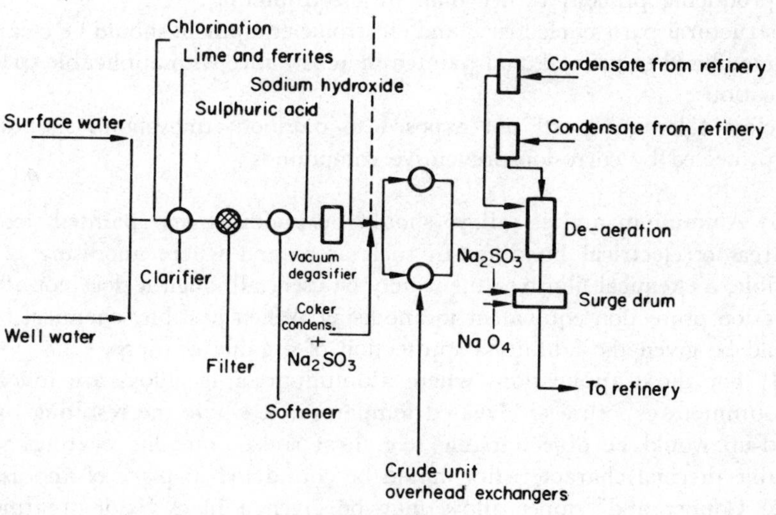

Figure 10.68

10.7 Electrical and Electronic Equipment

(1) The following techniques are mostly employed to prevent deterioration of electrical and electronic equipment:

(*a*) protective coatings;
(*b*) encapsulation;

(*c*) evacuation of air, followed by hermetical sealing (units);
(*d*) filtering of air—removal of gaseous pollutants and solid particles;
(*e*) exclusion or effective removal of moisture;
(*f*) air conditioning—continuous maintenance of optimum environmental conditions.

(2) Iron and mild steel used in electrical and electronic equipment is normally coated with cadmium, nickel or tin, except in the following circumstances:

(*a*) precision parts which are totally and continuously immersed in oil, grease, encapsulant or moisture-proof coating, or those contained in hermetically sealed units (i.e. sealed hermetically by fusion of metals, glass or ceramics), need not be given other protection; local application of oil or grease should not be taken for adequate corrosion protection;
(*b*) laminations used in magnetic circuits need not be plated if otherwise protected from corrosion;
(*c*) springs should, by preference, be protected by organic coatings or by metallic coatings applied by vacuum deposition or other non-hydrogen producing process, rather than by electroplating;
(*d*) structural parts of electrical and electronic equipment should be cleaned, metallised or primed, and painted in accordance with applicable specifications;
(*e*) close tolerance parts not exposed to outdoor atmospheres should be protected by corrosion-preventive compounds.

(3) Aluminium and its alloys should be anodised and painted, except in areas of electrical bonding. In such areas and where anodising is not possible, a chemical film treatment may be used, although it does not afford corrosion protection equivalent to anodising. When possible, chemical films should be given the additional protection of organic coatings.

(4) For those applications where aluminium or its alloys are involved in continuous exposure to elevated temperatures, where the resulting oxide build-up would be objectionable (e.g. heat sinks), metallic coatings with suitable thermal characteristics should be considered in place of anodising.

(5) Copper and copper alloys may be given a black oxide treatment, plated or painted.

(6) Magnesium, when its use is essential, should be protected as follows:

(*a*) rigid magnesium parts should be anodised—those subject to flexing should be chemically treated;
(*b*) all magnesium parts should then be given two coats of alkali-resistant primer, followed by one or more coats of a compatible top coat—magnesium parts for electronic applications may be given other moisture-

proofing coatings (epoxy, polyurethanes) in place of the primer and top coating.

(7) The noble metals, i.e. gold, platinum, palladium and rhodium, and the corrosion-resistant metals, i.e. chromium, nickel, tin, tin–lead solder and titanium, require no finish other than cleaning.

(8) Soldered joints should be protected with a moisture-proofing compound or coating.

(9) Corrosion-resistant (or treated to resist corrosion) minor devices (fasteners, etc.) should be used. Fasteners should be treated with zinc chromate, zinc chromate paste or graphite-free dry film anti-seize compound.

(10) Joint area of the electrical bond should be provided with a protective finish after bonding (organic coating, sealant, paint system).

(11) Metallic coatings should be selected for their suitability for the application involved, attention being given to the problems of ageing, diffusion and corrosion:

(*a*) cadmium or tin are used on metals which will be in contact with aluminium or magnesium;
(*b*) cadmium or tin are used as a pre-paint coating;
(*c*) rhodium over silver; gold over silver, copper or nickel; and nickel between copper and silver are applicable where tarnish prevention is required;
(*d*) heavy gold (0.03 mil (0.762 μm) thick) is used where subject to marine exposure;
(*e*) tin, gold or tin–lead are used for solderability;
(*f*) gold, rhodium or reflowed heavy tin are used for storage;
(*g*) chromium, nickel, rhodium or hard gold are used for wear;
(*h*) cadmium, nickel (in ferric chloride only), indium or tin are used for an easy etching;
(*i*) when base metals intended for inter-metallic contact form non-compatible couples these should be plated with those metals which will reduce the potential difference (see Chapter 6);
(*j*) heavy metal coatings should be used in preference to thin coatings;
(*k*) where practicable gold, platinum or tin–lead coatings should be used in preference to silver.

(12) Metallic coatings which may be applied by vacuum deposition to metallic or non-metallic surfaces for electrical conductivity should not be used for any mechanical application, due to their extreme thinness, fragility and susceptibility to damage.

(13) Use of cadmium plating for enclosed assemblies containing acids, ammonia, organic coatings, adhesives, plastics, varnishes or other organic materials subject to heat or their vapours should be avoided.

(14) Avoid use of palladium plating in enclosed assemblies containing organic materials to prevent polymerisation.

(15) Hot dip tinning should be used instead of electrodeposited tin; tinplating should be reheated to relieve stresses.

(16) The forming of copper oxide corrosion in pin-holes in metallic plating over copper should be prevented by interposing a layer of nickel between the copper and the top plating film.

(17) Silver plating should be protected from sulphurous fumes on storage and it should be cleaned immediately prior to soldering.

(18) Zinc plating should not be used.

(19) The thickness of gold plating should be sufficient to minimise porosity and provide complete corrosion protection. Recommended thicknesses:

(*a*) for tarnish prevention of silver, 0.05 mil (1.27 μm);
(*b*) for waveguides or contacts where a non-migrating material is required, 0.10 mil (2.54 μm);
(*c*) for general engineering use, 0.20 mil (5.1 μm);
(*d*) for resistance to extreme corrosion and wear, 0.30 mil (7.62 μm).

(20) Special care should be taken to prevent or retard the diffusion of substrate metals (silver, copper, chromium) into the electrodeposited gold under high temperature conditions. A suitable barrier to prevent diffusion is a thin nickel or palladium coating between the gold plating and the substrate.

(21) Soldering over gold should be avoided whenever possible. Where necessary, care should be taken to minimise the formation of brittle gold–solder compound by one or more of the following methods:

(*a*) using extremely pure (99.99 + %) gold;
(*b*) using thin plate;
(*c*) using minimum soldering time at minimum temperature.

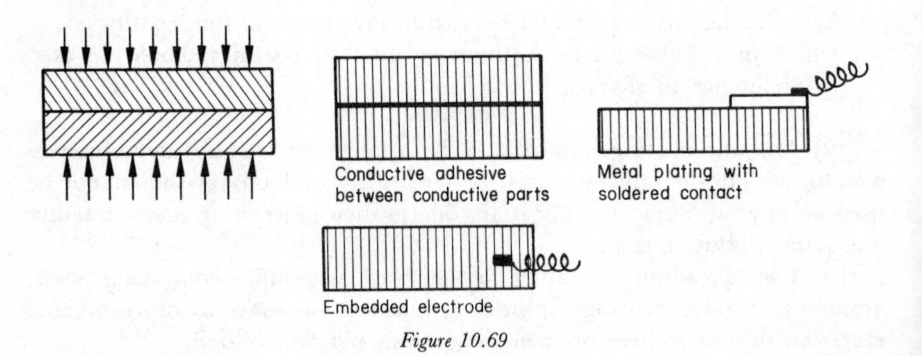

Figure 10.69

(22) Metallic coating may be applied to non-metals to provide a conductive surface. Although some problems (e.g. dissimilar metals corrosion) are thereby minimised or obviated, other corrosion reactions of the metal coating should be considered in the same manner as for plated or solid metals.

(23) Attachment of conductive plastics (filler types: silver flake, gold dust, pure carbon- or metal-plated types) for the purpose of establishing necessary electrical contact should be engineered to suit the form of the product (*Figure 10.69*).

(24) The ideal coating for an electrical circuit (jelly-type, foam-type, rigid, elastomeric, aerosol-type varnishes) shields it from environmental conditions mechanically, electrically and chemically without affecting the circuit's original characteristics. The coating should serve the following basic purposes:

(*a*) environmental protection to seal out moisture and other airborne contaminants, especially salts and sulphur compounds;
(*b*) handling protection to prevent damage from hand-borne contaminants (salts, oils, sulphur compounds) which are deposited on component during the final assembly or field testing;
(*c*) 'ruggedisation' against shock or vibration for protection during in-plant handling, testing and final use;
(*d*) insulation protection to maintain the electronic circuitry in a condition very close to its sterile design environment.

(25) Organic or inorganic coatings, when used, should be compatible with the substrate.

(26) Protective coatings should be applied after all punching, drilling, machining, forming and fabrication have been completed. Paint finishing systems, if required, may also be applied after metal deposition or over chemical film treatment.

(27) Corrosion-resistant steels should be passivated. No further finish is required to provide corrosion resistance to steel of the 300 series; where tarnish, rust or surface stain would be objectionable the 400 series and precipitation-hardening steels should be given additional protection by a suitable plating or, after passivation, they should receive one coat of zinc chromate primer followed by a suitable top coat.

(28) Soldered joints should be moisture proofed and the joint area of an electrical bond should, after bonding, be provided with a protective finish (organic coating, sealant, paint system). All contaminants should be removed from conductor surfaces.

(29) Direct corrosive environment away from critical equipment by judicious use of drainage and ventilation.

(30) Condensation should be precluded, by keeping components at temperatures above the dew-point.

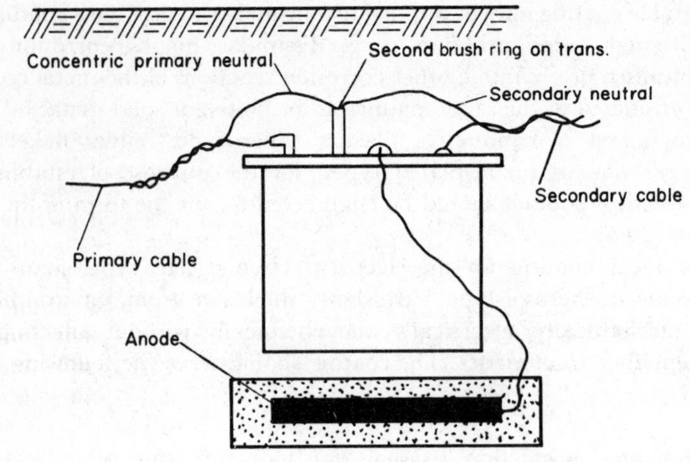

Figure 10.70

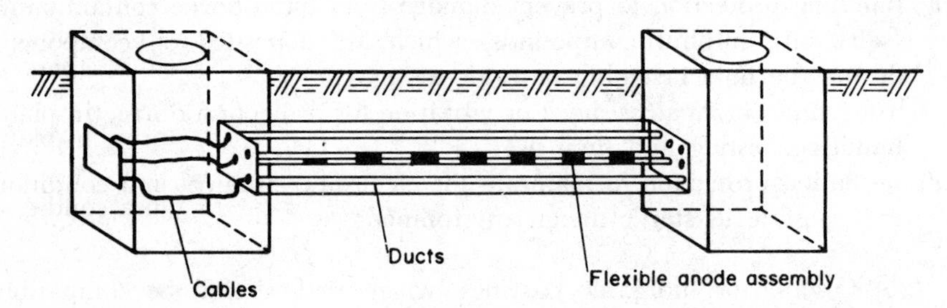

Figure 10.71

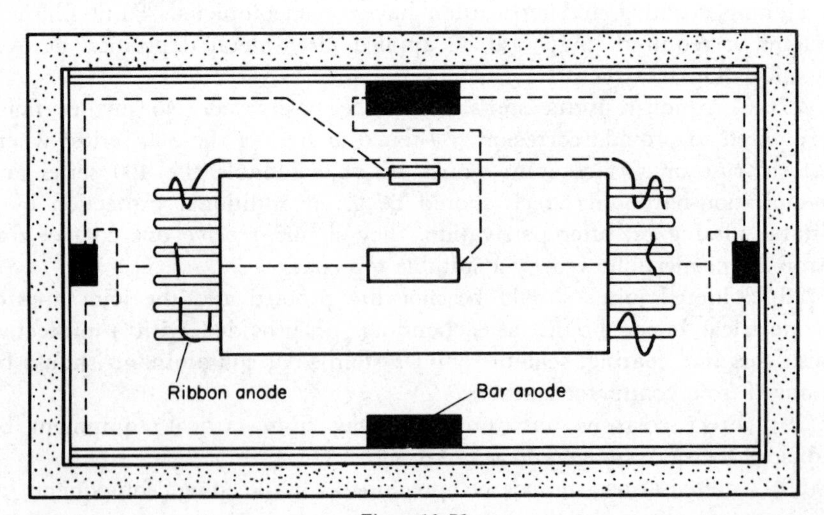

Figure 10.72

(31) Unless specified otherwise, the moisture level inside electronic equipment should be maintained below 30% RH at 68°F (20°C).

(32) Moisture should be further excluded from electronic devices by adequate housing, seals, gaskets and closures (see Chapter 7).

(33) Hygroscopic materials should be avoided. Desiccants should be used as little as possible and then these should not be in contact with unprotected metallic components.

(34) Contaminants should be removed from cooling air before it enters electronic equipment.

(35) Conductors should be provided with organic moisture barrier coatings where necessary.

(36) Typical arrangement of cathodic protection for buried electric underground residential systems equipment—direct buried type (*Figure 10.70*).

(37) Typical arrangement of cathodic protection by impressed currents in underground cable duct systems (*Figure 10.71*).

(38) Typical arrangement of cathodic protection for subsurface transformers by sacrificial anodes (*Figure 10.72*).

11 Maintainability

11.1 Introduction

It would be a truly ideal achievement to provide, in design, for an absolute functional reliability of a utility during its full operational life—from its conception to the termination of its usefulness. This, however, will be only very seldom possible in practice and the designed structure or equipment may fail as a result of random catastrophic failures of individual parts or by progressive degradation and deterioration of performance. Corrosion-control assessment should take account both of catastrophic and degraduation failures, with the understanding that each can be caused by a conglomerate of both mechanical and corrosion factors.

The designer can expect that during the utility's operational life its failures, including the failures of intrinsic corrosion control, will proceed in three stages. In the first phase, representing the debugging period, the failure rate will rapidly decrease from an initial peak, due to replacement of parts and correction of design and practical manufacturing and preservation errors.

In the second phase, a constant failure rate of lower power can be expected and this can further be reduced to the lowest possible frequency and potential by the diligent effort of designers and corrosion specialist. In this way the period of the utility's useful life can be extended to the most economic length of time, with the help of operational maintenance, when necessary.

In the third, or wear-out phase, the incidence of failures will again rise through a combination of catastrophic and deterioration failures. It is again in the hands of the mentioned group of specialists to postpone this period to the ultimate end of the utility's useful life.

It is obvious that whilst it is probable that failures will occur, the cost of major repairs (labour and materials) and replacements, the upkeep of the functional state, the danger of pollution of contents by corrosion products, the loss of productivity, use or profits and last but not least the danger of destruction of a utility itself and of human life shall, in most cases, enforce some sort of operational or preventive maintenance. Thus, as long as the equipment has to be maintained its maintainability must be ensured; that is, the equipment must preferably be so designed that the maintenance can be undertaken as a regular and economically feasible activity, unless it is of course entirely pre-empted during the suitably restricted lifetime of the utility. This applies to the functional as well as to the corrosion-control maintenance.

11.2 Scope

Every designer should acquire an eye for a design form which is economically maintainable, and also maintain a sympathetic understanding of the problems met by maintenance crews, who keep his creation alive. He should endeavour to make their task as simple and cheap as possible. The subsequent paragraphs indicate a possible relevant approach.

11.3 General

(1) All corrosion-control precautions included in design should be 'inspectable'. Their temporal state and efficiency should be open to observation under the conditions given by the selected geometry of the utility, its location and its arrangement and this should be possible in the ambient environment of production, operation or maintenance and with reasonable ease.

(2) All specified corrosion-control precautions should be 'repeatable and repairable' with reasonable ease in the given position and environment, as far as possible.

(3) Preference of replacing the whole utility (or its individual parts) instead of repetitive maintenance should be assessed on economic and operational conditions of the utility.

(4) Fundamental changes of the design concept may be necessary to avoid major impediments in maintenance, especially on complex structures and equipment in a corrosion-prone environment.

(5) Anti-corrosion maintenance should either require the shortest possible length of time to be executed or it should be as widely spaced as possible (e.g. in the case of structures and equipment or parts located of necessity in inaccessible positions). Design should be based on optimal length of corrosion prevention.

(6) The anti-corrosion maintenance should lend itself to a planned maintenance programme, which should be implemented with the least interference to the operation or use of the utility and be at the same time compatible with the corrosion incidence.

(7) Periodicity of maintenance should suit the operational cycles to cause the least possible inconvenience all round.

(8) All precautions should be taken in design to reduce the cost and degree of difficulty of subsequent maintenance, including dismantling and reassembling of structures and equipment, cleaning and preparation of surfaces *in situ*, preservation, etc., to a minimum. All such operations should be made easy and foolproof.

(9) Due consideration to the safety of the maintenance personnel should be given (e.g. easy and adequate access with the tools, provision of maintenance platforms, temporary ventilation, lighting, etc.).

(10) Fast corroding and critical parts shall not be located in inaccessible positions within a utility, which would involve maintenance personnel in major dismantling operations. Such parts or units should be easily removable.

Parts requiring difficult anti-corrosion maintenance should not be positioned in dark places.

(11) Materials and protective systems should be selected with an inherent economic length of resistance to corrosion.

(12) Anti-corrosion maintenance should be adjusted to suit problems arising from engineering maintenance procedures (e.g. physical damage, spillage of chemicals, etc.).

(13) Accessibility for inspection should be reconciled with the selected inspection and destructive or non-destructive testing procedures (e.g. X-rays, ultrasonic, visual, probes, instruments, etc.).

(14) Probable expertise and availability of future maintenance personnel may govern selected maintenance requirements.

(15) Use of modular assemblies tailored to the requirements of corrosion control is recommended.

11.4 Structures and Equipment (See also Chapter 7)

(1) Accessibility for maintenance of structures and equipment is a necessity (*Figure 11.1*).

(2) Obstructions to maintenance of structure or equipment should be subdivided, where necessary, and their width should be kept to a minimum (*Figure 11.2*).

(3) Optimum distance of an obstacle to the background structure or equipment depends on available and required maintenance equipment and technique (*Figure 11.3*).

(4) Excessively low inaccessible seatings for machinery and equipment should be avoided, especially if an entry or an accumulation of an unwelcome electrolyte within their frame is possible (*Figure 11.4*).

(5) Coverplates and other items which have to be removed for maintenance purposes shall be readily removable (*Figure 11.5*).

(6) Access space should be provided behind furniture, ventilation ducting, lockers and trunking, or such equipment should be incorporated and form part of the structure (*Figure 11.6*). Alternatively, any such equipment should be easily removable to obtain access to strength structures.

(7) Critical equipment subject to corrosive environment may require arrangement for accommodation of corrosion coupons, probes and corrosometers (*Figure 11.7*).

(8) Piping systems should be designed for cleaning on the stream, if possible and necessary (*Figure 11.8*).

(9) Critical parts or parts of systems subject to heavy corrosion should be separately dismantleable (*Figure 11.9*).

(10) Blind or non-inspectable areas shall be kept to a minimum in the fail-safe design and, where unavoidable, additional strength must be provided to prevent critical effect of corrosion fatigue (*Figure 11.10*).

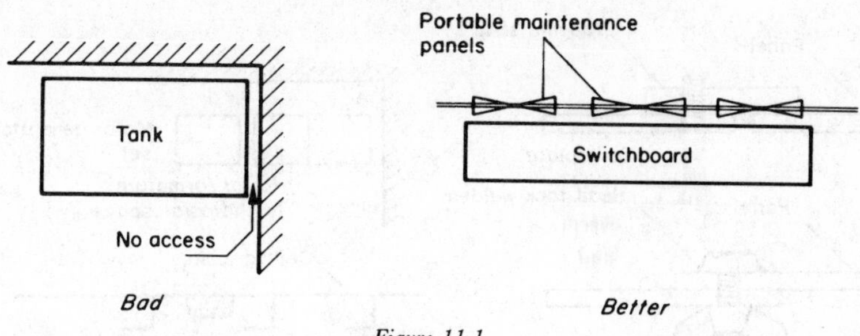

Figure 11.1

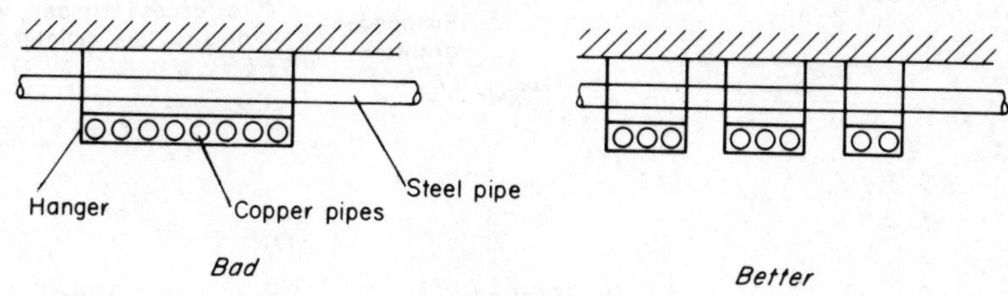

Figure 11.2

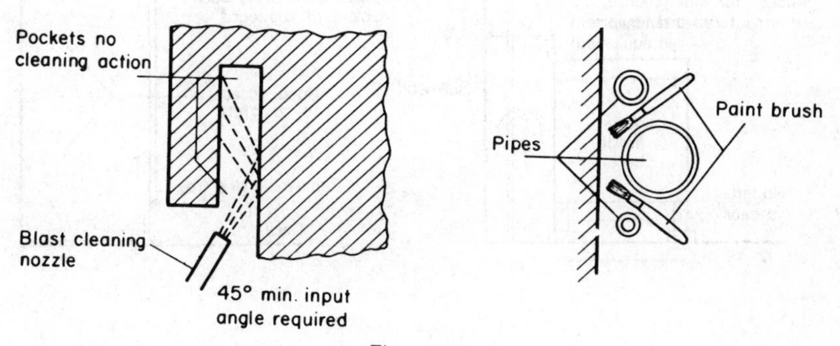

Figure 11.3

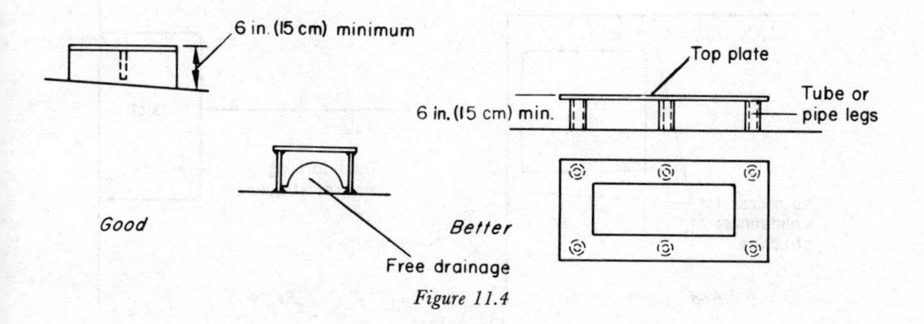

Figure 11.4

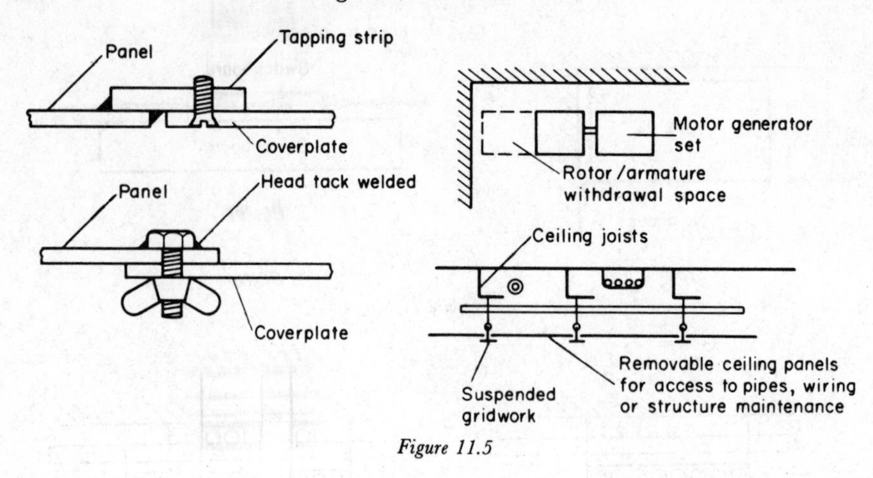

Figure 11.5

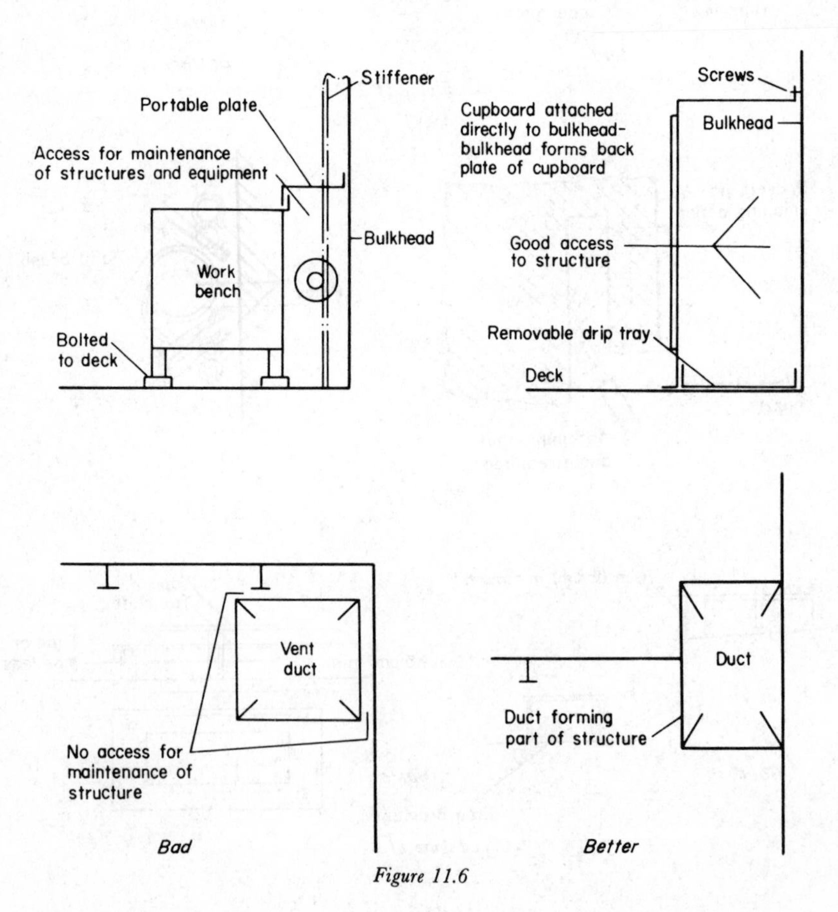

Figure 11.6

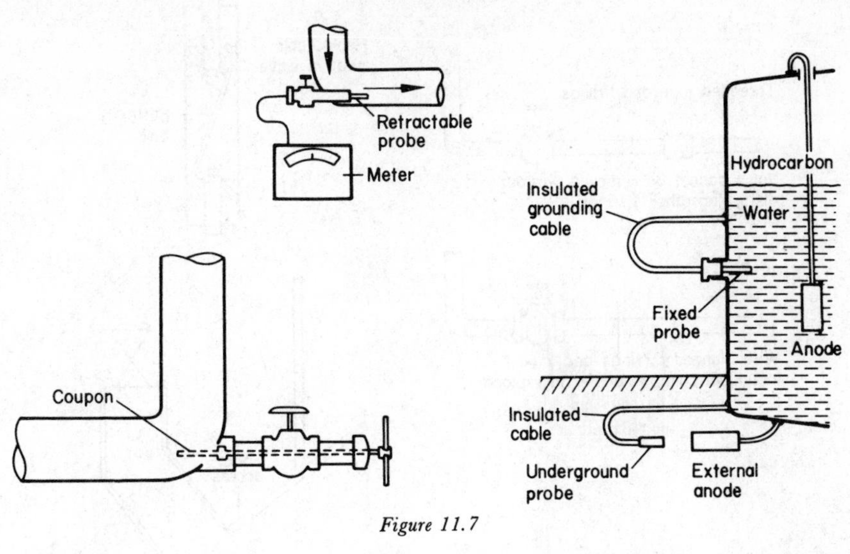

Figure 11.7

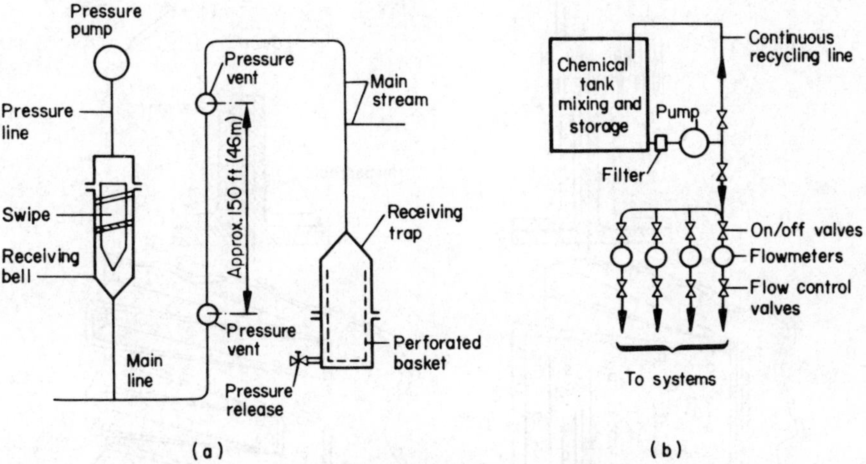

Figure 11.8. (a) Arrangement for resilient pipe-cleaning swipe; (b) arrangement of chemical cleaning system

11.5 Preservation and Protection

(1) Periodic replacement of protective devices of cathodic protection systems should be made easy and foolproof (*Figure 11.11*).

(2) Protection techniques, which can be repeated economically *in situ*, are preferred.

(3) Where blast cleaning *in situ* for maintenance will be necessary, either a form of inherent protection or auxiliary arrangements against ingress of abrasive should be built into the designed utility (*Figure 11.12*).

(4) Local and general geometry of the utility should be designed for repetitive cleaning and preservation (*Figure 11.13*).

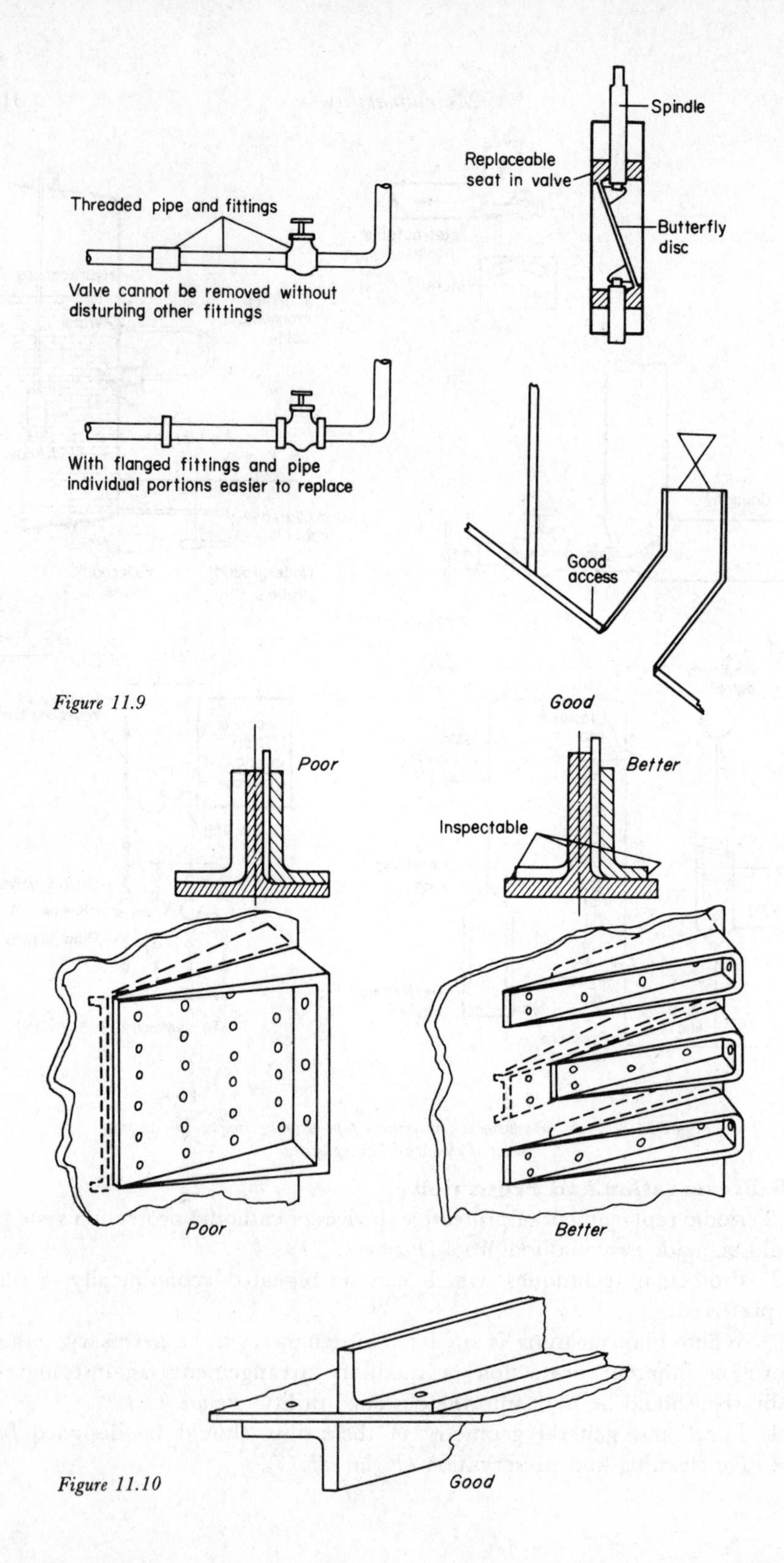

Figure 11.9

Figure 11.10

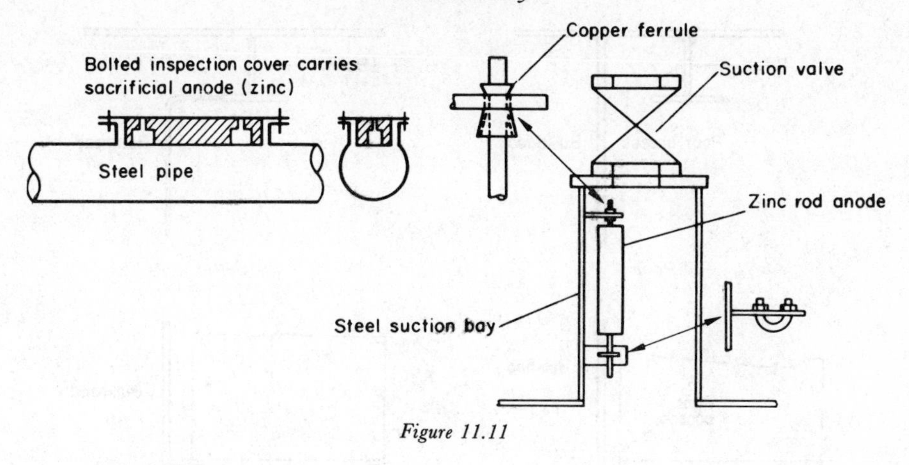

Figure 11.11

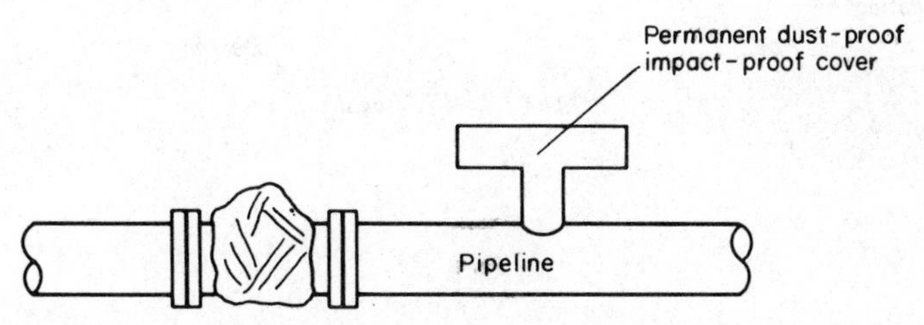

Figure 11.12. Arrangement for attachment of wrappers or protective boxes

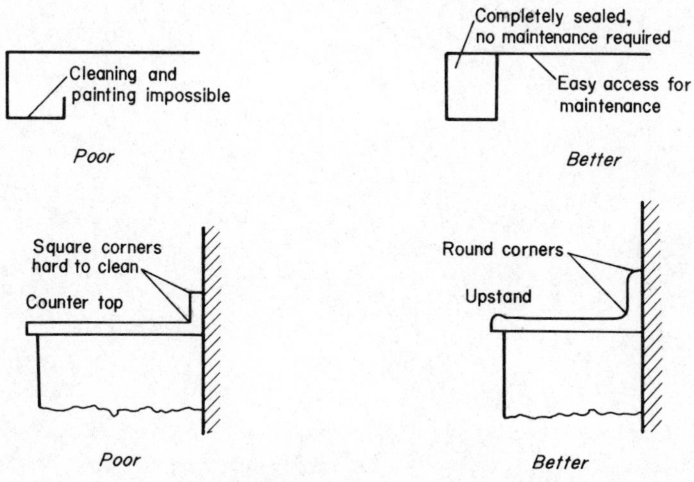

Figure 11.13

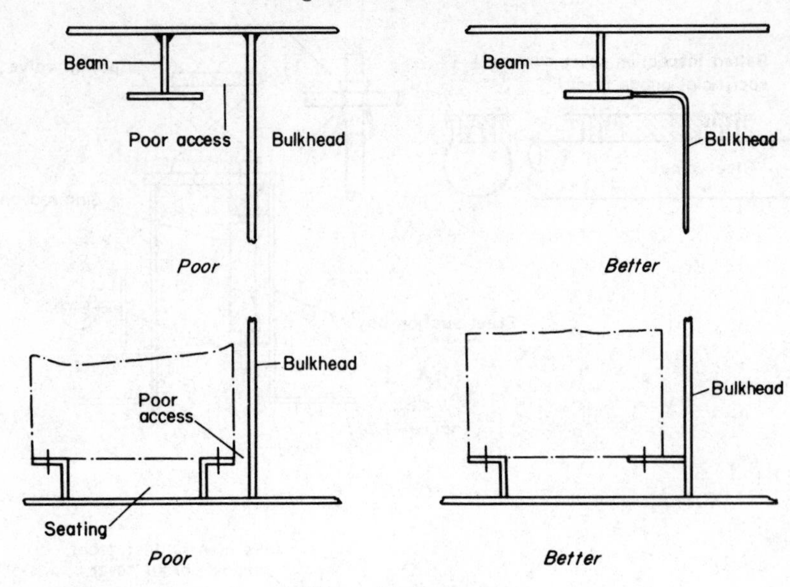

Figure 11.13 (*Contd.*)

12 Economics

12.1 Introduction

Whether corrosion-engineering projects are being initiated or reviewed, possibly the most searching question the responsible management can ask is how much the projects will cost them. Corrosion, as such, is basically an economic problem associated closely with the loss of capital assets and business profits; corrosion control should provide the most advantageous course of obviating such losses in the first case, whilst not neglecting the most important implications of safety, health, pollution of environment and products, which have a very close affinity with corrosion damage.

The corrosion specialist and the involved designer should primarily be concerned with achieving optimum allowable economy in each specific case on their drawing board. Current corrosion-control technology is usually able to offer a selection of acceptable solutions to any particular corrosion problems, although it may be difficult to select the best one without a common denominator. The selection of optimum solution cannot thus be based only on the understanding and knowledge of corrosion technology but the pertinent cost value must also be known. Selecting an ultimate in corrosion technology may sometimes prove the least profitable venture and a losing proposition.

Whilst basing his decision on a realistic appraisal of economy a designer can distinguish which combination of material, method, technique and system from the selective range is the best proposition, by trying to answer following questions:

(*a*) would more expenditure on corrosion control be economically advantageous;
(*b*) could greater savings be obtained, at the present level of expenditure, by better application of the money spent;
(*c*) does it pay to accept a given degree of corrosion rather than to spend more money on controlling it;
(*d*) what premium on expenditure will satisfy the public requirements for safety and healthy environments?

The fullest investigation is necessary to answer these questions.

12.2 Scope

This chapter provides the basic guidance and introduction to the art of economic evaluation of the particulate factors of corrosion-control appreciation, as well as the composite compromise of the complete, integrated corrosion control of the whole project. It is enriched by the NACE Standard RP-02-72, pages 326–343, in which the excellent presentation of the subject matter makes it superfluous to enlarge on the details of calculations of economic appraisal in this chapter independently.

12.3 General

(1) Where possible, and subject to availability at the time of analysis, all elements of costs should be included in the economic evaluation of corrosion control in design:

(*a*) Direct costs:
direct labour (original, service, replacement; operating time for tank washing and gas freeing, fresh water rinse);
direct materials (original, service, replacement);
over-design (part);
subcontracts (including ex works treatments);
deterioration and preventive measures;
travel and transport (e.g. dock fees, etc.);
contract labour;
consultants' services;
sampling and testing including NDT procedures;
pilot projects.

(*b*) Indirect costs:
indirect labour;
indirect materials and supplies (receiving, inspection, storage, handling, etc.);
over-design (part);
delays;
safety and insurance;
downtime (shut down);
efficiency losses;
product losses;
contamination of environment;
contamination and decontamination of product;
freight and cargo loss (due to the coating and rust scale formation);
freight and cargo loss during critical period;
loss or gain of turnaround time (improvement of drainage in rustless tanks and vice versa);
cost of potential catastrophic failure;
cost of potential fire;
repair of damage caused by corrosion to own and other properties.

(*c*) Overheads (prorated):

general and administrative (marketing, executive salaries, corporate expenses, etc.);
rental of machinery and equipment;
depreciation of machinery, equipment and facilities;
fringe benefits, sickness, vacation and holidays;
insurance;
taxes, less development subsidies;
heat and light;
utilities;
plant operating expenses;
repairs and maintenance;
research and development;
laboratory supplies and stationery.

(*d*) Profit or loss:
cost of borrowing (interest on loans);
cost of capital (return on investment);
recovery profits and salvage values;
inflation trends.

(2) Good and comprehensive specifications are essential to any accurate economic evaluation of corrosion-control precautions. The estimator needs to know exactly what is needed from the producer/erector/assembler and applicator in order to intelligently evaluate the whole job. The estimator is not a mind reader; he cannot assume details which might be important to the customer, that is if these are not written into the contract, instructions, specifications or drawings. Standard procedures should be noted by reference.

(3) Where exact instructions or specifications are not available, the contractor is going to bid and consequently furnish that product which is most economical for him. This may often be quite different from what is in the interests of good corrosion control and any preliminary economic analysis is illusory. For a meaningful economic evaluation the following information must be known:

(*a*) accurate description of the job; what is to be done and what is not to be done; where the job is to be done;
(*b*) avoidance of general terms of instructions; instructions to be referenced to standards or spelled out in exact technical terms and comprehensively described;
(*c*) production and application methods formulated and accurately described; machinery and tools described;
(*d*) corrosion-control systems exactly specified and conditions of their application stated;
(*e*) safety requirements and auxiliary equipment described;
(*f*) storage and materials or product movements, housekeeping, workman-

ship, weather limitations, flow of productions and assembly, decontamination and ventilation procedures, specific instruments by trade name and number, hiring charges, inspection requirements, etc., comprehensively described.

(4) Parallel to management's anxiety at the alarming increase of the cost of construction and investment, the main concern of the economics of corrosion control in design lies in the reduction of expenditure and the increase of profits. Economics being a social science and engineering an applied science, the economic evaluation of corrosion-control contents of design serves to bridge the gap between these two human activities, to interpret the intention of relevant engineering endeavour to the management in the meaningful language of money and to temper the flight of the engineering invention by the stark realities of commerce.

(5) Any unproductive deterioration of materials and facilities by corrosion damage should be recognised on the national level as an important loss of material and human resources, which could be otherwise used for a productive purpose.

(6) The degree of corrosion control applied in the designed product is a matter of sound economics. The designer's job is to achieve the desired balance of effect for a minimum in capital outlay, maintenance and replacement costs, cost of inconvenience and social depredations or the potential wastage of human life. However, more expensive materials or products than absolutely necessary should not be chosen, unless it is more economical in the long run, necessary for safety or for other important reasons.

(7) Sometimes a timely replacement of deteriorated components may be more economical than such precautions which require high initial expenditure or an extended maintenance. The cost of remedial maintenance or the cost of making do with the deteriorating structures and equipment should be known. Corroding components can then be replaced at a later date with new, used or improved materials or components; the design should be adjusted accordingly.

(8) To increase the utility of structures and equipment subject to corrosion one may look at the long-term factors influencing their degree of utilisation:

(*a*) amount of capital equipment per user;
(*b*) level of technology applied to the work;
(*c*) quality of labour force, depending on the education, training and motivation;
(*d*) skill, drive and intent of the executive branch of the involved organisation.

(9) The groundwork of correct economic evaluation and decision-making in corrosion control rests with comprehensive recording and periodical evaluation of the cost and effectiveness of materials (including their inter-

relation), systems, technologies and personnel. This preferably segregated for each given component, system, unit or grouping. The whole of a given project should be covered.

(10) Good records are better than a good memory in establishing the onset of the corrosion problem, the frequency of its occurrence and immediate past cost. Decision should be reached on all information available and recorded, using the best estimates of cost, life and other germane factors that can be obtained at that time.

(11) Effects of previous corrosion-control precautions or procedures should be classified as to whether these are constant problems or seasonal, whether they are progressive or static.

(12) Historically documented cases of corrosion behaviour of materials, structures, plant, equipment and the methods of corrosion control should be considered with caution; the response of materials, designs and preservation systems may vary considerably in the vast variety of corrosive conditions in existence.

(13) Where a considerable number of items of corrosion control import (e.g. corrosion failures) of a comparatively restricted and repetitive nature can be expected, electronic data processing and periodic analytical summaries may be very useful for the process of economic evaluation.

(14) Corrosion failures, especially those associated with stress corrosion cracking, fatigue corrosion and similar, are often, by mistake, classified under the costs of mechanical breakdowns.

(15) In any given project, life expectancy, frequency of recurrence, etc., should be estimated and evaluated.

(16) All estimates of life expectancy may be only an educated guess, based on a thorough study of a range of possibilities obtained by the averaging of statistical data; these do not always agree with actual experience.

(17) Service life of existing materials, even when correlated with recorded corrosion test data, should be weighed in the light of the costs (first cost, average yearly expense of replacement, average yearly expense of routine maintenance, average yearly expense of operating costs, average yearly expense of down-time, etc.).

(18) Information on the service life pattern of selective materials is based on the following data:

(*a*) information on wastage limits in a given environment and in a particular location, based on the mean corrosion rate plus the standard or observed deviation;

(*b*) information on the critical thickness or strength of materials to secure the integrity of the appreciated structure or equipment;

(*c*) information on the expected materials renewal periods in the particular locations for various structural members or groupings;

(*d*) information on the distribution of materials amongst various structural members, or groupings by logical divisions in corrosion patterns;
(*e*) information on miscellaneous material repairs of structures and groupings;
(*f*) information on possible benefits of corrosion.

(19) Both the external and internal corrosions should be included in the estimation of the service life in the progressive corrosion-rate figures.

(20) For comparative evaluation of various technological solutions of the given corrosion problem the results of individual calculations should be converted to a common denominator, e.g. cost per day, cost per year, cost per life or to any other suitable financial base.

(21) Selection of the design, system or method which will make the greatest reduction in overall operating cost and provide the prerequisite safety accounts for the first application of corrosion economics in corrosion-safe design.

(22) For estimation of corrosion-control economy the estimator should make every effort to project the immediate past experience into the future with a help of projection graphs, where possible.

(23) The annual corrosion rate of steel has little or no bearing on paint selection; the main concern is with the service life of the paint system in destructive exposures and a possible occurrence of rust-creep. Neither does it affect the selection of surface preparation nor the selection of a cathodic protection system.

(24) The ultimate paint system should be restricted for economic reasons to the aggressive areas and the less expensive systems used for less destructive exposures.

(25) In the economic evaluation of painting preservation the cost of materials is not indicative on its own—the cost of labour (approximately 80%), auxiliary provisions (scaffolding, plant, ventilation, drying, lighting and cleaning), accommodation and transport of personnel must be included.

(26) Surface estimating aids presented by many paint suppliers are not valid for economic evaluation of paint systems—the true surface areas form the base for estimates (texture, undulation, fasteners, flanges, etc.).

(27) For appropriate evaluation of a particular paint system the most important factor is its maintenance cost per square metre per year, and this consideration should be coupled with the serviceability of the paint system in variable conditions (e.g. change of cargo in a ship or similar).

(28) The present taxation system, in most cases, does not seem to favour an efficient preventive corrosion-control effort. The short-term deductibility of approximately 50% of the annual maintenance cost in a form of tax credit, when compared with the long-term depreciation write-off of the capital expenditure on preventive measures, offers a premium on inefficiency. In

the interest of national economy responsible governments should support the policy of preventive corrosion control by preferential taxation.

12.4 Methods of Appraisal

Having decided upon the integrated technical concept of corrosion control the corrosion specialist must be prepared to obtain optimum economy from his conclusions and to present his supporting arguments to his principals in terms that are readily comprehensible. He has thus to secure their backing not only for the preferred corrosion control that is technically sound but also offers the greatest economic advantage and increases their profits. Basically he is faced with three possible types of decision:

(1) Selection of optimum corrosion control for a brand new process or untested project. No money value can be consigned to corrosion control.

(2) Clear-cut decision; concept A costs less money than concept B and has the same life expectancy or costs the same and has a longer life—normally both lives being shorter than one year.

(3) Comparison between specific concepts of different costs and different anticipated lives, where not only the financial cost of the new project but also the subsequent upkeep of its utility during its economic life has to be compared.

The systematic analysis of corrosion control in design is primarily concerned with the last type of financial comparison. There are several different techniques of calculating the economic appraisal of corrosion control at hand:

(*a*) comparison of money outflow for short-time projects;
(*b*) life versus cost comparison—value of money and possible cost of production losses not being considered;
(*c*) pay-out period comparison of the cost of an item of corrosion control, divided by the income or savings it produces per year—taxes and time value of money not being considered;
(*d*) return on investment comparison, which is reciprocal to the pay-out period system—the results can be compared directly in terms of interest:
(*e*) annual cost comparison; the cost is multiplied by capital recovery factors for comparison of items of corrosion control with different service life, or for comparison of an item requiring major expense with another one in need of recurring expenses;
(*f*) discounted cash flow comparison—present worth and present worth after taxes—where the cost of money is also being considered.

In consideration of the need for an overall simplified and standard approach to the economic appraisal calculations, the National Association of Corrosion Engineers, Houston, Texas, USA published their Standard RP-02-72,

'Direct Calculation of Economic Appraisals of Corrosion Control Measures', which is based on the method of discounted cash flow analysis. This is aptly suitable to complement the chosen system of analysis of corrosion-control parameters in this book in a significant number of cases.

The author of this book received the kind permission of the National Association of Corrosion Engineers to include here the relevant contents of the above-mentioned Standard, to illustrate a procedure of an economic evaluation of corrosion-control measures which is easily applicable by the corrosion specialist or designer and comprehensible to the economist. This privilege is acknowledged with sincere thanks.

Recommended Practice

Direct Calculation of Economic Appraisals of Corrosion Control Measures

The following Standard (RP–02–72) is issued by the National Association of Corrosion Engineers, Houston, Texas, USA and is reproduced here with their permission.

The National Association of Corrosion Engineers issues this Standard in conformity to the best current technology regarding the specific subject. This Standard represents minimum requirements and should in no way be interpreted as a restriction on the use of better procedures or materials. Neither is this Standard intended to apply in any all cases relating to the subject. Unpredictable circumstances may negate the usefulness of this Standard in specific instances.

This Standard may be used in whole or in part by any party without prejudice if recognition of the source is included. The National Association of Corrosion Engineers assumes no responsibility for the interpretation or use of this Standard.

Nothing contained in this Standard of the National Association of Corrosion Engineers is to be construed as granting any right, by implication or otherwise, for manufacture, sale, or use in connection with any method, apparatus, or product covered by Letters Patent, nor as indemnifying or protecting anyone against liability for infringement of Letters Patent.

Section 1: General

1.1 In common with other engineering problems the choice between alternative corrosion control measures must consider economic factors as well as technical aspects.

1.2 The need has long existed for an overall, standard approach to aid in the selection between alternatives. Standardisation, in the form of a Recommended Practice, will contribute significantly to a broader understanding of the different approaches possible within industry and to an effective exchange of ideas.

1.3 This Standard (Recommended Practice) is comprehensive in its consideration of the many factors involved, simplified for clarity and quick application by the working engineer, and mathematically expressed to facilitate calculation of unknown parameters.

1.4 The Standard embodies accepted economic terminology in use at the accounting and managerial levels, so that a corrosion engineer's judgment can be properly expressed to and understood by management.

1.5 Standard methods herein presented are precise calculations within the parameters defined. Although individual company parameters may vary, differences should be minor and should permit application of these methods as useful tools of engineering and management decision making.

Section 2: Appraisal Techniques

2.1 Several different techniques exist for economic appraisal of alternative corrosion control measures. Among these are the concepts of 'Return on Investment', 'Pay-Out Period', and 'Discounted Cash Flow'. Some of these appraisal techniques lack adequate sophistication; others are unduly complex and do not lend themselves readily to comprehension and use, especially as calculations.

2.2 A standardised approach is offered in this NACE Recommended Practice which comprises a mathematical expression of a 'Discounted Cash Flow' analysis. This permits rapid calculation of 'Present Worth After Taxes', and equivalent uniform 'Annual Cost'. The formula also facilitates solution for unknown parameters.

2.3 Calculation of specific factors for the effect of tax rate, depreciation, and accounting schedules permits the correlation of these formulas with the requirements of individual businesses.

Section 3: Definitions

3.1 Cash Flow: A cash flow (a) is designated as plus or minus ($\pm$ a), for the following reasons:

3.1.1 Negative Cash Flow, or outlay, is the actual cost of any proposal, which may include: components or materials; labor required for fabrication, installation and maintenance; periodic, regular, or irregular expenses; working capital; salvage value; and overhead.

3.1.2 Positive Cash Flow, or income, is the income which can be derived from sales, depletion allowances, salvage value, and depreciation or other tax credits.

3.2 Life: This is the anticipated duration of a cash flow, expressed in years. Life may constitute actual duration of the proposal (n), or an artificial period of depreciation or depletion calculated for accounting purposes (n').

3.3 Interest: Interest is the rate of payment of money paid for the use of money, expressed as a decimal percentage per unit of time. For example, an interest rate of 6% is expressed as $i = .06$, with the added requirement that the element of time (i.e., whether the interest rate is compounded annually, semi-annually, etc.) be expressed (e.g., 6% per year).

The interest rate of return after taxes (r) is related to (i) as follows: $r = i(1 - t)$ where t is the tax rate, also expressed as a decimal.

3.4 Tax Rate: The tax rate, (t), expressed as a decimal, is the rate of Federal Income Tax extant. It enters into calculations of r and into certain discounted cash flows.

3.5 Depreciation: Depreciation is the systematic scheduled rate of decrease in the book value of property. It is a method of accounting for wear, deterioration, or obsolescence. Accounting procedures recognise a variety of methods for depreciating investment: Straight Line (SL); Sum of Digit (SOD); Constant Percentage (of the declining balance); Constant Percentage based on life such as Double Declining Balance; Sinking Fund, etc. NACE Unit Committee T-3C has elected to illustrate the first three listed. The calculations for Double Declining Balance are identical with Constant Percentage. The calculations presented can be readily adapted or revised as required by the users.

3.6 Year Zero Concept: The current year, or 'now', or year considered the starting point in the evaluation is considered as 'year zero' and cash flows occurring during that year are not discounted. For example, if 1969 is the now or zero year, then cash flows in 1969 are not discounted and cash flows in 1970 are discounted 1 year's interest. This is equivalent to considering all cash flows during a year as occurring at the beginning of that year.

3.7 Future Worth (FW): Because money can earn interest, its value—or worth—today is different from the future. Thus, by calculating the amount of interest it could earn, the 'future worth' of a 'today' dollar can be determined. Similarly, the value of a dollar spent or earned in the future is decreased by a similar amount 'today'. (This should not be confused with 'inflation'.)

3.8 Present Worth (PW): The value 'now' of any cash flow or flows ($\pm$ a), or of any summation of cash flows now or in the future, constitutes Present Worth.

3.9 Present Worth After Taxes (PWAT): The 'now' value or Present Worth After Taxes is obtained by correcting any cash flow or flows now or in the future, for all taxes and tax allowances.

Section 4: Symbol Description

4.1 a: Any cash flow, dollars.

4.2 n: Life of project in years.

4.3 n′: Write-off or depreciation period, years.

4.4 n″: Time to occurrence of cash flow, years.

4.5 i: Interest rate, as a decimal.

4.6 r: Rate of return after taxes, as a decimal.

4.7 t: Tax rate, as a decimal.

4.8 F: Variant form of Capital Recovery Factor. Cf. paragraphs 6.2 and 6.3.

4.9 d: Tax and depreciation factor equal to the PWAT of $1.00 depreciated at the applicable schedule and taxed at the applicable rate, expressed in cents on the dollar (or decimal). d may be calculated if on a regular basis. Otherwise, it is derived by DCF analysis (See 4.15).

4.10 N: The denominator for Sum-of-Digit depreciation, equal to $(n'^2 + n')/2$.

4.11 P: The depreciation rate for constant percentage depreciation.

4.12 PW: Present Worth.

4.13 FW: Future Worth.

4.14 A: Annual Cost. This is the equivalent uniform annual cost beginning one year hence or 'end of year'. A may relate to either PW or PWAT.

4.15 PWAT: Present Worth After Taxes (sometimes called Net Present Value, or NPV).

4.16 DCF: Discounted Cash Flow, the formulation of the PWAT of all cash flows in a project.

4.17 C: Capital outlay.

4.18 X: Expense outlay.

4.19 S: Annual Sales or Income.

4.20 K: Annual Expenses.

4.21 m: Years between regular recurring cash flows (e.g., 2 for a biennial item).

4.22 m′: Total number of years over which regularly recurring cash flows occur (e.g., for 3 biennial costs, m′ = 6).

4.23 WC: Working Capital.

4.24 V: Salvage Value.

4.25 y: Tax credit on capital items, as a decimal.

Section 5: Present Worth and Future Worth Relationships

5.1 Present Worth (PW) is the value 'now' of any cash flow ($\pm$ a), or of any summation of cash flows.

5.2 The relationship of Present Worth to Future Worth (FW) is governed by the compound interest formula:

$$\text{(a) } PW\,(1+i)^n = FW \text{ or (b) } PW = FW/(1+i)^n$$

5.3 The (b) relationship of the formula in 5.2 defines 'discounting', which states that any Future Worth can be discounted to Present Worth simply by dividing Future Worth by the compound interest factor for i and n years.

5.4 Future Worth can be obtained by either compounding a given Present Worth for n years at interest rate i, or by setting aside a fixed amount annually A at the end of each year. This is the so-called "annuity" formula (see Appendix for derivation):

$$FW = A\,\frac{(1+i)^n - 1}{i}$$

Section 6: Annual Costs

6.1 Annual costs can be figured by combining the compound interest formula and the annuity formula:

$$PW\,(1+i)^n = \frac{A\,(1+i)^n - 1}{i}$$

6.2 The annual cost of any cash flow 'now' can be expressed as:

$$A = PW \cdot i \cdot F_n \text{ where } F_n = \frac{(1+i)^n}{(1+i)^n - 1}$$

For any cash flow now (also called beginning of year and year zero) the first A occurs at the end of the year or in year one. Thus F_n converts the product of Present Worth and interest rate to an equivalent series of n annual cash flows (Annual Cost) beginning one year hence.

6.3 F_n is a variant form of the Capital Recovery factor in the Standard Financial Tables. F_n is derived from the interest rate (i) if taxes are not being considered and from rate of return after taxes (r) if taxes are being considered.

Thus: $A \text{ (before taxes)} = PW \cdot i \cdot F_n \text{ where } F_n = \frac{(1+i)^n}{(1+i)^n - 1}$

and: $A \text{ (after taxes)} = PWAT \cdot r \cdot F_n \text{ where } F_n = \frac{(1+r)^n}{(1+r)^n - 1}$

(This item will be developed fully in Section 8.) See references at the end of this standard.

Section 7: Present Worth After Taxes from Discounted Cash Flow

7.1 One of the most sophisticated economic analysis methods, capable of taking into consideration all types of present or future cash flows of income or disbursement, arrives at Present Worth After Taxes (PWAT) by a tabulation known as Discounted Cash Flow (DCF).

7.2 The formal procedure of a DCF analysis involves the designation of years of occurrence of cash flows, depreciation schedule, and the minimum required interest rate after taxes (r).

7.3 A hypothetical model of a DCF analysis for a capital expenditure of − a dollars, for a three-year project, is illustrated. The typical major business tax rate of 48% is used, money worth 10% after taxes, starting in year 'one'.

1. Year	1	2	3
2. Capital	− a	± 0	± 0
3. Depreciation, SL	+ a/3	+ a/3	+ a/3
SOD	$\frac{+3.a}{6}$	$\frac{+2.a}{6}$	$\frac{+1.a}{6}$
4. Tax Credit	0.48 Depr	0.48 Depr	0.48 Depr
5. Cash Flow*, SL	(− a + 0.16a)	+ 0.16a	+ 0.16a
SOD	(− a + 0.24a)	+ 0.16a	+ 0.08a
6. DCF, SL	$\frac{(-0.84a)}{(1+r)^1}$	$\frac{+0.16a}{(1+r)^2}$	$\frac{+0.16a}{(1+r)^3}$
DCF, SOD	$\frac{(-0.76a)}{(1+r)^1}$	$\frac{+0.16a}{(1+r)^2}$	$\frac{+0.08a}{(1+r)^3}$

7. PWAT = Algebraic sum of DCF's: − .511a (SL) or − .499a (SOD).

*Cash Flow is the algebraic sum of lines 2 and 4.

7.4 A simple example of a capital outlay of $1.00 for a three-year project, calling the first year 'zero' (i.e., not discounting the first year) follows. Depreciation is Sum-of-Digit, tax rate 48%, money worth 10% after taxes.

1. Year	0	1	2
2. Capital	− $1.00	± 0	± 0
3. Depreciation, SOD	+ 0.50	+ 0.333	+ 0.167
4. 48% Tax, Credit	+ 0.240	+ 0.160	+ 0.080
5. Cash Flow (2 + 4)	− 0.760	+ 0.160	+ 0.080
6. DCF @ 10 %	− 0.760	+ 0.145	+ 0.066
7. PWAT	− $0.549		

7.5 Paragraph 7.3 has illustrated that PWAT is equal to the cash flow (a) multiplied by a factor.

Paragraph 7.4 has illustrated the calculation of the factor directly by making the cash flow equal to \$1.00. The PWAT for a capital outlay of \$1.00 for any duration of project, accounting schedule, depreciation schedule, tax rate, and rate of return can be calculated in a similar manner to give a factor for tax and depreciation (dn'). Thereafter, the PWAT for any capital outlay of n' years depreciation period (not necessarily equal to the project life, n) is:

$$\text{PWAT} = \pm a \cdot d_{n'}/(1+r)^{n''} - n'' \text{ is the time in years to occurrence of the cash flow.}$$

7.6 If the accounting schedule follows a regular pattern of depreciation as described in Section 11, these factors for tax and depreciation can be calculated directly.

7.6.1 For regular Straight Line (SL) depreciation of n' years wifh first year as year zero (not discounted):

$$d_{n'} = \text{PWAT of } \$1.00 = \$1.00 - \frac{\$1.00t}{n'} \cdot \frac{F_1}{F_{n'}}$$

7.6.2 For regular Sum of Digits (SOD) depreciation of n' years with first year as year zero (not discounted):

$$d_{n'} = \text{PWAT of } \$1.00 = \$1.00 - \$1.00t\,(1+r)\left[\frac{F_1 - 1}{N}\right]n' - \left[\frac{F_1 - 1}{F_{n'}}\right]$$

7.6.3. For regular Constant Percentage depreciations of P constant percentage depreciation and with the first year as year zero (not discounted):

$$d = 1 - \frac{tP}{r+P}$$

7.6.4 If the accounting schedule follows some arbitrary and irregular pattern of depreciation, the $d_{n'}$ factors are best derived by DCF calculation.

Section 8: Present Worth After Taxes and Uniform Annual Costs

8.1 To revise the Annual Cost equation (See paragraph 6.2) for inclusion of taxes and depreciation, Present Worth must be replaced by Present Worth After Taxes (PWAT) where:

$$\text{PWAT} = \text{PW} \cdot d_{n'}$$

8.2 The calculation of ($d_{n'}$) was given in Section 7.

8.3 The subscript (n') refers to the period over which depreciation is allowed. (Capital items have varying periods; expense or maintenance items are written off in one year under present tax rules.)

8.4 The Interest (i) must be replaced by interest after taxes (r), which is thus equated as:

$$A = \text{PW} \cdot d_{n'} \cdot r \cdot F_n \text{ where } F_n = \frac{(1+r)^n}{(1+r)^n - 1}$$

8.5 A general form of this equation utilizes either Present Worth or Future Worth (the actual cash flow a 'now' or at some future date), and expressed as:

$$A = \frac{a}{(1+r)^{n''}} \cdot d_{n'} \cdot r \cdot F_n \text{ where } F_n = \frac{(1+r)^n}{(1+r)^n - 1}$$

This expression allows the direct calculation of the Uniform Annual Cost of any cash flow from the answers to the following questions:

8.5.1 How much is the cash flow (a)?

8.5.2 What is the 'write-off' period (n')? ($d_{n'}$ from Table 1, 2, or 3 or d from Table 4).

8.5.3 What is the rate of return, or interest, after taxes (r)?

8.5.4 What is the total life of the project (n)? ($F_{n'}$ from Table 1, 2, or 3).

8.5.5. When will the cash flow occur (n")? $\left(\frac{1}{(1+r)^{n''}} \text{ from Table 1, 2, or 3}\right)$

8.6 Alternatively, the Present Worth After Taxes can be calculated:

$$\text{PWAT} = \frac{A}{r\,F_n} = \frac{a\,d_{n'}}{(1+r)^{n''}}$$

8.7 The choice between calculation as PWAT and subsequent reduction to annual cost (or vice versa) may be based upon ease of formulation, clarity, or personal preference.

8.8 Alternatives expressed in terms of PWAT can be directly compared only when project lives are equal or, if unequal, the projects are calculated on a length of time equal to the least common denominator of the project lives (e.g., 35 years for alternatives of 5 and 7 years duration).

8.9 Alternatives of different lives can be directly compared by reducing PWAT to annual cost.

Section 9: Expressions for Various Cash Flows

9.1 From the relationships developed in Section 8, various types of cash flows can be expressed as follows (the detailed derivation is given in Section 14 and the Appendix):

TABLE 1—Tax and Depreciation Factors

1. Irregular sum-of-digit depreciation.
2. Tax rate = 48%.
3. Money is worth 10% after taxes.
4. First year is year zero.

Years, n	$(1+r)^n$	d_n	F_n
1	1.100	.520	11.000
2	1.210	.556	5.762
3	1.331	.583	4.021
4	1.464	.603	3.155
5	1.611	.617	2.637
6	1.771	.627	2.297
7	1.948	.633	2.055
8	2.145	.636	1.873
9	2.358	.639	1.736
10	2.595	.639	1.627
11	2.855	.640	1.539
12	3.14		1.467
13	3.46		1.407
14	3.80		1.357
15	4.18		1.314
20	6.72		1.175
25	10.82		1.102

TABLE 2—Tax and Depreciation Factors

1. Regular straight line depreciation.
2. Tax rate = 48%.
3. Money worth 6% after taxes.
4. First year is year zero.

Years, n	$d_{n'}$	F_n	$(1+r)^n$
1	0.520	17.666	1.060
2	0.534	9.091	1.124
3	0.546	6.235	1.191
4	0.558	4.810	1.262
5	0.572	3.957	1.338
6	0.583	3.389	1.418
7	0.594	2.986	1.504
8	0.606	2.684	1.594
9	0.617	2.450	1.689
10	0.626	2.264	1.791

TABLE 3—Tax and Depreciation Factors

n	$d_{n(1-5)}$	$d_{n(1-10)}$	$d_{n(1-15)}$	$d_{n(1-20)}$	$d_{n(1-25)}$	$d_{n(1-30)}$	$d_{n(1-40)}$	$(1+r)^n$	$\frac{1}{(1+r)^n}$	F_n	n
1	0.5000	0.5000	0.5000	0.5000	0.5000	0.5000	0.5000	1.100	0.9091	11.0000	1
2	0.5364	0.5409	0.5424	0.5432	0.5436	0.5439	0.5443	1.210	0.8264	5.7619	2
3	0.5612	0.5740	0.5782	0.5804	0.5817	0.5825	0.5836	1.331	0.7513	4.0211	3
4	0.5762	0.6003	0.6083	0.6123	0.6147	0.6163	0.6183	1.464	0.6830	3.1547	4
5	0.5830	0.6208	0.6333	0.6396	0.6434	0.6459	0.6491	1.611	0.6209	2.6380	5
6		0.6363	0.6540	0.6629	0.6682	0.6718	0.6762	1.772	0.5645	2.2961	6
7		0.6476	0.6710	0.6827	0.6897	0.6944	0.7002	1.949	0.5132	2.0541	7
8		0.6553	0.6846	0.6993	0.7082	0.7140	0.7214	2.144	0.4665	1.8744	8
9		0.6599	0.6955	0.7133	0.7240	0.7311	0.7400	2.358	0.4241	1.7364	9
10		0.6620	0.7040	0.7250	0.7376	0.7460	0.7565	2.594	0.3855	1.6275	10
11			0.7104	0.7346	0.7492	0.7588	0.7709	2.853	0.3505	1.5396	11
12			0.7151	0.7425	0.7590	0.7699	0.7836	3.138	0.3186	1.4676	12
13			0.7183	0.7489	0.7673	0.7795	0.7948	3.452	0.2897	1.4078	13
14			0.7202	0.7540	0.7742	0.7877	0.8046	3.797	0.2633	1.3575	14
15			0.7211	0.7579	0.7800	0.7947	0.8131	4.177	0.2394	1.3147	15
16				0.7609	0.7848	0.8007	0.8206	4.595	0.2176	1.2782	16
17				0.7631	0.7887	0.8058	0.8271	5.054	0.1978	1.2466	17
18				0.7646	0.7919	0.8101	0.8328	5.560	0.1799	1.2193	18
19				0.7655	0.7944	0.8137	0.8378	6.116	0.1635	1.1955	19
20				0.7659	0.7964	0.8167	0.8421	6.727	0.1486	1.1746	20
21					0.7978	0.8191	0.8458	7.400	0.1351	1.1562	21
22					0.7989	0.8212	0.8490	8.140	0.1228	1.1401	22
23					0.7997	0.8228	0.8518	8.954	0.1117	1.1257	23
24					0.8001	0.8241	0.8541	9.850	0.1015	1.1130	24
25					0.8003	0.8251	0.8562	10.835	0.0923	1.1017	25
26						0.8259	0.8579	11.918	0.0839	1.0916	26
27						0.8265	0.8594	13.110	0.0763	1.0826	27
28						0.8268	0.8606	14.421	0.0693	1.0745	28
29						0.8271	0.8616	15.863	0.0630	1.0673	29
30						0.8272	0.8625	17.449	0.0573	1.0608	30
31							0.8632	19.194	0.0521	1.0550	31
32							0.8638	21.114	0.0474	1.0497	32
33							0.8643	23.225	0.0431	1.0450	33
34							0.8647	25.548	0.0391	1.0407	34
35							0.8649	28.102	0.0356	1.0369	35
36							0.8652	30.913	0.0323	1.0330	36
37							0.8653	34.004	0.0294	1.0300	37
38							0.8654	37.404	0.0267	1.0270	38
39							0.8655	41.145	0.0243	1.0250	39
40							0.8655	45.259	0.0221	1.0226	40

	PWAT	A
9.1.1 Capital Expenditure	$-C \cdot d_{n'}$	$-C \cdot d_{n'} \cdot r \cdot F_n$
9.1.2 Expenses	$-X \cdot d_1$	$-X \cdot d_1 \cdot r \cdot F_n$

	PWAT	A
9.1.3 Regular Annual Items	$(S-K)\cdot d_1 \cdot F_1/F_n$	$(S-K)\cdot d_1 \cdot r \cdot F_1$
9.1.4 Regular Recurring Expenses or Savings	$\pm X \cdot d_1 \cdot F_m/F_{m'}$	$\pm X \cdot d_1 \cdot F_m/F_{m'} \cdot r \cdot F_n$
9.1.5 Working Capital	$-WC + \frac{WC}{(1+r)^n}$	$\left(-WC + \frac{WC}{(1+r)^n}\right) r \cdot F_n$
9.1.6 Salvage Value	$+V\cdot(1-t)/(1+r)^n$	$\frac{+V\cdot(1-t)\cdot r \cdot F_n}{(1+r)^n}$

9.1.6.1 The treatment as shown in Section 9.1.6 is valid when salvage value is expected to be less than 10% of the capital investment; the salvage value is treated as income at the end of the project. Otherwise, it is treated as working capital, thus:

	PWAT	A
Capital	$-(C-V)d_{n'}$	$-(C-V)d_{n'} \cdot r \cdot F_n$
Salvage	$-V+\frac{V}{(1+r)^n}$	$\left(-V+\frac{V}{(1+r)^n}\right) r \cdot F_n$

9.1.7 Tax Credit $+y \cdot C$ $+y \cdot C \cdot r \cdot F_n$

TABLE 4—Tax and Depreciation Factors

1. Regular Constant Percentage depreciation.
2. Tax rate = 50%.
3. First year is year zero.

Depreciation Rate Percent	Money Worth After Taxes—Percent 6	8	10	15	20
5	.7728	.8077	.8334	.8750	.9000
6	.7500	.7858	.8125	.8572	.8847
8	.7143	.7500	.7778	.8261	.8572
10	.6875	.7223	.7500	.8000	.8334
12.5	.6622	.6952	.7223	.7728	.8077
15	.6429	.6740	.7000	.7500	.7858
20	.6154	.6429	.6667	.7143	.7500
25	.5968	.6213	.6429	.6875	.7223
30	.5834	.6053	.6250	.6667	.7000
40	.5653	.5834	.6000	.6364	.6667

9.2 Delayed Cash Flows

9.2.1 The formulae must be divided by $(1+r)^{n''}$ for delayed cash flows, where n'' is equal to the total number of fully completed time periods prior to the cash flow. For example, any time during the third year of a project, $n''=2$; at the close of the third year, $n''=3$.

Section 10: Examples and Applications

10.1 Process Equipment (r = 10%, t = 48%, irregular SOD depreciation per Table 1).

10.1.1 A new heat exchanger is required in conjunction with re-arrangement of existing facilities. Due to corrosion, the expected life of a steel exchanger is five years. The installed cost is $9500. It is proposed to substitute a Type 316 item, costing $26,500 installed, with an estimated life of 15 years, written off in 11 years. Which is the economical choice, both items being capitalized?

Annual costs:

$$A_{Steel} = -C \cdot d_5 \cdot r \cdot F_5$$

$$= -\$9500 \cdot (.617)(.1)(2.637) = -\$1546$$

$$A_{316} = -C' \cdot d_{11} \cdot r \cdot F_{15}$$

$$= -\$26{,}500 \cdot (.640)(.1)(1.314) = -\$2229$$

Note that the steel is more economical despite the higher ratio of cost to life (\$1900/year vs \$1767/year).

10.1.2 If the steel item were to require \$2000 yearly maintenance (e.g., painting, inhibition, cathodic protection, etc.), would it still be economical? (the yearly maintenance costs are treated as expense items with a one-year life, one year hence).

Now:

$$A_{Steel} = -C \cdot d_5 \cdot r \cdot F_5 - K \cdot d_1 \cdot r \cdot F_1/(1+r)^1$$
$$= -\$1546 - \$2000\,(.52)\,(.1) \cdot 11/1.1 = -\$2586$$

The Type 316 item is now more economical.

10.1.3 Under the conditions described in 10.1.1 above, if it is not certain that a 5-year life would be attained for steel, at what life is it economically equivalent to the Type 316?

Equating

$$A_{316} = A_{Steel}$$

$$-\$2246 = -\$9500 \cdot d_x \cdot r \cdot F_x$$

$0.234 = d_x \cdot r \cdot F_x$ @3 yrs (.583) (.1) (4.021 per Table 1)

10.1.4 Under the conditions described in 10.1.1 above, how much product loss could be tolerated after two of the five years life (e.g., from roll leaks or a few tube failures) before we could have justified the Type 316 condenser?

(Note: Production losses are expense, two years hence, accruing once in five years).

Equating

$$A_{316} = A_{Steel} + A_{prod.\ loss}$$

$$-\$2246 = -1546 - \frac{X \cdot d_1 \cdot r \cdot F_5}{(1+r)^2}$$

$$-\$2246 = -\$1546 - X(.113)$$

$$-\$680 = -X(.113)$$

$$\$6018 = X$$

If production losses were to exceed this amount, the Type 316 item could have been justified.

10.2 Plant Painting ($r = 10\%$, $T = 48\%$, irregular SOD depreciation per Table 1).

10.2.1 A present paint system costs \$0.38/sq ft applied and total failure has occurred after four years. If this condition persists for a 12-year period, what is the Annual Cost? (The first application is capitalized, those in the 4th and 8th years are expense.)

$$PWAT = -\$0.38 \cdot d_{11} - \frac{\$0.38\,(d_1) \cdot F_4/F_8}{(1+r)^4}$$

$= -\$0.243 - \$0.228 = -\$0.471$

$A \quad = PWAT \cdot r \cdot F_{12} = -\$0.471\,(.1)\,(1.467) = -\$0.069/sq\ ft.$

10.2.2 Total failure could be eliminated by biennial touch-up (wire-brush, spot-primer and top-coat). What is the most that can be spent on this preventive maintenance?

Equating $\qquad PWAT = PWAT'$

$$-\$0.471 = -\$0.38\ d_{11} - \frac{X \cdot d_1 \cdot F_2/F_{10}}{(1.1)^2}$$

$$-\$0.471 = -\$0.243 - X\,(1.5215)$$

$$-\quad .228 = -1.5215\ X$$

$$.15/sq\ ft = X$$

The biennial touch-up should cost no more than \$0.15/sq ft.

10.3 Gas Cracking ($r = 10\%$, $t = 48\%$, irregular SOD depreciation per Table 1).

10.3.1 A chrome-nickel heat resistant casting is used to fabricate a gas furnace cracking set. Each set costs \$3300 (expense) and has a 2-year life, provided it is maintained at a yearly cost of \$1550. It is proposed to substitute a more expensive alloy costing \$4178 (expense), having no anticipated repair costs. How long must the proposed alloy last?

Equating $\qquad A_1 = A_2$

$$-\$3300 \cdot d_1 \cdot r \cdot F_2 - \$1550 \cdot d_1 \cdot r \cdot F_1 = -\$4178 \cdot d_1 \cdot r \cdot F_n$$

$$-\$1875 = -\$217\ F_n$$

$$8.6 = F_n \qquad 1.5\ \text{years}$$

10.4 Heat Exchanger ($r = 6\%$, $t = 48\%$, regular straight line depreciation per Table 2).

10.4.1 A compressor intercooler costs \$2500 installed, the tubes having a 6-year life and the shell and tubesheet a combined salvage value of \$1000. What is the annual cost?

$$PWAT = -(2500 - 1000)\ d_6 - 1000 + 1000/(1 + r)^6$$

$$= -1500\,(.583) - 1000 + 705$$

$$= -875 - 1000 + 705 = -1170$$

$$\text{and } A = -1170 \cdot r \cdot F_6 = -1170\,(.06)\,3.389 = -\$238$$

10.5 Cathodic Protection ($r = 10\%$, $t = 48\%$, irregular SOD depreciation per Table 1).

10.5.1 It is proposed to cathodically protect an underground piping installation. Three proposals have been made, as follows:

No. 1: Anodes, 4250 @ \$100,000 installed, 10 year life.

No. 2: 30 Rectifiers with groundbeds @ \$100,000 installed, 20 year life plus yearly operation and maintenance of \$5900.

No. 3: A 'mixed' system involving rectifiers (20 years and \$18,000 + \$1200 yearly) and anodes (10 years and \$82,000).

The Annual Costs are as follows:

$$A_1 = -\$100{,}000 \cdot d_{10} \cdot r \cdot F_{10} = \$10{,}400$$

$$A_2 = -\$100{,}000 \cdot d_{11} \cdot r \cdot F_{20} - \$5900 \cdot d_1 \cdot r \cdot F_1 = -\$7520 - \$3375 = -\$10{,}895$$

$$A_3 = -\$18{,}000 \cdot d_{11} \cdot r \cdot F_{20} - \$82{,}000 \cdot d_{10} \cdot r \cdot F_{10} - \$1200 \cdot d_1 \cdot r \cdot F_1 = -\$1354 - \$8528 - \$686 = -\$10{,}568$$

10.5.2 A pipeline installed 25 years ago will require either: (1) 50% replacement at a cost of \$690,000 to last another 25 years or; (2) cathodic protection to effect the same life. How much can be spent now for a rectifier installation if operating and maintenance costs are 6% yearly?

Solution: A repair of this magnitude would have to be capitalised. Its PWAT would be:

$-\$690{,}000\,(.64) = -\$442{,}112$ Equating

$$-442{,}112 = -C\,(.64) - .06\ C \cdot d_1F_1/F_{25}$$

$$-442{,}112 = -C\,(.64) - .311\ C = -.951\ C$$

$$C = \$464{,}892$$

10.5.3 In 10.5.2 above, it develops that cathodic protection can be applied for an installed cost of \$34,613 with yearly expenses of \$840. What annual savings are expected compared with repairing the line? Again, PWAT of repair is − \$442,112.

$$PWAT' = -\$34{,}613\ d_{11} - \$840 \cdot d_1 \cdot F_1/F_{25}$$

$$= -\$22{,}152 - \$427(9.982)$$

$$= -\$26{,}514$$

The difference in PWAT of the alternatives is (\$442,112 − 26,515) = \$415,598 which gives an annual savings by installing cathodic protection of:

$$A = PWAT \cdot r \cdot F_n$$

$$= (\$415{,}598)\,(.1)\,F_{25}$$

$$A = \$45{,}799$$

10.6 Truck Rental (r = 10%, t = 50%, regular constant percentage depreciation of 15% per Table 4). Tax regulations do not allow equipment that is sold to be withdrawn from the capital account.

10.6.1 A truck may be purchased for \$9865 and sold at the end of 5 years for \$2135. A new set of tires will be needed after 3 years costing \$1520 and an overhaul will be required every two years costing \$625 each. Minor tune ups are expected to cost \$215 annually and collision insurance will be \$185 annually. What is the Annual Cost of operating this truck exclusive of gas, oil, servicing, and liability insurance?

$$A_{Capital} = -C \cdot d_{10,15} \cdot r \cdot F_5 = (-9865)\,(.7000)\,(2.6380) = -1822$$

$A_{Salvage}$ $= +V(1-t)\cdot r\cdot F_n/(1+r)^5 = (+2135)(.5000)(.1)(2.6380)/1.611 = +175$

A_{Tires} $= -X\cdot d_1\cdot r\cdot F_n/(1+r)^3 = (-520)(.5000)(.1)(2.6380)/1.331 = -151$

$A_{Overhauls}$ $= -X\cdot d_1\cdot F_m/F_m{}'r\cdot F_n/(1+r)^2 = (-625)(.5000)\dfrac{5.7619}{3.1547}(.1)(2.6380)/1.210$
$= -124$

$A_{Insurance\text{-}Tune\text{-}Ups}$ $= -K\cdot d_1\cdot r\cdot F_1 = -(400)(.5000)(.1)(11.0000) = -220$

A_{Total} $= -\$2142$

10.6.2 A rental Company will assume all of the obligation of owning the vehicle for \$350/month. Is rental attractive?

$A_{Rental} = -K\cdot d_1\cdot r\cdot F_1 = -(350)(12)(.5000)(.1)(11.000) = -\$2310$

Rental is not attractive at this price.

10.6.3 If the truck were wrecked after 2 years and the trade-in cost \$4000 which was not compensated by insurance and the salvage value increased to \$4000, would rental then be attractive?

$A_{Capital} = (-4000)(.7000)(.1)(2.6380)/1.210 = -610$

$A_{Salvage} = (+1865)(.5000)(.1)(2.6380)/1.611 = +153$

A_{Tires} = delete – not necessary now (+ 151)

$A_{Total} = -\$2142 - 610 + 153 + 151 = \$2448$

Rental is now attractive.

Section 11: Depreciation Derivation Schedules

11.1 Straight Line (SL) depreciation schedules the diminution in value in unit fractions of the life (n); i.e., each year's depreciation of a cash flow $(-a)$ is equal to $(+a).1/n$:

11.1.1 An investment of – \$1000 having a 10-year life will depreciate (\$1000).1/10 = + \$100/year. The investment is said to be 'written off' in 10 years.

11.2 Sum-of-Digit (SOD) depreciation writes off an investment in fractions of a number (N) representing the sum of the digits comprising the life (n).

11.2.1 The general expression is: $N = (n^2 + n)/2$

11.2.2 The depreciation schedule is: $n/N + (n-1)/N \ldots \ldots \dfrac{+n-(n-1)}{N}$

11.2.3 An item having a three-year life $(3 + 2 + 1 = 6)$ would be written off 3/6 in the first year; 2/6 in the second year; and 1/6 in the third year.

11.3 Constant Percentage (P) (of the declining balance) depreciation writes off an investment at some percentage of the remaining balance. The percentage may be related to life as a Double Declining Balance or it may be some arbitrary percentage based on class of equipment.

11.3.1 The general expression for $P = 20\%$ is:

$$.2a + .2(a - .2a) + .2(a - .2a - .16a) \text{ etc.}$$

11.3.2 For Double Declining Balance:

5-year life P = 40%
10-year life P = 20%
20-year life P = 10%

11.4 If such schedules are actually used by the accountants, the SL or SOD depreciation is said to be 'regular'.

11.5 It is frequently elected to set up an artificial or 'irregular' schedule (e.g., 11 years for process equipment). Then, the depreciation is set up on the basis of 11 years, the residual being taken in the last year if the life of equipment is less than 11 years.

11.5.1 An item of 4-years life would be depreciated:

$$\text{SL} \ldots \frac{1}{11}+\frac{1}{11}+\frac{1}{11}+\frac{8}{11}$$

$$\text{SOD} \ldots \frac{11}{66}+\frac{10}{66}+\frac{9}{66}+\frac{36}{66}$$

11.5.2 Constant Percentage is most frequently used on a 'regular' basis with all items of a certain percentage in a pool. In some cases, rules allow given items to be removed from the pool and written off immediately. Such situations will rarely be foreseen when making economic comparisons, however.

Section 12: Tables

12.1 Table 1 is a table of factors for tax and depreciation based on an eleven year, irregular SOD schedule (d factors for less than 11 years have the remaining depreciation taken in the final year, d_{11} is the d factor for a regular eleven year schedule). Tax rate is 48%, money worth 10% after taxes, and the first year is year zero.

12.2 Table 2 is a table of factors for tax and depreciation based on regular straight line depreciation for 1 through 10 years. The tax rate is 48%, money is worth 6% after taxes, and the first year is year zero.

12.3 Table 3 is a table of factors for tax and depreciation based on irregular straight line depreciation schedules of 5, 10, 15, 20, 25, 30 and 40 years (d factors for less than the full schedule have the remaining depreciation taken in the final year; d factors for the final year in each schedule are the regular SL depreciation factors for that schedule). The tax rate is 50%, money is worth 10% after taxes, and the first year is year zero.

12.4 Table 4 is a table of d factors for regular Constant Percentage schedules of 5, 6, 8, 10, 12.5, 15, 20, 25, 30 and 40%. The tax rate is 50% and money is worth 6, 8, 10, 15 or 20% after taxes; the first year is year zero.

12.5 A similar set of factors may be calculated for any given set of circumstances. Thereafter, PWAT and Annual Cost may be directly calculated as shown in Section 8.

Section 13: Factors for Various Industries

13.1 Process Industries.

13.1.1 The process industries are most likely to use Table 1 or similarly derived factors.

13.1.2 In the process industries, manufacturing equipment (e.g., reactors, stills, heat exchangers) is frequently depreciated on an eleven year, irregular SOD basis. If the equipment lasts less than eleven years, the remaining depreciation is taken the year in which failure occurs (i.e., a capital item having a three year life would be depreciated 11/66 + 10/66, + 45/66).

13.2 Utilities

13.2.1 Utilities are most likely to use Table 2 or similarly derived factors.

13.2.2 Utilities are usually regulated at a rate of return relating to their size and complexity of operation; their method of depreciation is generally controlled as part of this regulation.

13.3 Production Industries

13.3.1 Production industries are most likely to use Table 3 or similarly derived factors.

13.3.2 The production industries frequently use depreciation based on depletion of reserves so, the schedule period varies from project to project. Irregular SL depreciation is frequently used for economic comparison purposes. If the equipment does not last as long as the schedule period, the remaining depreciation is taken in the year in which failure occurs (e.g., a three year life item in a five year schedule would be depreciated 1/5 + 1/5 + 3/5).

13.4 Canadian Industries

13.4.1 Canadian industries are required to depreciate capital items at a Constant Percentage such as in Table 4.

13.5 Petroleum Refining

13.5.1 Petroleum refining and similar industries are more likely to use Table 4 or similarly derived factors.

Section 14: Types of Cash Flows

14.1 The major types of cash flows involved in economic assessments are as follows (summarized in Section 9):

14.1.1 Capital Items—A capital expenditure is a negative cash flow (C) corrected for tax and depreciation over a specific write-off period of n′ years. The write-off period may or may not equate with project life, n years.

Then

$$PWAT = -C \cdot d_{n'} \text{ or } A = -C \cdot d_{n'} \cdot r \cdot F_n$$

where $n' \geqq n$

14.1.2 Expense Items—An expense item (X) is written off in the year in which it occurs (i.e., in one year).

$$PWAT = -X \cdot d_1 \text{ or } A = -X \cdot d_1 \cdot r \cdot F_n$$

14.1.3 Annual Items—(cf. Appendix 1)—The common regularly recurring annual items are both negative (e.g., maintenance costs or $-K$) and positive (e.g., sales income or $+S$). If uniform, they may be algebraically expressed as

$$PWAT = (S - K) \cdot d_1 \cdot F_1/F_7 \text{ or } A = (S - K)\, d_1 \cdot r \cdot F_1$$

where F_1/F_n sums the series of uniform annual cash flows starting in the year zero, as below:

$$\frac{(S-K)}{(1+r)^0} + \frac{(S-K)}{(1+r)^1} + \frac{(S-K)}{(1+r)^2} \cdots \frac{(S-K)}{(1+r)^{n-1}}$$

$$\frac{(1+r)^1}{(1+r)^1} - 1 = (S-K) \cdot F_1/F_n$$

$$= (S-K).\ \frac{(1+r)^n}{(1+r)^n - 1}$$

14.1.4 Other Regularly Recurring Cash Flows (cf. Appendix 1)—Any regularly recurring cash flows can be expressed as PWAT by discounting, correcting for tax and depreciation (either capitalized or expense) and summing as to the incidence of repetition within the total number of years involved. For example, an expense occurring at the end of the 2nd, 4th and 6th (but not in the final) year of the eight-year project is

$$PWAT = \frac{-X \cdot d_1}{(1+r)^2} \cdot F_2/F_6 \quad \text{or } A = \frac{-X \cdot d_1}{(1+r)^2} \cdot F_2/F_6 \cdot r \cdot F_8$$

assuming the schedule starts in the year zero.

14.1.5 Working Capital – Working Capital (WC) consists of funds put aside to operate a project. It is analogous to funds in a checking (rather than savings) account. As such, it is not taxable and is returned at the end of the project *without interest*. Then

$$PWAT = -WC + \frac{WC}{(1+r)^n}$$

$$A = \left[-WC + \frac{WC}{(1+r)^n} \right] r\,F_n$$

Note: The ultimate return is discounted because one loses the interest it could have earned applied as an investment.

14.1.6 Salvage Value—When the salvage value (V) of equipment is less than 10% of capital investment it is not separated from capital, thus it constitutes taxable income at the end of the project.

$$PWAT = \frac{+V(1-t)}{(1+r)^n} \quad \text{or} \quad A = \frac{+V \cdot (1-t) \cdot r \cdot F_n}{(1+r)^n}$$

When the salvage value is 10 or more percent of the capital, it is separated from the capital as shown:

depreciable capital $= -(C - V)$

salvage value: $PWAT = -V + \dfrac{V}{(1+r)^n}$

$$A = \left[-V + \frac{V}{(1+r)^n} \right] r\,F_n$$

14.1.7 Investment Tax Credit—Where and when applicable, an investment tax credit per cent is a simple decimal fraction (y) *now* applied to the capital investment. Then,

$$PWAT = +yC \quad \text{and } A = +yC \cdot r \cdot F_n$$

In general, the effect of a tax credit on capital investment is to add incentive to such an investment by treating the capital investment more like an expense.

References

1. "Perry's Chemical Engineers' Handbook." 4th Edition, Tables 1–14, pp. 1–32 to 1–38.
2. G. A. Taylor, "Managerial and Engineering Economy—Economic Decision-Making." (D. Van Nostrand Co., Inc.) 1964.
3. C. G. Edge, "A Practical Manual on the Appraisal of Capital Expenditure," (The Society of Industrial and Cost Accountants of Canada) 1970.
4. F. J. Jelen, "Cost and Optimisation Engineering" (McGraw-Hill Book Co.) 1970.

APPENDIX 1

Mathematical Derivations

A series of regular annual cash flows ($\pm$ a) at the *end* of each year sums as follows in terms of Present Worth:

$$\sum_1^n = a\,\frac{1}{(1+i)^1} + \frac{1}{(1+i)^2} + \frac{1}{(1+i)^3} \ldots + \frac{1}{(1+i)^n}$$

Multiplying by $(1+i)^n$ gives:

$$\sum_1^n = a\,\frac{(1+i)^{n-1} + (1+i)^{n-2} + (1+i)^{n-3} \ldots + 1}{(1+i)^n} = a\,\frac{(1+i)^n - 1}{i\,(1+i)^n}$$

PW of a series of annual end-of-year flows $. = \pm\, a/i\; F_n$ where $F_n = \dfrac{(1+i)^n}{(1+i)^n - 1}$

Multiplying by $(1+i)^1$

$$\sum = \frac{a}{(1+i)^1} \cdot \frac{(1+i)^1}{i} \cdot \frac{1}{F_n}$$

since $i = (1+i)^1 - 1$

$$= \frac{a}{(1+i)^1} \cdot \frac{\dfrac{(1+i)^1}{(1+i)^1 - 1}}{F_n}$$

$$\sum_1^n = \frac{a}{(1+i)^1} \cdot F_1/F_n$$ When first year is 'Year 1' (i.e., for end of year annual flows)

Then the PW of series of annual flows starting now (beginning of year)

$$PW = a \cdot F_1/F_n$$

In effect the 1st year is treated as Zero (i.e., 'now', it is not discounted) whereas n = actual number of years (e.g., 8, which is 'year 7' when first year is treated as 'Zero').

A regularly recurring series every m years starting now can be summed as follows:

$\sum = a \cdot F_m/F_{m'}$ where m' = m times number of instances of cash flow. e.g., biennial flows for a total six years (3 flows).

$\sum = \pm\, a \cdot F_2/F_6$

Increasing annual cash flows may result from inflation, expanded sales, etc. The increase may be arithmetical ('gradient') or geometrical.

(a) Regular series starting with year zero, the regular annual cash flow might be:

n	1	2	3	4	....	n
year	0	1	2	3	......	n − 1
Flow	+ a	+ a	+ a	+ a	......	+ a

and

$$PW = +a \cdot F_1/F_n$$

(b) With geometric increases (e.g., Sales increasing 10% annually, or cost-of-living increasing 5% annually):

n	1	2	3	4	......	n
year	0	1	2	3	......	n − 1
Flow	+ a	+ ab	+ ab^2	+ ab^3	......	+ ab^{n-1}

and

$$PW = +a \frac{1 - \dfrac{b^n}{(1+i)^n}}{1 - \dfrac{b}{(1+i)^1}}$$

The PW of projects of different duration, to be comparable, must be run out to a number of years equal to the least common denominator of n_1 and n_2. Then where $n_1 = 5$ and $n_2 = 7$

$$PW_1 = a \cdot F_5/F_{35}$$

and

$$PW_2 = a \cdot F_7/F_{35}$$

or if run to infinity $F_n = 1$ and

$PW_1 = a \cdot F_5$ infinite number of replacements of 'PCI'

$PW_2 = a \cdot F_7$

but multiply by i

$i\,PW_1 = a \cdot i\,F_5 = A_1$ from Section 6

and

$i\,PW_2 = a \cdot i\,F_7 = A_2$

which reduces PW to annual cost

12.5 Structures

(1) The cost of structural steel design is very closely related to the weight of steel used in the design; the cost of fabrication, shipping, erection, aesthetics, maintenance and corrosion control normally increases with the increase of weight of steel in the structure. Amongst the other savings which accrue from lighter steel weights one may also count the occasional profit resulting from greater payloads.

(2) Reduction in weight of structures permissible by smaller loss of strength integrity through improved corrosion control stands for direct savings in costs. The reduction of material cost is the major consideration; fabrication cost including processing and others should also be calculated.

(3) Trade-in between improvement of corrosion control and reduction of weight or vice versa should be considered multilaterally.

(4) Some special considerations, such as improvement of notch toughness and other precautions against stress corrosion cracking at low temperatures for cryogenic tanks, ships and submarine design, are specifically relevant in individual cases of economic evaluation.

(5) Corrosion control which allows planned elimination or reduction of labour motions or shop operations can reduce costs:

(*a*) rolled shapes of any grade of steel in a structure are in most cases more economical than any built-up shapes of the same size and grade of steel;
(*b*) rolled shapes of a higher strength may be used to replace lower strength steel shapes requiring flange plates;
(*c*) thinner plates of higher strength steels can replace thicker plates of lower strength steels (*note*: consider reduction in cost of welding and danger of stress corrosion cracking and fatigue corrosion).

(6) The equipment used during erection of structures and the time of erection is not necessarily reduced simply by reduction of the weight of the handled components—*note* for calculation of costs.

(7) A reduction in weight caused by improved corrosion control may accentuate the aesthetics of the structure and, by its pleasing outline, it may be inducive to a reduction in fabrication and erection costs.

(8) Use of smaller size structural members, due to effective corrosion control, reduces the maintenance costs, e.g. paint and painting time.

(9) Simplification and standardisation of structural members, which are conducive to improved corrosion control, also reduce the cost of maintenance and enhance maintainability.

(10) Use of corrosion-wise safer steel in structures may also provide improved economy in tension, in bending, in compression and economy in framing and plating.

(11) Permanent structures designed for a service life of more than five years should be protected from deterioration with the best coating system available. Other corrosion-control measures should also be considered. The cost limit in each case is individual and should not be arbitrarily dictated.

(12) The economics of buying quality materials or techniques for treatment of new construction surfaces can be negated through incomplete specifications, poor application or lack of inspection.

(13) The sequence of economic rating of protective coatings for steel permanent structures:

(*a*) hot dip galvanising;
(*b*) hot metal spray or zinc-rich coatings;
(*c*) high built organic coatings—where zinc or other sacrificial metals cannot resist certain types of corrosion attack and as a sealer for anodic metals containing coatings in places where such metals are easily corroded.

(14) Some economic guidelines for application of various protective coatings for standard components:

(*a*) below 5/8 in (16 mm) flange thickness the galvanising is more economic—above this thickness hot metal spray or spray and brush applied coatings are preferred;
(*b*) galvanising on the inside and outside of steel piping is more economic up to 12 in (30 cm) OD—above this diameter hot metal spray or spray and brush applied coatings are preferred;
(*c*) galvanising is more economic on fabricated vessels up to 4 ft (1.2 m) diameter—above this diameter hot metal spray or brush and spray applied coatings are preferred;
(*d*) zinc-rich paints (e.g. prefabrication primers) are more economic for plate sections of large field fabricated tanks than other coatings.

12.6 Equipment and Pipe Systems

(1) The service life of materials contained in any equipment, even when correlated with the corrosion test data, should be primarily weighed in the light of the actual cost.

(2) Both the damage and the cost of external and internal corrosion should be considered for evaluation of the service life of hollow equipment and pipe systems.

(3) The effect of occasional and periodical variations of internal and external environmental conditions should also be evaluated for the true extent of service life of relevant equipment.

(4) Whilst it may not always be possible to estimate the actual money cost of equipment it is possible to make the relevant economic decision with the help of the relative costs for comparison. This applies especially to such equipment for which the price quotations vary so greatly and are influenced by so many factors. Thus it is more meaningful to examine the ways in which the choice of materials may affect primarily the first cost and then the future expenses.

(5) The number of renewals forecast or necessary in the normal life-span of the equipment should be determined in the economic analysis.

(6) The degree of difficulty of the replacement of equipment or pipeline in a particular location will affect their cost.

(7) The sum of the renewals of equipment is to be multiplied by the sum of labour costs which will accrue through removing and re-installing the equipment *in situ* and by the sum of material costs, this to obtain a base for the calculations.

(8) Selections providing a safe reduction of weight should be credited with the side effects of their weight reduction, e.g. lower power needed to haul given payloads, additional earning power through extra cargo capacity at the same consumption of fuel, etc.

(9) Some factors to be considered in the economic evaluation of condensers:

(*a*) comparative cost of several materials;
(*b*) size and geometry of condenser to suit the variety of materials;
(*c*) comparative heat transfer factor (area of surface for a given wall thickness);
(*d*) cleanliness factor;
(*e*) compensation for corrosion scales formed on service;
(*f*) tolerance for velocity of water to suit variety of materials, etc.

(10) For fabricated vessels, organic coatings become more economical at a diameter of 4 ft (1.2 m) or more.

(11) Zinc-rich paints are more economical for the preservation of plate sections of large field-fabricated tanks, water towers, etc.

(12) The comparative throughput of transferred liquids should be used as a base of economic evaluation instead of the outside diameter of the pipes.

(13) There is no set figure for unscheduled out-of-service time for all equipment within a utility. For example, whilst a leak in a sanitary or service piping system, although undesirable, will not stop the normal operation of a ship, a leak in a tube of a main condenser will stop the ship for cooling down, opening up, plugging the leaking tube and the close down.

(14) It is more economical to galvanise both the inside and outside of pipes of up to 12 in (30 cm) OD.

(15) Longer life expectancy can be expected from larger diameter pipes than from the smaller diameter ones.

(16) Life expectancy of tubes and pipes cannot be judged on their nominal pipe thickness. Reasons: tolerances in nominal thickness; metal removed in threading will considerably reduce the wall thickness which a pit must penetrate.

(17) Life expectancy of pipes cannot be judged solely on the available pitting data; these tend to vary a lot. Rate of penetration is not affected by pitting only; impingement attack, erosion, etc., should be taken into account.

(18) Pitting and preferential corrosion in the vicinity of welds make the available weight loss data of some austenitic stainless steels meaningless; note the loss in local pits or in adjacent welds.

(19) Note, in life expectancy calculations of condenser tube materials, that the test data on their specimen do not ordinarily reflect the hot-wall effect.

(20) Economic evaluation of any equipment should not be based on any single portion of this equipment but on a balanced investigation of the whole system.

13 Plan of Action

13.1 Introduction

Once the individual corrosion-control fundamentals of a particular utility have been duly appraised, evaluated and culled in each separate division of the set procedures, and prior to the reconciliation of the corrosion-control complex with its functional counterpart in design, a final summary, concept or a corrosion-control plan is the next logical step. In this plan each individual product of the appraisal procedures is adjusted and rationalised to obtain an optimum compromise of corrosion-control precautions, answering to the anticipated production, environmental conditions and sound economy; that is, as near as is obtainable within the boundaries of the corrosion-control domain.

The corrosion-control decision and subsequently formulated instructions, accurately defined in a succinct technical form and free from ambiguities, are then utilised either by a corrosion specialist for his advisory activities, or by a designer as a mainstay of his effort to capture the logic of corrosion control in his functional design and to formulate the participating specifications. It is also used by the management executive to assess the value and utility of particular corrosion-control precautions.

13.2 Scope

Corrosion-control thinking cannot remain solely the vested property of corrosion-control experts. The speed of to-day's work and the interdependence of all facets of design enforces the closely knit co-operation of all men engaged in producing an economically sound product and securing the continuity of its desired effect. It is then to the advantage of all members of such a team to develop, as far as is practicable, a common system of corrosion-control thinking, a common system of communicating and analysing corrosion-control problems and, furthermore, a common home base for their individual and personally divergent interests. The plan of action should be such an expression of corrosion-control know-how, addressed to the whole design team at large, and should be written in a language understood by all—the initiated ones and the unconverted ones.

To provide an accurate recipe for such a work of art and diplomacy is a most difficult task to attempt in a chapter of this nature. One can try only to indicate but not to prescribe; it must therefore be left to the kind offices of each individual specialist, or generalist who wishes to try, to do his best and present his solution for a soft sale and the common good of the project in hand.

13.3 General

A properly designed structure or equipment should serve its purpose with an acceptable degree of efficiency, should last as long as it is to the advantage of the proprietor or user and should be as cheap a possible. This means that it does not necessarily need to last for ever, unless it is an object of irreplaceable intrinsic value. As long as corrosion does not limit life to a markedly shorter period than originally planned, the result of corrosion does not vitally interrupt its general and particular utility, and the corrosion damage does not cause danger to user's life or limb, then only economically reasonable precautions should be incorporated in the corrosion-control plan. An over-design should be avoided as much as humanly possible and, consistent with the product's function, design should be as simple, straightforward and cheap as can be reconciled with the utilitarian value of the product during its lifetime.

When anticipating the influence of corrosion control on the designed product, the analyst and compiler of the corrosion-control plan should be aware of the constantly changing pattern of corrosion, of the new techniques and the new chemical substances which can exert their influence on the forecast environment at a later date and also of the failure of the human mind to grasp, at the planning stage, the full implications of all aspects of the practical employment of the anti-corrosion measures taken. The best corrosion prevention is not good enough if it cannot be employed, repeated, improved and also implemented in maintenance.

Whilst trade-off between individual anti-corrosion measures and between corrosion prevention and over-design is possible, neither a neglect nor gross underestimating of all corrosion probabilities can be tolerated. Once a logical conclusion has been reached, that by a law of probability a certain corrosion damage can occur within the economic lifetime of the utility, then the appropriate preventive measures should be taken. This is especially critical with new and not fully tested materials and components which may sometimes be marketed under false pretences. Also a designer can misunderstand the full implications of the supplied descriptive literature.

There is no excuse for neglecting to stop corrosion on critically loaded structures. Sometimes even a negligible corrosion can bring about a major catastrophe. Corrosion can initiate and cause mechanical breakdown; thus all corrosion should be stopped at all costs, within economic limits.

Designing obsolescence into any product by breeding corrosion or increasing corrosion potential by selection and treatment of materials, or by aggravating the geometry of a structure or equipment, gives proof of reckless venality and shortsightedness of the producer for which all of us will have to pay sooner or later. These policies should therefore be strongly opposed. One should, on the other hand, be warned against excessively idealistic application of acquired corrosion knowledge in the design practice. Well-balanced design, with corrosion and functional requirements complementary to each other and economically reasonable in costing, may be suggested as a sound policy.

Corrosion or excessive deterioration of electrical and electronic parts can change the electrical properties of equipment and affect its performance. It can also considerably affect the safety of operating the equipment. Provisions arising from corrosion-control analysis of electrical and electronic design apparent in the plan of action should apply to housings, chassis, hardware and similar items forming part of the equipment, as well as the electrical, electronic and electromechanical components, and also to the associated conduits, fittings and fixtures. The plan of action should establish the minimum requirements for procedures, materials and systems and for protecting such systems from adverse environments; it should also provide for a well-engineered electrical bonding and grounding of the equipment.

13.4 Inherent Reliability

The main purpose of the corrosion-control plan is to give the relevant utility its inherent reliability at economic costs. This inherent reliability, however, does not necessarily need to conform to the utility's operational reliability and it may, in a general sense, be regarded as a corrosion-control maturity of the design together with its specifications. After all, when all factors have been considered and the best available corrosion-control and design experience utilised, the ultimate inherent corrosion-control reliability will depend upon the reliability of each individual part separately and combined.

In practice, any designed equipment or structure may fail in its function during its operational life as a result of random catastrophic failures of individual parts or by progressive deterioration or degradation of its performance. Corrosion-control assessment should therefore take account of both the catastrophic and the degradation failures, and it should be understood that each one can be caused by multiple mechanical and corrosion factors conjoint.

As discussed at the start of Chapter 11, during the operational life of a utility one can expect that the failures of its corrosion control will occur in three stages. In the first phase, representing the debugging period, the failure rate should rapidly decrease from its initial peak by a timely replacement of parts and correction of practical manufacturers' errors.

In the second phase a constant failure rate of a lower level can be expected, and it is mostly in the power of the combined team of operational maintenance crews and corrosion specialists or technologists to reduce the failure rate during this useful period of utility to the lowest possible level and to extend it as much as it suits the economy of the organisation. During this phase catastrophic failures occur only at random and may be suspected of being initiated or aggravated by the corrosion.

In the third, or wear-out phase, the incidence of failures will again rise through a combination of catastrophic and deterioration failures. The commencement and duration of this period is very much affected by the effectiveness and continuity of such precautions which have been taken on the drawing board.

In consideration of reliability, one should bear in mind that the presumptions and conclusions made in the corrosion-control plan of action and eventually in the functional design itself have been based mostly on laboratory tests, past experiences of failures and successes and are therefore more or less an educated guess which may not need to conform to the actual manufacturing and operational characteristics of the particular utility being designed. Variations in actual quality of parts or treatments made or applied by different manufacturers, varying environmental conditions, as well as the thoroughness of the operational maintenance team, will profoundly influence the rate of failures and thus the useful life of the utility.

It is not conceivable that one can calculate the actual rates of failure—only the corrosion hazards. The accuracy of the calculation of the rate of failures will be very much influenced by the accuracy and comprehensiveness of records of each organisation, by their correctness of averaging and by their future projection, coupled with the experience of each evaluator. Absolute accuracy is hardly possible.

13.5 Objectives of Integrated Corrosion Control

The basic objective of adopting integrated corrosion control as a whole is the rational utilisation of design methods that can provide technically the most effective and economically the most advantageous means of stopping corrosion of structures and equipment.

The objective of the *control of materials* is to include in the product materials of construction that are suitable for their function and also for their ability to maintain their positive function for the required length of time at a reasonable cost. The resistance to corrosion in the given environment, the tendency to specific types of corrosion, requirements of specific treatments and joining methods and the adjustability of material to a producible form that gives the best chance to the product to resist corrosion, are appraised.

The objective of the *control of compatibility* is to exclude from the product all inter-material influences that can locally or within the structure or system as a whole adversely affect the degradation of materials and include those which have a propitious effect. Such influences may be caused by direct contact between dissimilar metals, by changes of polarity, transfer of electrolysis, by the carry of metallic particles in the stream, stray currents or by any other derogatory effect arising from the near proximity of various materials.

The objective of the *control of geometry* is to include, in the product, a selection of such shapes, component forms and configurations of structures and equipment that these can be kept clean and free from destructive corrosion in all stages of fabrication, assembly and operation without excessive effort.

The objective of the *control of mechanics* is to include in the product optimal upkeep of mechanical strength of materials when under stress loading, so that their calculated strength will not be considerably and critically reduced by corrosion.

The objective of the *control of surfaces* is to include, in the product, optimum relations between the chosen materials and their environmental interfaces; optimal configuration of surfaces, preparation and cleaning of surfaces, texture, pretreatment of surfaces and maintenance of the electrochemical balance of surfaces should be noted.

The objective of the *control of protection* is to include in the product the optimal complex of anti-corrosion measures applied to the surfaces of structures and equipment for the purpose of preventing corrosion, this by separation of material surfaces from environment, by anodic or cathodic protection or by adjustment of environment, each individually or combined.

The objective of the *control of maintainability* is to include in the product precautions which can reduce the cost of detecting the occurrence of corrosion and the cost of repetitive maintenance, by making it easy, safe and foolproof.

The objective of the *control of economics* is to cost, evaluate and select corrosion-control precautions as to their optimum economy during the lifetime of the product, this including the cost of any probable repetitive anti-corrosion maintenance, necessary dismantling and assembly of adjoining and interfering obstacles and other associated workload.

Summarily, the objective of the complete integrated corrosion-control programme contained in the corrosion-control plan is to enhance the ultimate practical corrosion-control reliability of the product and the components in the forecast environmental conditions, as much as it is possible to achieve by selection, reconciliation, elimination and optimum compromise of all known corrosion-control factors.

13.6 Checking the Design

To facilitate optimum corrosion reliability, the design should be checked to ensure that optimum corrosion-control requirements contained in the first

Table 13.1 CORROSION-CONTROL DESIGN CHECK-OFF LIST

(*a*) *Materials*

Are the materials appropriate to the environments met on all surfaces of the product?

Were all necessary precautions to improve resistance of materials to corrosion effects incorporated or specified?

Was the corrosion rate of materials added or do compensating protective measures allow reduction of weight?

Table 13.1 (*Contd.*)

Are the chosen materials appropriate to avoid contamination of contents by their corrosion products?
Is the choice of materials economically sound?

(*b*) *Compatibility*
Is the use of dissimilar metals in the unit or system necessary?
Were all possible precautions for reduction of galvanic corrosion taken?
Was the right type of connection between dissimilar metals used?
Were precautions to avoid galvanic action between off-standing dissimilar metals in the conductive medium taken?
Are the fasteners compatible to joined metals and, if not, were necessary precautions taken?
Is the dc electrical equipment well insulated to prevent emission of stray currents or were necessary precautions taken?
Was the right composition of welding rods, brazing metal, solder or adhesive specified?

(*c*) *Geometry*
Is the design reasonably simple and protectable?
Is the design well drained; were precautions for effective drying taken (in and outside)?
Are the pockets for accumulation of dirt and debris eliminated (inside and outside); are solids (liquids) removed from the stream?
Is the design streamlined (inside and outside) and are turbulence, surging, impingement and bubble formation reduced or eliminated?
Are hot spots eliminated; is the design evenly heated, cooled or corrosion loaded?
Has the design only necessary joints; were precautions against crevice corrosion taken?
Can the design be kept rustless and clean at all stages of fabrication and operation without excessive effort?
Is the relative position of the product in the arrangement corrosion-wise sound?

(*d*) *Mechanics*
Is localised static or cyclic stress reduced or eliminated; were precautions against stress or fatigue corrosion taken?
Were precautions for reduction of initiation and transfer of shock and vibration taken?
Was the best joining method for reduction of localised stress used?
Can the designed product be stress relieved?
Was the corrodant adjusted to reduce stress?
Were precautions against fretting taken?
Is the lubrication arrangement effective and accessible?

(*e*) *Surfaces*
Is the design electrochemically stable?
Is effective protection of the cathodic surface specified?
Is the surface of the design smooth in contour and continuous, inside and outside?
Were rounded surfaces, corners and edges incorporated in design; were precautions for removal of burrs, dents and notches specified?
Are the welds accessible for cleaning?
Are the surfaces self-cleaning; can they be kept dust-free?
Is efficient surface preparation cleaning specified; is the surface texture optimal?

(*f*) *Protection*
Are protective coatings indicated and correctly specified?
Are sacrificial pieces, cathodic protection and control indicated and correctly specified?
Is environmental control indicated and correctly specified?
Can the surfaces be treated on pre-production or pre-assembly and assembled without major damage?
Can the protective system be renewed *in situ*?
Are the protective systems compatible throughout?
Is the access of condensation and airborne pollution to metal prevented?

Table 13.1 (*Contd.*)

(*g*) *Maintainability*
Is the design easily accessible for inspection, cleaning and maintenance preservation?
How will corrosion affect utility and safety?

(*h*) *Economics*

Cost per year: £. . . .
Date
Costing Clerk

Summary

Date
Corrosion Officer

Action taken

Date
Divisional Chief

case in the corrosion-control plan have been duly incorporated; this generally in accordance with the design check-off list in *Table 13.1.* It is not, however, inferred that all points must be considered in each design, nor is it imperative to follow exactly the given example word by word or in the exact style for every type of design work; only those points or parts appertaining to the particular design under consideration may be called upon. The essence of the check-off list is to ensure that corrosion-control requirements receive adequate consideration in the work of engineering designers and draughtsmen. Note that all temporarily effective corrosion-control measures should be checked for their easy repetition or curative control *in situ*.

Index